2021 International Conference on Architectural Research and Design and 21st Sustainable Environment and Architecture

IOP Conference Series: Earth and Environmental Science
Volume 1007

Online
2-3 November 2021

Editors:

Arina Hayati
Sri Nastiti
Nugrahani Ekasiwi
Kirami Bararatin

ISBN: 978-1-7138-5449-4
ISSN: 1755-1307

Printed from e-media with permission by:

Curran Associates, Inc.
57 Morehouse Lane
Red Hook, NY 12571

Some format issues inherent in the e-media version may also appear in this print version.

Institute of Physics
Dirac House, Temple Back
Bristol BS1 6BE UK

Phone: 44 1 17 929 7481
Fax: 44 1 17 920 0979

techtracking@iop.org

Additional copies of this publication are available from:

Curran Associates, Inc.
57 Morehouse Lane
Red Hook, NY 12571 USA
Phone: 845-758-0400
Fax: 845-758-2633
Email: curran@proceedings.com
Web: www.proceedings.com

TABLE OF CONTENTS

KEYNOTE PAPERS

THEME I: SUSTAINABLE BUILDINGS AND TECHNOLOGY

THEME II: THEORY AND HISTORIOGRAPHY OF ARCHITECTURE AND SUSTAINABILITY

THEME III: SUSTAINABLE HOUSING AND SETTLEMENT

THEME IV: SUSTAINABILITY IN CITY AND URBAN ENVIRONMENT

THEME V: SUSTAINABLE ARCHITECTURAL DESIGN AND PRACTICE

Preface

Tackling climate change has attracted many built environment researchers devoted to reducing environmental problems in dangerous ways. The phenomenon contributes to the cause of extreme weather events worldwide, posing an immediate risk to human survival, especially in urban areas. This situation has challenged designers and urban planners to carry out studies. The results need to be elaborated and enriched to a broader practicality and community participation level. As a response, many efforts addressed minimised threats posed by climate change. The UN launched Sustainable Urban Agenda in 2015, including Goals 6 of The UN's Sustainable Development Goals (SDGs), to reduce the impact of climate change on cities and urban areas. The agenda is signed by countries worldwide committed to climate change mitigation and limiting the increase of global average temperatures.

Architecture needs to enable its roles to mitigate climate change, lower urban heat islands, reduce carbon emissions, and preserve biodiversity and favourable human habitation. Humans as users and actors, nature as a context and resources, and technology as means of realising objectives must be synchronised and synergised in creating a sustainable built environment. All elements and stakeholders should play their roles by strengthening partnerships with long-standing collaboration and cooperation. Having all in the same plate will ensure the health and safety of the user and the environment.

The conference theme is "Sustainable Built Environment: Bridging the role of Human, Nature, and Technology." Issues like sustainable cities, sustainable building and settlement, sustainable building and technology, theory of sustainability and historiography of architecture, and sustainable architectural design and practice are among essential topics related to the role of humans, nature, and technology. The conference acts as a reminder to all of us to contribute our knowledge and expertise, bringing humans, nature, and technology development to a better world.

International Conference on Architecture Research and Design (AR+DC) is an annual international event organised by the Department of Architecture, Faculty of Civil Engineering, Planning and Geo-Engineering, Institut Teknologi Sepuluh Nopember (ITS). This event was first initiated in 2009 bearing different names and, in 2014 has been constantly called AR+DC, inviting academics, researchers, and professionals from a wide range of expertise and fields. For years, AR+DC's topics have been focused on architecture and its related subjects, such as Architecture design, Architectural theory, history, and criticism, Environmental architecture, Housing and settlement, and Urban design.

Whereas SENVAR (Annual International Conference in Sustainable Environment Architecture) is an annual gathering for scientists, scholars, and architects that focuses on nature and built environment issues like urban planning, building and architecture, outdoor and indoor comfort, local and global green, and sustainability. SENVAR is a yearly event hosted voluntarily by various universities in Indonesia and outside Indonesia, based on their availability and updated issues. SENVAR was first initiated and started by the late Professor Mas Santosa from the Department of Architecture, Institut Teknologi Sepuluh Nopember (ITS) in 2000 as a seminar on environmental architecture; in 2011, it changed the name to Sustainable Environment and Architecture. We have the privilege to conduct the event every ten years (decennary). This year is a special moment for all of the members of the SENVAR community because it is the third time SENVAR international meeting has been held again in ITS.

Along with that privilege, in 2021, we also celebrate the 55th lustrum anniversary of the Department of Architecture ITS. Therefore, we initiate to merge AR+DC and SENVAR to define a broader topic in an integrated understanding of sustainability in the context of the built environment.

Due to the pandemic of COVID 19 situation, AR+DC SENVAR 2021 is conducted online on November 2-3, 2021, through the Zoom Meeting platform. We promote the event through our official website (https://www.events-arch-its.org/ardcsenvar2021) that more than 33 countries have accessed. The conference has successfully gathered 222 participants from 10 countries: Indonesia, Japan, Malaysia, Netherland, Australia, New Zealand, Jordan, China, Madagascar, and Slovakia. The conference has invited 7 Keynote Speakers from Indonesia, China, Japan, Netherland, and New Zealand. We accepted 37 papers that have been blinded double-review and are presented in 2 days of the presentation session. Each day consisted of 4 breakout rooms divided based on the conference themes. Each presenter had 10 minutes to present their papers then continued with QnA or discussion session. We also provided

IOP Conf. Series: Earth and Environmental Science 1007 (2022) 011001 doi:10.1088/1755-1315/1007/1/011001

the pre-recorded video sent by the author on the previous day in case of technical difficulties.

We would like to express the highest appreciation to all organisers for their hard work in preparing, managing, and organising this successful event. We also thank authors, reviewers, and participants who have contributed to the success of AR+DC SENVAR 2021.

Organising Committee

Advisory Board

Dr. Ir. Murni Rachmawati M.T.	Dean of Faculty of Civil, Planningand Geo-Engineering, Institut Teknologi Sepuluh Nopember, Indonesia
Dewi Septanti, S.Pd., S.T., M.T., Ph.D	Head of Department of Architecture, Institut Teknologi Sepuluh Nopember, Indonesia
Dr. Eka Sediadi	American Univ of Ras Al Khaimah, United Arab Emirate
Prof. Prasasto Satwiko	Universitas Atmajaya Yogjakarta, Indonesia
Prof. Dr. Mohd Hamdan	University Technology Malaysia
Dr. Eng. Ir. Sri Nastiti Nugrahani Ekasiwi, M.T.	Institut Teknologi Sepuluh Nopember, Indonesia
Prof. Happy Ratna Sumartinah, Ph.D	Institut Teknologi Sepuluh Nopember, Indonesia
Prof. Dr.Ir. V.Totok Noerwasito, M.T.	Institut Teknologi Sepuluh Nopember, Indonesia

Scientific Committee

Dr. Eka Sediadi	American Univ of Ras Al Khaimah, United Arab Emirate
Dr. Tetsu Kubota	Hiroshima University, Japan
Prof. Dr. Mohd Hamdan	University Technology Malaysia

Prof. Yandi Andri Yatmo, S.T., Dip.Arch., M.Arch., Ph.D	Universitas Indonesia
Prof. Paramita Atmodiwirjo, S.T., M.Arch., Ph.D	Universitas Indonesia
Dr. Eng. Beta Paramita, S.T., M.T.	Universitas Pendidikan Indonesia
Dr. Eng. Donny Kurniawan	Institut Teknologi Bandung, Indonesia
Prof. Ir. Prasasto Satwiko, M.B.Sc., Ph.D	Universitas Atmajaya Yogjakarta, Indonesia
Dr. Ir. Arif Kusumawanto, M.T., IAI., IPU	Universitas Gajahmada, Indonesia
Prof. Dr. Ir. Sangkertadi, DEA	Universitas Sam Ratulangi, Indonesia
Prof. Dr. Ir. Erni Setyowati, M.T.	Universitas Diponegoro, Indonesia
Dr. Ing. Ir. Bambang Soemardiono	Institut Teknologi Sepuluh Nopember, Indonesia
Dr. Ir. Asri Dinapradipta,M.B.Env	Institut Teknologi Sepuluh Nopember, Indonesia
Ir. I Gusti Ngurah Antaryama, Ph.D	Institut Teknologi Sepuluh Nopember, Indonesia
Dr. Ima Defiana, ST. M.T.	Institut Teknologi Sepuluh Nopember, Indonesia
FX Teddy Badai Samodra, S.T., M.T., Ph.D	Institut Teknologi Sepuluh Nopember, Indonesia
Sarah Cahyadini, ST., M.T., Ph.D	Institut Teknologi Sepuluh Nopember, Indonesia
Dr. Eng. Didit Novianto S.T. M.Eng.	Institut Teknologi Sepuluh Nopember, Indonesia

Organizing Committee

Collinthia Erwindi, S.T., M.T.	Conference chair
Adinda Sih P. R. U., S.T., M.T.	Secretary
Angger Sukma Mahendra, ST., M.T.	Bursar
Wahyu Setyawan, S.T., M.T.	Members
Iwan Adi Indrawan, S.T., M.Ars.	
Fenty Ratna Indarti, S.T., M.Arch	
Nurfahmi Muchlis, S.T., M.Ars.	
Defry Agatha A. S.T., M.T.	
Endy Yudho P. S.T., M.T.	
Irvansyah, S.T., M.T.	

IOP Conf. Series: Earth and Environmental Science 1007 (2022) 011002 doi:10.1088/1755-1315/1007/1/011002

Peer Review Statement

All papers published in this volume have been reviewed through processes administered by the Editors. Reviews were conducted by expert referees to the professional and scientific standards expected of a proceedings journal published by IOP Publishing.

1. Type of peer review: Double Anonymous
2. Conference submission management system: Morressier
3. Number of submissions received: 42
4. Number of submissions sent for review: 36
5. Number of submissions accepted: 25
6. Acceptance Rate (Submissions Accepted / Submissions Received × 100): 69,4 %
7. Average number of reviews per paper: 2.00
8. Total number of reviewers involved: 19
9. Contact person for queries:
   Name: Arina Hayati
   Affiliation: Department of Architecture, Institut Teknologi Sepuluh Nopember (ITS)
   Email: arina_h@arch.its.ac.id

* means value has been edited

3D scanning as an architectural tool

R Potangaroa[1, 2]

[1]School of Architecture, Victoria University of Wellington, New Zealand

Abstract. Laser scanners are gaining acceptance as a tool for three-dimensional modelling of existing buildings, but not much more than that. The idea that a digital model constructed from hundreds of thousands of measured laser points having 'soulful' applications remains foreign to Architects. This paper presents the work that has been ongoing for over 5 years at the School of Architecture at Victoria University of Wellington in New Zealand and the 'soulful' experiences we have encountered in that work.

1. Why talk about 3D scanning as an architectural tool?

The first reason to discuss this issue is that it is not recognised as a potential tool, other than making a digital record of it [1]. While on the other hand, equipment such as the iPad Pro 12[2] and more recently the iPhone 13[3] is making access to the hardware more accessible. But the second reason is that 3D scanning as an architectural tool opens up research opportunities that have not been readily available. And that lack of prior access is now preventing future adoption and the realisation of future opportunities. Finally, it does allow more operational based research that deals with the 'here and now' of spatial design.

2. The 3D scanning process

2.1. Hardware

The scanner is the core piece of equipment for this work. In the first instance we used a Matterport scanner [4] and more recently changing over to a Leica BLK360 [5] scanner. The increased hosting costs and the inability to use the scans outside of Matterport made it unworkable in the medium to long term for research. And while this has pushed us into using the more open-sourced Leica machine the issues of portability of files still remains.

2.2. Software

Part of the issue has been (and remains) is the proprietary nature of the associated software. MatterPort for example, is online and scans are uploaded, and the stitched model is normally available next day. Such a model is ideally suited for estate and property sales but is not so useful for research. The models cannot be used outside of the Matterport site and where they can, models are deliberated downgraded. Nonetheless, the turnaround service is appreciated though now costs for hosting models has made it an expensive option for research-based organisations. On the other hand, the Leica model while producing interchangeable file types (such as e57 and PTX) generates huge file sizes. Several of the PTX files for the case studies below are 37GB.

And while the software is available to handle these files (slowly) there are associated subscription costs and complications in moving files from the scanner to your computer and finally via UNITY [6]

[2] Corresponding author: regan.potangaroa@vuw.ac.nz

IOP Conf. Series: Earth and Environmental Science 1007 (2022) 012001 doi:10.1088/1755-1315/1007/1/012001

or UNREAL [7] to create the virtual reality environment that is the platform for the research work we do.

However, the meshing step; where we take the point clouds generated by the scanner and produce meshes as an 'intermediate' step before UNITY/UNREAL installation, has proved to be the fickle part of this process. We have been able to get 'workable' (but not great) meshed models from RECAP PRO [8], VRmesh [9] and CloudCompare [10]. There has been a long list of software that we have tested, but the gap between what the advertising says the software can do and what you can get it to do remains. And the goal of a photo realistic model (other than via MatterPort) remains elusive.

3. The 3D scanning process

The perspective of the paper is from a practitioner rather than an academic and while there is not sufficient material to statistical draw conclusions, there is evidence to move forward, as a practitioner would. Hence, this paper centres around 4 case studies as outlined below:

Case Study 1 : Heritage-Old Wooden Government Building
Case Study 2 : Retail- 'Good as Gold'
Case Study 3a : Housing-Refugee
Case Study 3b : Housing- Maori
Case study 4 : Office-CMIC office

Case studies link theory and practice according to Yin (2014) [11]. His stated approach is as a "linear but iterative process" which these case studies have built up over the 5+ years of work by different researchers and in different contexts. The 3D scanning process being the common thread.

4. Case study

4.1. Case study 1: Heritage-Old Wooden Government Building OWGB [12]

The first case study was work on a large 4 storey timber heritage building in Wellington's CBD (across the road from the NZ Parliament, the Beehive). It was built in 1876, is a remarkable example of timber construction and one of New Zealand's most important historic buildings. It was designed by New Zealand's first and only Colonial Architect, William Clayton, in the classically derived Italian Renaissance revival style, much in favour for government buildings throughout the British Empire. Despite shortages in timber, and considerable cost overruns, the building was finished in 22 months, a remarkable achievement at the time. When it opened, the building was the largest in the country and housed all Wellington-based civil servants and ministers of the Crown. The building included two staircases, eight vaults, 143 rooms, 126 fireplaces, 22 chimneys, two hydraulic lifts, 64 toilets, eight verandah and seven porticos. It is currently used by the Law Faculty of Victoria University but owned and maintained by Heritage New Zealand HNZ.

Figure 1. The OWGB c1880
(https://www.heritage.org.nz/the-list/details/37)

In 2019, HNZ needed to decide as to whether the roof should be replaced or not and the School of Architecture took on the project as part of their Summer Scholar Program. And the parts of that work relevant to this paper were the 3D scanning of the interior space of the building plus the external 3D point cloud scanning using drones. The idea being to bolt the two data sets together.

What did we find out from that work?

We found that 3D scanning is an excellent way to record and digitise spaces, facades and roofs (as expected). The interior roof scans when cut horizontally (just above the top chord of the roof trusses) gave HNZ their first accurate plans of the roof structure. That was not expected; but useful. As was the experience of a fly through the roof space. But the joining of the drone 3D data and the Leica BLK360 (ground based) scanner was not successful. The plan was to join the data sets through common points at the two roof access points. In the future this would require the location of control points at both access locations plus 5 of the chimneys to get a tie in of the two data sets. In a similar way the scans of the front facades of the OWGB would also need control points across the face and at various levels to pin scan data against. Nonetheless, the outside mapping of the OWGB from the drone point clouds using PIX4D [13] was successful.

Hence a level of accuracy which would have been difficult/impossible any other way was achieved, but more work in terms of control points is required.

Figure 2. External BLK 360 scans of the OWGB

Figure 3. External drone scanning of the OWGB

IOP Conf. Series: Earth and Environmental Science 1007 (2022) 012001 doi:10.1088/1755-1315/1007/1/012001

4.2. Case study 2: retail- 'Good as Gold' [14]
What we had shown thus far was what was expected, namely that 3D scanning was an excellent way to record existing buildings. But how or can we make it 'soulful'? And what would soulful mean and do?

Retail design has always been much more than selling products. It is a way for the brand to express themselves, in what they stand for through their brand by selling through their physical space. However, it is much more. It does allow consumers to connect with the brand but also creates an experience for the consumer that sets the brand apart from their competitors. The interior architecture bridges the gap between the brand and consumers, envisaging the brand in an architectural form or narrative. However, the birth of e-commerce has created a virtual shop, a non-physical space and this relationship is being scrutinised.

This case study explored how this virtual shop and its associated virtual reality (VR) can be incorporated for the e-consumer retail market and used to enhance the experience. This was expected to involve integrating the retail and brand into a virtual environment; but adding a new layer of interaction from the traditional 2D web interface and an experience that would make it more immersive and engaging for the consumer.

The way we interact with an online platform is limited, often requiring scrolling through website pages of data. A 3D scan (using a Matterport scanner) of the shop was inputted into UNITY and 15 participants experienced both the physical shop and its Virtual 3D scanned version. The results were recorded in the form of questionnaires which were then analysed to give an insight into the differences and the potential improvements.

We approached shops that we felt had 'brand' nearby to the School of Architecture. The school is in the Cuba Precinct of wellington which is seen as an alternative and 'creative' area in the city and found the Good as Gold Shop [15]. The shop was scanned using a Matterport 3D scanner, a virtual model of the shop was developed, and 15 participants were then taken through both the shop and the model and then debriefed/interviewed.

When asked to describe the atmosphere in the actual shop participants used adjectives such as warm (5 participants) and relaxed (4). They noted the natural aesthetic of the shop, the 'cabin' feels, its spaciousness, high ceilings and squeaky wooden floors in the midst of a busy urban context. They noted the lighting, the touch of the wood and the sound of the background music. The clothes were displayed against a strong timber aesthetic but one that spoke to them about being urban (5), relaxed/casual (6) and being 'cabin' (4) or getting away from it. Hence the actual store was strategic and strongly focused.

The 3D experience was done in the VR lab area of the school. Participants suggested that the Virtual experience was lighter/brighter (7), faster/visually stimulating (3) but lacked sound (4) and perhaps surprisingly other people (2). It had good light and colour (3) but lacked tactile/sound (4) and thermal (2). Architectural details such as the stairs (2), the cupboards (2) and the timber aesthetic of the shop (2) were enhanced but unfortunately could not be touched in the Virtual environment

Interestingly, the VR experience did not shift the perception of the Good as Gold Shop; and participants commented that that the VR experience was urban (4), relaxed/casual (3) and being 'cabin' (4), which are similar numbers to the actual shop.

Figure 4. Actual shop and 3D scanned model side-by side

However, at least 2 commented that the Virtual was not as clear or had lost the actual shop brand. Nonetheless, 8 participants would have bought from the Virtual shop, 3 would not and the remainder were in between. Moreover, 7 confirmed they could/would shop online in a similar environment with the remainder saying they conditional could/would. With the three characteristics of touch (4), sound (7) and thermal/warmth (2) being most notably absent from the Virtual experience. Furthermore, the study did support the 'influential nature of vividness' and the immersive nature of the Virtual despite participants being aware of the context they were experiencing the shop.

The Virtual Retail Strategy that was emerging was that the Virtual site needs to reflect the strategy of the Good as Gold shop, that reflection should/could include vivid colours and light (the desire for music, touch and other people would be held back because of computer coding issues), the materiality of elements should be explicit, and the use of architectural features (like the stair) should be developed in the online site.

Up to this point the focus of the research study followed what was in the literature that the shopping experience was the driver for the Virtual online shop. But what was actually happening from the interviews and the enthusiasm/insistence of the shop owner with the 'idea' of a Virtual Shop that it was now the driver for the shopping experience. We had done a 180-degree turn. One that was not reflected in the literature. The goal from the literature was how can the Virtual Reality 'mimic' reality rather than how it can be used as a design tool.

Consequently, the Virtual shop design took on a different complexion and began to drive how the shopping experience in ways that while aligning with the previous strategy, extended it in ways that were not otherwise possible. The 3D scanning had become an architectural tool. And the full design impact of the Virtual shop can be seen in the published thesis.

4.3. Case study 3: housing [16]

Home is a universal experience of our human condition. Yet for marginalised groups understanding what makes 'home' is difficult. This case study set out to propose an innovative methodology, using immersive virtual 3D scanning processes, to facilitate an understanding of the meaning of home for both designers and marginalised groups. Cooper (1974) [17] notes "a violation of the self (house) is perhaps one of man's most deep-seated and universal fears" highlighting the importance we place on the house as an extension of ourselves.

However, New Zealand's architectural vernacular of housing is changing. Driven by a growing population and increasing unaffordability, the demand for medium and high-density mass housing has pushed the traditional single unit family house beyond young families. Thus, a disconnection exists between architectural designers and end-users that affects the ability for housing to meet the needs of inhabitants.

The impact is more so for marginalised communities. Hence, a refugee family from Myanmar living in a 3-storey housing block and a Maori family from Ngati Kahungunu ki Wairarapa [18] living in their own single storey timber house (both living in Wellington) were selected as case studies.

4.3.1. Case study 3a: housing-refugee. The initial key reaction when entering a 3D scan of their home for both the Refugee and Maori families was shocked laughter. For example, the refugee mother noted that her house was messy while the father noted the broken balustrade on the stair and the front door handle. Both were details they were living and engaging with daily, but only noticed in the 3D scanned environment.

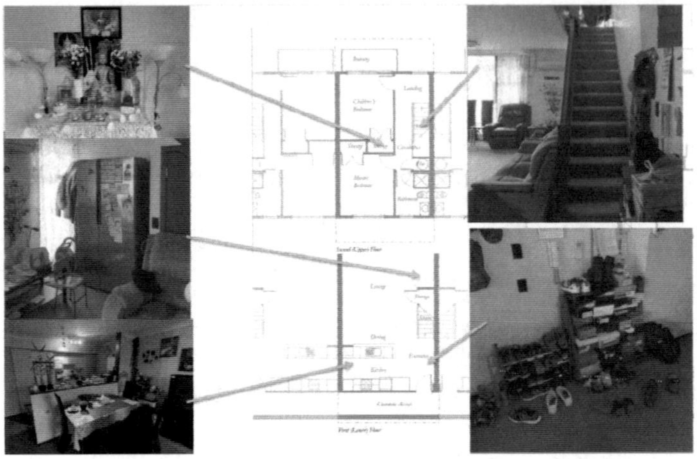

Figure 5. The refugee house unit plan and photographs

The responses from the refugee family (mum and dad) besides the laughter when in the virtual house was:

- All were concerned and had intense interest in the details around the family shrine and it was noted to make this a key area for their 'home'
- A sense of safety in the communal corridor
- Both felt their house was small but safe
- But that the virtual house was 'messy'

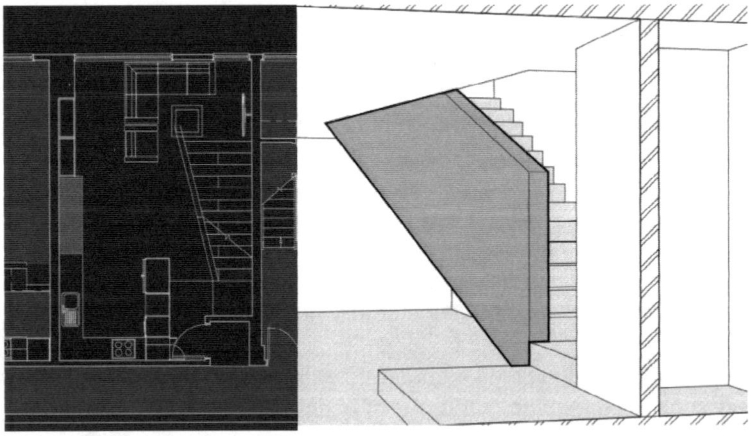

Figure 6. The proposed staircase alteration

The key design aspects given that their two-storey apartment was part of a block that emerged from the above observations, the later interviews, and our observations of how the family lived zeroed in on the staircase, the front entry, storage space and display areas, the link and flows between the kitchen and lounge, the dining area and finally the bedroom spaces upstairs.

The central aspect was the staircase, and how to use and create a place that one could occupy in addition to access the upper level; thus, creating additional space by optimising the space that was already there. The final plan for it is included above and incorporates a closed rather than an open balustrading, widening the top of it, included display and seating along it so the children of the house and those that visit could occupy the stair and be in contact with what was happening with the adults in the kitchen/lounge area while at the same time be separate from them. The display space and the storage underneath addressed the 'messy' evident in the Virtual.

4.3.2. Case study 3b: housing - Maori [16]. The family, six permanent occupants spanning three generations, live at the house. It consists of two teenagers, and four adults. One adult is retired. The household has strong community engagement and family network evidence by the large whanau wall of photographs in the house. There is a steady flow of guests and family friends plus a strong Maori cultural connection.

There was large variation in the time participants spent in the virtual house. Occupant Three spent more than thirty minutes in the experience while Occupants Five and Six were immersed for less than ten. However, Occupant Five suggested this was because she was unwell at the time. But there is a connection between time in the Virtual and data quantity. Nonetheless, the level of detail provided from observational commentary was much higher than expected. Therefore, as opposed to individual interviews the debriefing was done as a family unit as everyone was already talking about the Virtual and what they had found.

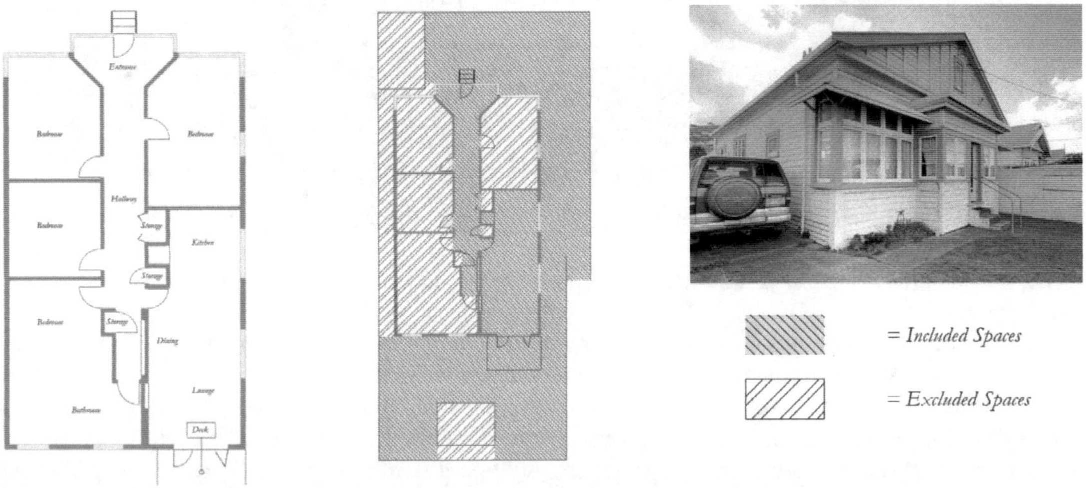

Figure 7. House plans and scanned areas in the Virtual House. (Note that only family parts of the house were used for the Virtual)

The occupants have extensive/personal knowledge of the spatial details of their home. Hence, any difference in geometric and textural would be noticed. And in this case study we used two different head sets a HTC Vive system and a Samsung Gear VR (which was compatible with the Matterport scanner used for this house). Occupant Three quickly identified that the HTC virtual house lacks overall realism. However, he suggests that because the model was 'not quite right', recognisable parts became important as a means for contextualising himself in the environment. Thus, many details of the house stood out to him. The participant also appeared to be considerably more investigative when experiencing the virtual house using the HTC Vive. This is observed as acts of crouching to examine objects and moving closer to different areas and objects within the virtual house. Our impression was that photo realistic scans should be the standard to aim for but that all is not lost if there are some areas of missed scans [16].

This was in part because we had scanned the outside of this house to capture the inside/outside nature of the house as we wanted to gauge the impact that would have in the Virtual. No one went outside and we are rethinking this approach moving forward.

Figure 8. Scan from within the house (Note the Virtual House was in colour)

The results are interesting; the grandmother was attached to her bedroom. Home was specific for her. While the son-in-law noted the timber floors, doors, and a wrinkle in the couch. These were parts of that house that he and his father-in-law built. The comment that 'home is where the whanau (family) is' summed up this situation. Family photos were (strangely) not a point of engagement or presence. Not mentioned by anyone? And the level of engagement at photo-realistic texture is right for 'presence' and spatially correct geometries means that participants do not have to 'work' to understand the environment. For example, spaces behind chairs. This does however require more scanning points and more work to create the Virtual.

We were sceptical at the start of this case study as to what we would be able to ascertain from the Virtual model that we would not be able to find through talking to the occupants and observing both houses. Could we uncover and explore new ideas and emotions for the user? Interestingly, that was the view of the 3 Master's thesis reviewers of this work when it was originally suggested, that there would be nothing to be found in a digital copy that could not be found using 'traditional' methods of inquiry. That was not the case. And watching this family rediscover their house and home in front of us was the evidence. For example, the idea that there was not the connection to the photographs on the walls of the home would have been impossible to discussion or discern by traditional methods. It still requires more validation but would have been an automatic conclusion going into the house/home as can be seen in the earlier scans of the house/home. What is clear so far is that a 3D scan and virtual representation of a space is a digital copy of that space as shown in case study 1 but is much more than that alone.

4.4. Case study 4: office-CMIC office [19]

The Architect's role in distinguishing the spatial requirements of an office have traditionally been that of a non-human centred design approach. Staff turnover are a feature of office life, one which the office culture sets out to instead establish company loyalty and organisational sustainability. Yet in attempting to do that, companies lack the knowledge and understanding of their own workplace, resulting in a disconnected understanding of what spatial qualities benefit employee well-being.

Figure 9. The CMIC office 3D scanned model

Hence, the research question for this case study was 'What is the potential immersive virtual reality technologies have when designing to improve the spatial quality and well-being for the end users within New Zealand offices: Using 3D scanning and virtual reality as an applied tool during the design process'.

The chosen office for the case study, Victoria Universities Computational Media Innovation Centre (CMIC), provided an opportunity to assess a methodology that was influenced and based upon previous successful case studies performed through Victoria University of Wellington.

It sought to extract connections between users and space at both an emotional and technical level. As such, the resulting data informed the design proposals, creating solutions that are cultured by the emotional connections extracted between user and office, as well as the spatial qualities needed to promote wellbeing.

The open plan model office and variations of this layout are believed to be the most dominant office model. However, it is debated as to whether it is an adequate office environment for contemporary work with Covid-19 influencing the majority of the globe into forced telework; show casing the possibility for a higher demand of agile workspaces / flexible workspaces. This was an issue for the CMIC office

Table 1. Participant responses in the 3D scanned virtual office

Researchers notes	Participants comments	Observations
Participant 1 Began VR within the office area	1. Initially began giggling 2. Noticed the rendering quality and the buffering of the visualization and asked, "so what's this?"	1. Begins to move around immediately 2. Participant seemed disappointed with the quality of the rendering and the speed of the buffering within VR
Participant 2 Began VR within the office area	1. "My desk is a mess" 2. "Surprising Quality" 3. "Control Method is much more intuitive than I anticipated it to be"	1. Participant was very calm and seemed used to the control of movement within VR 2. Participant immediately went to their own desk and was interested in the items on their desk
Participant 3 Began VR within the office area	1. "Once I'm close to objects I can see individual points" 2. "I think the detail is general good" 3. Colors seem a bit strange...Very purple"	1. Participant was very observant of the point cloud information and was critical of the visual representation 2. Was observant of the walls and writings captured on the white boards
Participant 4	1. "Much better than what we would normally have in the main campus"	1. Participant seemed confused and began talking about the previous office design that they had worked in

Began VR within the office area	2. "Detail is generally good" 3. Noticed the hue of the colours were more saturated	2. Participant spent a lot of time within the office floor area and not much time exploring the other areas
Participant 5 Began VR outside of the model	1. Whoa... It's moving very fast" 2. "It's pretty cool, you recognize the space" 3. Noticed that the scale felt a bit strange	1. Immediately went to their desk and spent a significant amount of time looking at the desk and surrounding desks 2. Was slightly thrown off from the controls and moving within VR fast 3. Noticed trivial items such as a "party hat" left on the desk within the meeting room
Participant 6 Began VR within the office area	1. "It's really nice"	1. Was very interested in the point cloud data was looking around the space a lot 2. Was drawn to the surfaces of objects and the floor
Participant 7 Began VR within the office area	1. "Very interesting" 2. "Pretty realistic" 3. "Scan is very nice, dense I can see a lot of detail"	1. Was drawn to pathways that they would not normally take when walking through the office and moved through different areas suggesting "I'll take this way because I can" 2. Spent a lot of time observing the floor
Participant 8 Began VR within the office area	1. "Taken during covid times, so very interesting" - In terms of the scan 2. Artefacts really stood out "the small subtle things"	1. Noticed a lot of details in the kitchen 2. Drawn to the details of the artefacts

IOP Conf. Series: Earth and Environmental Science 1007 (2022) 012001 doi:10.1088/1755-1315/1007/1/012001

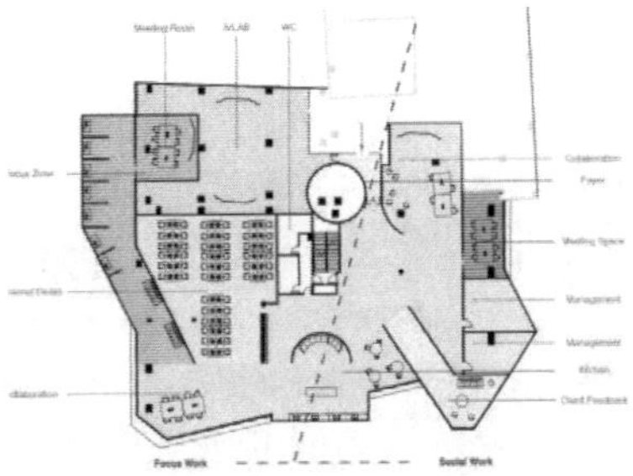

Perspective of Expanded Office Plan

Expanded Office Plan

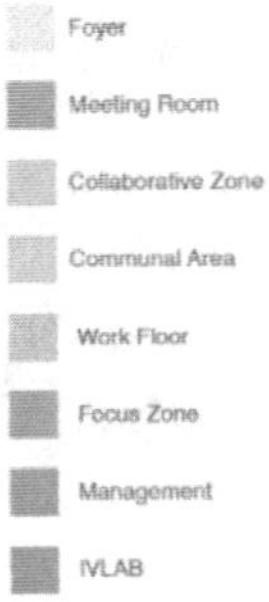

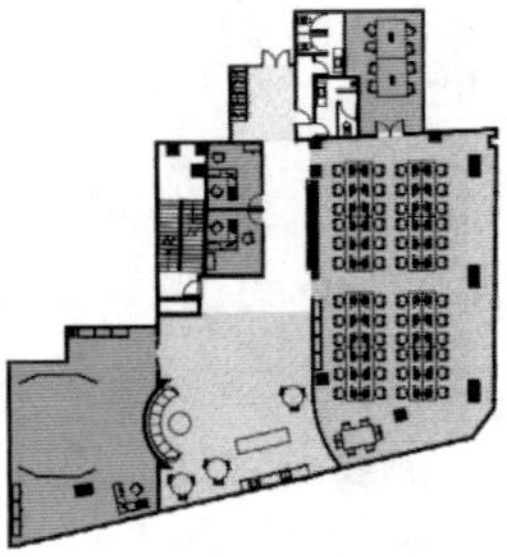

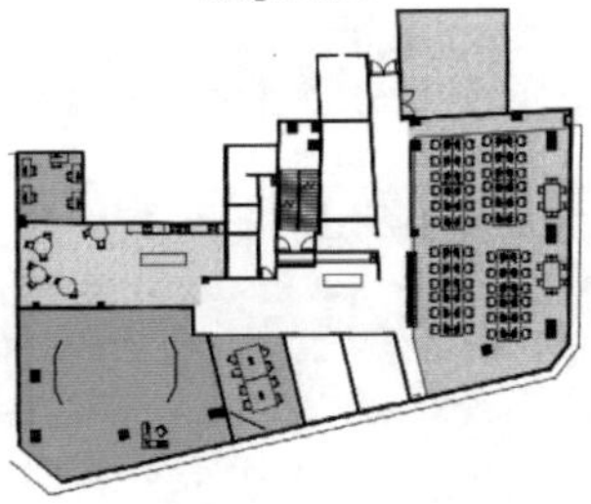

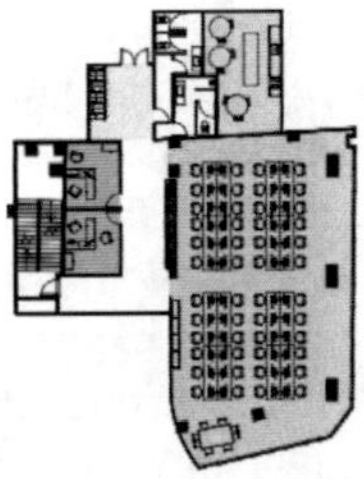

Existing Office Plan

Two Reduced Office Plans

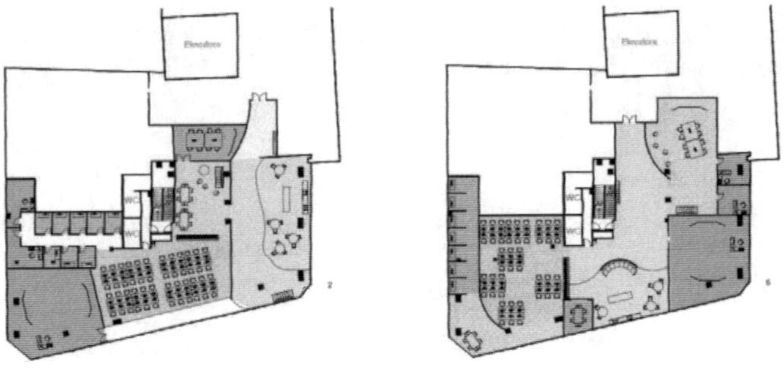

Same Floor Area: Two New Layout Plans

Figure 10. Proposed plans developed from the 3D scanned data of the existing office

Moving forward, the synthesize data has resulted in a focus being placed on the worker's personal desks and the ability to shift this system from a social working zone to a focused working zone. The connectivity within the office will also be a focus, with the need to reduce connectivity through uncomfortable and non-stimulating areas and promote pathways between comfortable and more stimulating areas. An opportunity for added programs to better fit the working style of the end users was apparent, plus a need for greater control of privacy. The interesting aspect of details, lighting, and colours to create a more comprehensive interior to strengthen the new architectural design.

Ultimately the results found from the virtual participant study alongside the information gathered were successful and principal drivers proceeding into design. This show cases the initial potential, 3D immersive technologies have, when used as an applied tool during the design process.

5. Findings [20]

As indicated at the start, this paper sets out to address the question of whether 3D scanning could be an architectural tool from a practitioner's perspective using a case study approach. The sense from that work is that 3D scanning is and can be an architectural tool for documenting and recording buildings, such as case study 1 of a heritage example. It does that well. But we wanted to go further and explore whether it could be soulful as described by Chris Milk; could it be the ultimate empathy machine? That is much harder and is where the profession stands back. Yet in an 'arcade' and gaming sense one has to wonder why? Reference 22 above is the well-known roller coaster experience, and it is difficult not to move and respond to those rollers coasters by just watching your screen. How much more for an immersive headset and an environment that you know well. Or at least thought you did. The shocked/laughter response from those engaging with a 3D scanned version of their home presented in a virtual form, was telling. Despite living in the house there were details and aspects and parts of it that were suddenly new, in spite of its familiarity. In the same but different way for the office in case study 4. We picked up patterns that would not have been identified by other tradition interviews and engagement used by designers. Even videotaping would not have readily identified the characteristics that then enabled 3 different office design options to have been developed. But case study 2 was for us the most significant because it was when we as a team realised that 3D scanning was much more than a digital copy. I remember one participant when in the virtual shop tapping the concrete floor of the school of architecture because the Good as Gold shop has a timber floor. It seems that the virtual timber may have won out on the actual concrete floor in this case.

Photo realistic is the goal for the 3D scanned virtual environment, and preferably doing the virtual on location, so that participants first don't have the above 'floor' difference issue; but more importantly in getting to the site experience the usual 'pre-experience' experience of occupying the space of concern.

There are many factors at play in this study and the paper does not set out to solve all of them scientifically but uses a designer's eye that practitioners are accustomed to work with.

6. Conclusion

The conclusion from this body of work using a case study approach is that 3D scanning can is a powerful architectural tool. There are caveats to ensure its best use which are touched on in this paper. But it opens up other research possibilities in addition to design, that cannot be realised within the existing toolbox.

7. References

[1] Chris Milk, TED 2015. How virtual reality can create the ultimate empathy machine. He uses cutting edge technology to produce astonishing films that delight and enchant. But for Milk, the human story is the driving force behind everything he does. https://www.ted.com/talks/chris_milk_how_virtual_reality_can_create_the_ultimate_empathy_machine?language=en

[2] Buy iPad Pro Free delivery. https://www.apple.com/nz/shop/buy-ipad/ipad-pro

[3] iPhone 13 Pro Oh. So. Pro. https://www.apple.com/nz/iphone-13-pro/

[4] Meet the world's leading spatial data platform. https://matterport.com/

[5] Leica BLK360 Imaging Laser Scanner The smallest and lightest imaging laser scanner available https://leica-geosystems.com/products/laser-scanners/scanners/blk360

[6] More than an engine. Create and grow more with Unity. Unity is the leading video game engine – and so much more. Discover solutions to help you at every stage of the game development lifecycle, from big idea to big success. https://unity.com/

[7] The world's most open and advanced real-time 3D creation tool https://www.unrealengine.com/en-US/
[8] Autodesk® ReCap™ Pro – 3D Scanning Software for Intelligent Model Creation https://www.youtube.com/watch?v=Oo70OwoZLa0
[9] VRmesh: Point cloud & mesh processing software. https://www.vrmesh.com/
[10] Welcome to the official website of the Cloud Compare project. https://www.cloudcompare.org/main.html
[11] Yin R 2014 Case study research design and methods (5th ed.) (Thousand Oaks, CA: Sage). 282 pages. ISBN 978-1-4522-4256-9
[12] Ermenyi T 2020 Using Laser Scanners & Drones to Document Heritage Buildings: Old Government Building 55 Lambton Quay, Wellington, NZ Summer Scholar Research Report Victoria University of Wellington
[13] Measure from images A unique photogrammetry software suite for mobile and drone mapping https://www.pix4d.com/
[14] Misty P 2017 Virtual Retail Master's Thesis Victoria University of Wellington
[15] Good as Gold; Shop online https://goodasgoldshop.com/
[16] Musson M 2019 Virtually Home; An innovative methodology for identifying the housing needs of marginalised groups Master Thesis Victoria University of Wellington
[17] Cooper C. 1974 The House as Symbol of Self. In Lang, J., & Burnett, C., & Moleski, W., & Vachon, D. (Eds), Designing for Human Behaviour: Architecture and the Behavioral Sciences (pp. 120-146). Stroudsburg, PA: Dowden, Hutchinson & Ross
[18] Ngāti Kahungunu from Wikipedia, the free encyclopedia. https://en.wikipedia.org/wiki/Ng%C4%81ti_Kahungunu
[19] Han R 2020 Occupation Immersive: An Innovative methodology to improve the spatial quality, well-being and understanding of contemporary office spaces and their users Master Thesis Victoria University of Wellington
[20] VR 360 Video of Top 5 Roller Coaster Rides 4K Virtual Reality https://www.youtube.com/watch?v=hNAbQYU0wpg

IOP Conf. Series: Earth and Environmental Science 1007 (2022) 012002 doi:10.1088/1755-1315/1007/1/012002

Sawdust and soil block as wall in simple buildings to gain the optimum heat and embodied energy in humid tropical area

V T Noerwasito[1,2]

[1] Architecture Department, Faculty of Civil, Planning, and Geo-Engineering, Institut Teknologi Sepuluh Nopember. Kampus ITS, Sukolilo, Surabaya 60111, Indonesia

Abstract. Red bricks are widely used in Indonesia. The process of making red brick itself causes many problems such as the combustion process and the creation of air pollutant during the production process. On the other hand, waste from timber industry in Indonesia, which is called sawdust, becomes a problem that need to be solved by the woodcraft industry. Therefore, this study aims to substitute the use of red brick as wall material as well as reducing the waste of woodcraft industry by mixing sawdust waste with clay and cement as an alternative. The purpose of this study is investigating the performance of sawdust as wall material in terms of heat transfer and embodied energy in buildings compared to red brick. This study aims to analyse the optimum composition of sawdust as wall material to gain a balanced heat and embodied energy in simple housing. The study investigates sawdust wall in building with the size of 36m2 by using simulation software called Archipak. The embodied energy calculation uses standard embodied energy per unit. The research variable is the thickness of sawdust walls. The result of this study is the thickness of sawdust wall with minimum heat transfer and embodied energy.

1. Introduction

Most wood factories in Probolinggo produce a lot of waste in the form of sawdust. It is really difficult for them to dispose the waste. Recent days, they have not found solutions for the waste problem. The only thing they do to reduce the waste is burning the waste that becomes air pollutant. The burning process degrades the environment condition and reduces the quality of life.

Potentially, wood factory's waste is able to become basic material to make brick for simple house. To make the brick, the sawdust is mixed with soil to make it more resistant toward water. Unlike traditional red brick, the sawdust brick does not need combustion process during the production, but it needs to be compressed and dried in the outdoors. In addition, cement is added to the mixture to make it more solid.

Brick made from sawdust and soil does not create pollutant during the production. Moreover, it helps to reduce the amount of waste from wood factory, which is thrown away to the surrounding area of the factory. There are two kinds of important energy in simple house, such as operational energy and embodied energy. Operational energy is energy that relates to the heat inside the building. It consists of cooling 1energy and heating energy. In humid tropical areas such as Indonesia, the cooling energy is majorly needed. Cooling energy is needed to overcome or avoid overheating in the building. On the other hand, embodied energy is the energy consumed (or needed) during the production process.

[2]Corresponding author: totoknoerwasito@yahoo.com

In the building, cooling energy and embodied energy should be in an efficient way. Wall of the building influences the efficiency of those energy. The thickness of sawdust wall affects the value of the cooling energy and embodied energy in the building. This research investigated the efficient condition of cooling energy and embodied energy of the building. The investigation studies a simple house with the area of 36m2 since this house is very popular domestic building in Indonesia.

The research question of this study was how thick the sawdust wall was to achieve the optimum efficiency of cooling energy and embodied energy. This study took several projects in Surabaya as case study. The benefit of this research is providing wall material alternative coming from industrial waste that relates to local material.

2. Research methods

In the building model, the sawdust wall had varied thickness depends on the building function. In tropical areas the mitigating the heat becomes the major concern as well as the use of material energy. The energy that relates to the heat is cooling energy. Meanwhile, the energy that relates to the production of the building material is called embodied energy.

Table 1. Research variables

No	Building condition	Code
1	Wall sawdust thickness 10 cm	Saw 10
2	Wall sawdust thickness 15 cm	Saw 15
3	Wall sawdust thickness 20 cm	Saw 20
4	Wall sawdust thickness 10 cm with panel	Sawpan 10
5	Wall sawdust thickness 15 cm with panel	Sawpan 15
6	Wall sawdust thickness 20 cm panel	Sawpan 20

The weakness of sawdust wall was the low resistant towards excessive water. Because of that, the outer part of the wall was protected by Calcium Silicate panel. This panel also replaced the use of wall plaster. This research investigated the building with area of 36 m2, which is a common simple house for low-income people in Indonesia. Here are the table of wall variable that is mentioned before (Table 1).

This research analysed the optimum energy used in the building based on wall variable, such as annual cooling energy, building material embodied energy and annual overheating energy. Simulation of simple building with area of 36m2 is used to investigate the research question. The calculation of annual cooling energy and overheating energy used Archipak software. The climatic condition utilises the local climate of Surabaya and thermal properties sawdust [1]. Moreover, the calculation of embodied energy used the value of embodied energy per unit of the building volume or area [2]. The optimisation result of heating energy and cooling energy was generated from the Gradient diagram.

3. Discussion and results

Cooling energy and overheating are one same value but have different meanings. Cooling energy relates to energy for cooling the room, while overheating refers to the frequency of the conditions when the room temperature above or under the comfort standard. Both cooling energy and overheating were generated from Archipak software. Below is the result of the cooling energy and overheating in the sawdust wall calculation with varied wall thickness on 36m2 building type (Figure 1). The thickness can be seen in Table 1.

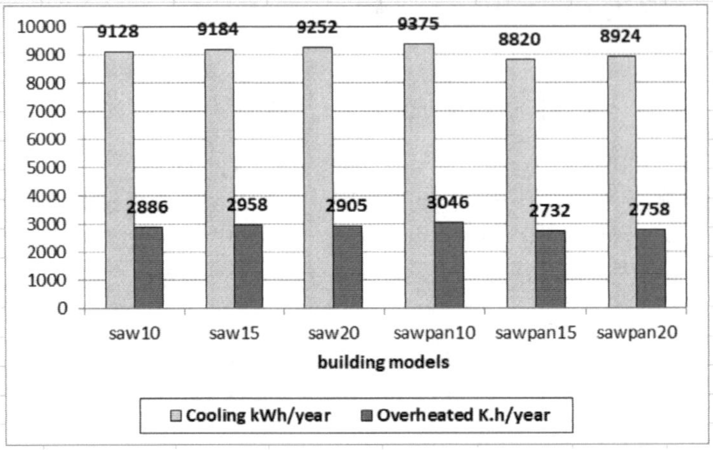

Figure 1. Annual cooling energy and overheating bar graph.

Figure 1 illustrates the small differences on annual cooling energy value and overheating on all wall variation. The lowest value is generated from wall thickness 15 cm with calcium silicate panel. In this type of wall, the panel helped to reduce the cooling energy. But the use of panel did not affect the 10 cm sawdust wall. It explains that the use of panel is very effective for the 15 cm and 20 cm wall, which is about 3.5-4% compares to the same thickness wall without panel.

The simulation shows that in the building uses panel with 15 cm and 20 cm wall have cooling energy and overheating value lower than other walls. The value of cooling energy and overheating illustrate the same trend among all variables. That is why the value used for the next section is overheating.

4. Embodied energy

The embodied energy value on 36 building type that uses sawdust wall with varied wall conditions was calculated based on the building material type and the used material area or volume. The selected material used for this study consists of sawdust wall, clay tile roof, floor plaster, glass window and timber door. Each material has different embodied energy value per unit.

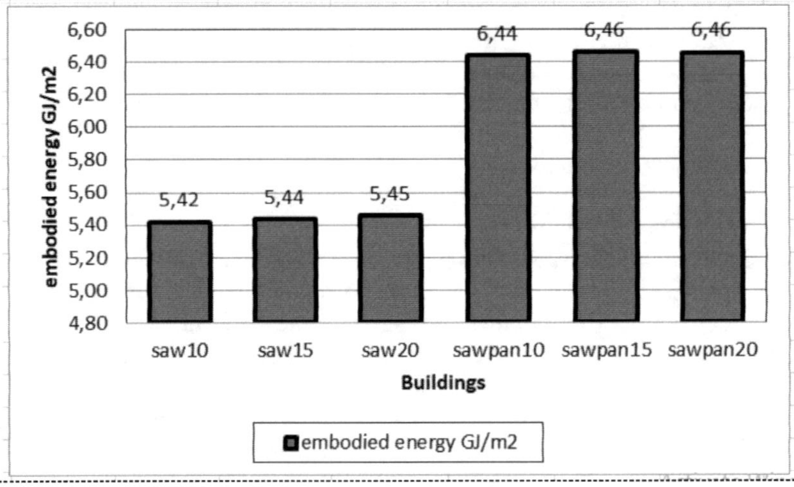

Figure 2. Embodied energy/m2 graphic

This research used building with area of 36 m2. The volume and building material area on every building was the same except the wall due to the variable of wall thickness and the used panel. The value of embodied energy used per unit is 51.88 MJ/m3. Figure 2 illustrates no differences between the value of embodied energy on among the wall without panel as well as among building with panel. The value of embodied energy per metre square of the wall with panel is higher than wall without panel. It indicates the use of panel gives a huge effect towards the inclination of embodied energy per metre square, which is 15.5%.

5. Optimisation

All overheating and embodied energy value are optimised in all variables. The conditions of each variable can be seen in Figure 3. Since cooling energy, overheating and embodied energy per metre square have various values, the optimisation is needed for this research. Due to the small differences of cooling energy and overheating, the annual overheating was compared to the embodied energy.

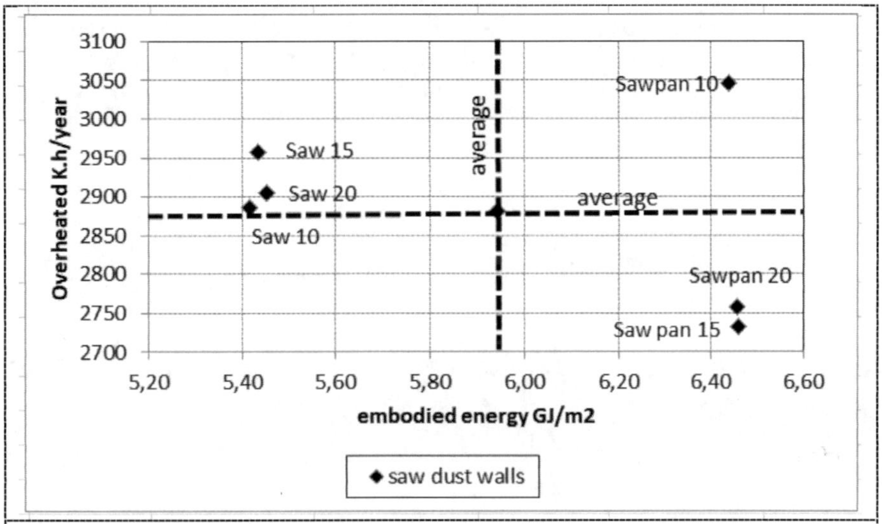

Figure 3. The optimisation of annual overheating and embodied energy/m2

There are 4 zones illustrated in Figure 3. Zone 1 and zone 3 represents building which has the lowest and the highest overheating and embodied energy. Zone 2 and 4 illustrates the zone where the building has the highest overheating and the lowest embodied energy and vice versa. On the figure, the optimum embodied energy is represented by Sawdust 10, which has 10 cm wall without panel. Meanwhile, the optimum overheating and embodied energy is indicated by sawpan 15 with 15 cm wall with panel. Wall with calcium silicate panel has the lowest cooling energy and overheating. The overheating comparison between 10 cm wall with calcium silicate panel is.

6. Conclusion

In this research the thinnest sawdust wall has the lowest overheating and cooling energy, which is 10 cm. The addition of calcium silicate panel in the outer wall affects cooling energy and overheating. Based on the cooling energy value, the most optimum wall configuration is the 15 cm wall covered by panel in the outer side. The application for simple building adopts the wall with calcium silicate on the outer layer because it does not need rough plaster. This action leads to the minimum use of cement. In addition, the wall configuration helps to reduce the heat energy inside the building. In this research, it is found that the use of calcium silicate panel has a big impact in the declination of cooling energy about 3.5-6.3% but the embodied energy is increased about 18.2-19.2%.

7. References

[1] Sam A R M, et al. (2017) *Mechanical and thermal properties of sawdust concrete* Jurnal Teknologi (Sciences & Engineering) 79:6 pp 23–27.
[2] Bejo L. (2017) *Operational vs. Embodied Energy: A Case for Wood Construction* (DRVNA INDUSTRIJA) 68 (2) pp163-172.
[3] Sawant A, et al. (2018) *Partial Replacement of Sand with Sawdust in Concrete* International Research Journal of Engineering and Technology (IRJET) Vol. 05 Issue: 03 pp 3098- 3101.
[4] Toukourou C A, et al. (2016) *Thermomechanical Characterisation of Compressed Earth Blocks Added with Sawdust* (Physical Science International Journal) 2(4) pp1-9.
[5] Christoforou E, et al. (2015) *Cradle to life cycle assesment (LCA) of adobe brick* (Journal of Cleaner Production) Xxx pp 1-10.
[6] Memon R P, et al. (2017) *Mechanical and thermal properties of sawdust concrete* (Jurnal Teknologi) 79:6 pp23–27.
[7] Kamel S, et al. (2008) *Mechanical Properties and Water Absorption of Low-Density Polyethylene/ Sawdust Composites* (Journal of Applied Polymer Science) Vol. 107 1337–1342
[8] Hafed S A. (2017) *Study of thermal insulation and some mechanical properties for hybrid composites (cement – wood sawdust)* (International Journal of Computation and Applied Sciences IJOCAAS) Vol. 3 Issue 2 October 2017 pp 212-216.

IOP Conf. Series: Earth and Environmental Science 1007 (2022) 012003 doi:10.1088/1755-1315/1007/1/012003

Traditional town houses in Kyoto, Japan: present and future

S Hokoi[1,2,4], **C Iba**[3]

[1]Kyoto University, Nishikyo-ku, Kyoto, Japan
[2]School of Architecture, Southeast University, Nanjing, China.
[3]Department of Architecture and Architectural Engineering. Kyoto University, Nishikyo-ku, Kyoto, Japan

Abstract. Climate change is a hot topic that requires urgent countermeasures; heat islands in urban areas significantly influence our daily lives such as health problems represented by heat stroke and insomnia. The health problems caused by abnormally hot weather cannot be solved only through the passive design of buildings but require artificial measures for cooling by air conditioners. The electricity consumed by air conditioners in Japan has consistently increased annually, resulting in an increase in the consumption of fossil fuels. The heat discharged due to the cooling worsens the environmental conditions around us. From these viewpoints, energy-saving standards have been implemented as one countermeasure. In this paper, these issues and the countermeasures to be taken are investigated regarding "Kyo-machiya," a traditional townhouse in Kyoto, Japan. Field survey on indoor environment was carried out both in summer and winter. As an example, indoor air flow is explained in detail. It is important from both thermal comfort (cooling in summer) and thermal discomfort (cold draft in winter) points of view. It is highly related to the building design such as atrium and fittings such as sliding doors used for partitioning off a house. It is also important for infection by virus. Finally, eco-friendly residences and life are discussed in connection with the Sustainable Development Goals (SDGs) through a new form of "Kyo-machiya," the "Heisei-no-Kyo-machiya."

1. Introduction

Climate change is a hot topic that requires urgent countermeasures; heat islands in urban areas significantly influence our daily lives, such as health problems represented by heat stroke on a hot and sunny day and insomnia during a hot and humid night.

The health problems caused by abnormally hot weather cannot be solved only through the passive design of buildings but require artificial measures for cooling by air conditioners. The electricity consumed by air conditioners in Japan has consistently increased annually, resulting in an increase in the consumption of fossil fuels. The heat discharged due to the cooling worsens the environmental conditions around us, aggravating the heat island and, ultimately, the global environment.

Therefore, the buildings we use and our lives themselves have enormous impacts on the heat island and global environment, causing a vicious cycle. The built environment has been claimed to contribute to this phenomenon. In this respect, from a building environment perspective, energy consumption must be minimalized under the requirement of a healthy indoors.

From this viewpoint, energy efficiency standards have been implemented worldwide as one countermeasure and satisfying them is regarded as a minimum requirement.

[4] Corresponding author: hokoi@maia.eonet.ne.jp

In this paper, these issues and the countermeasures to be taken are investigated regarding "Kyo-machiya," a traditional townhouse in Kyoto, Japan, as a studied subject. Furthermore, eco-friendly residences and life are discussed in connection with the Sustainable Development Goals (SDGs) through a new form of "Kyo-machiya," the "Heisei-no-Kyo-machiya."

2. Energy efficiency standards in Japan

Energy efficiency standards were first enacted in Japan in 1980 under the "Act on the Rational Use of Energy." This was one of the main countermeasures against the oil crisis in 1973, and the industrial sector, consuming a large fraction of the total energy, was the main target to be regulated. Office buildings were also regulated, and building performance, centered on thermal insulation, airtightness, and solar control, was regulated, along with the rational use of energy in building systems, including heating-cooling and lighting systems, electric appliances, and hot water systems. The indoor environment's wastage of energy, particularly the minimum requirements of the air exchange rate, was also taken into consideration. Residential buildings were also asked to improve their building performance, but only as a guideline.

Energy efficiency standards of residential buildings were enacted in 1980. The standard was improved under the "New Energy Efficiency Standard" in 1992, and further strengthened under the "Next Generation Energy Efficiency Standard" in 1999 to respond to the Kyoto Protocol issued in 1997 (Figure 1). Almost the same level of energy efficiency is adopted, even now, regarding building performance.

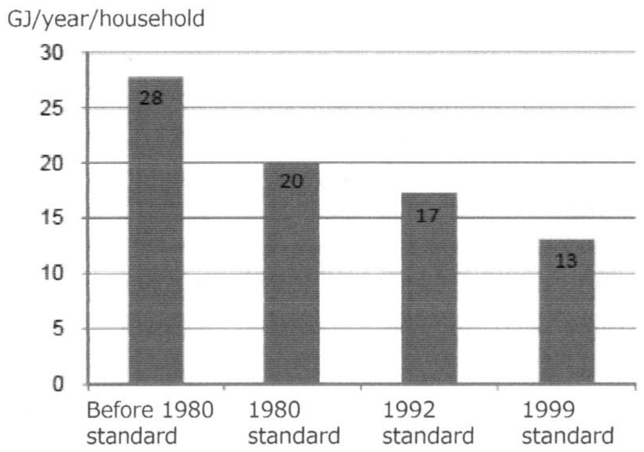

Figure 1. Annual heating-cooling load changes under the standards [1]

The total energy consumed in a building depends not only on the thermal performance of the building but also on the energy used for heating/cooling, lighting, hot water, etc., as well as on the energy production by solar panels. From this viewpoint, an integrated energy efficiency concept was introduced in 2013, and the evaluation of energy-saving performance was changed based on the primary energy consumption, resulted in the enactment of the "Act on Improvement of Energy Consumption Performance of Buildings" in 2015.

Furthermore, to respond to the Paris Agreement in 2015, the "Act on Improvement of Energy Consumption Performance of Buildings" was amended in 2019 to extend target buildings and changes from suggested (induction by a subsidiary) to obligatory.

3. Indoor environmental control in traditional residential buildings in Kyoto, "Kyo-machiya"

3.1. Japanese climate and building design

In this section, the present situation and issues related to energy-saving are explained with respect to "Kyo-machiya" as a study subject. Since Japan has four seasons, summer cooling and winter heating are required. Thus, energy-saving measures and building design must consider not only the hot season but also the cold outdoor environment, which is different from the situation in Southeast Asian countries (Figure 2). Furthermore, the intermediate seasons (spring and autumn), when heating or cooling might be dispensed with, must be carefully considered from an energy-saving point of view.

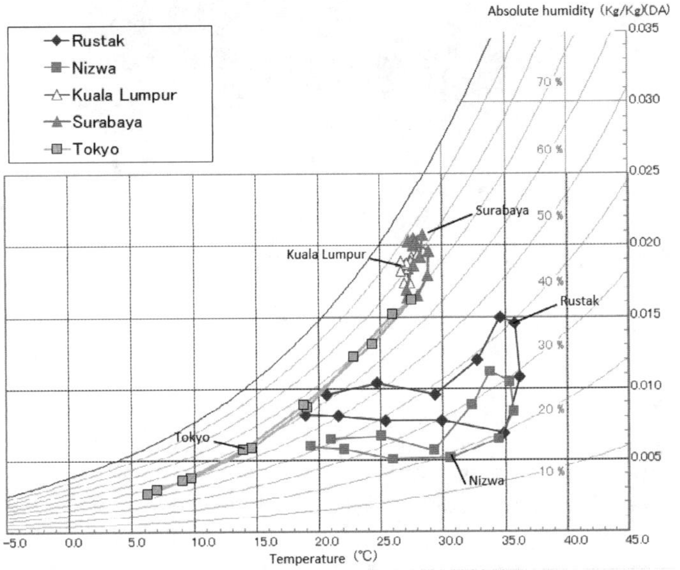

Figure 2. Comparison of climatic conditions in Japan and Indonesia [2]

Passive techniques, such as thermal insulation and airtightness, are the main techniques effective for energy-saving in a residential building. Solar radiation (heat) should be avoided in summer, although it is useful and should be introduced indoors in winter. Ventilation is an effective cooling technique in summer without consuming energy, but suitable means of avoiding cold drafts and air leakage are required in winter. From the viewpoint of air conditioning, spring and autumn are called intermediate seasons, meaning there is no need for cooling or heating. The term of this season can be extended by suitably controlling the gain/removal of solar heat and adjusting the ventilation air volume.

As the western parts of Japan have hot summers and mild winters, the people there historically emphasized measures to stand the heat of summer while comparatively neglecting the coldness of winter. As a result, it has become common sense that a residential building should be designed to be open, and the openings should be kept open to ensure sufficient ventilation. This seems to have become fixed in the lifestyle of Japanese people. The fact that most Japanese residential buildings are made of wood with poor airtightness may have helped to cement this common sense. The design of controllable airtightness should be a key factor in innovative energy-saving design.

Such a lifestyle seems to influence not only energy consumption but also the residents' health. This is a health issue closely related to residents' lifestyles; in some sense, similar situations occur in southeastern countries, where cooling for a long time under very low air temperatures is common with the rapid and widespread use of air conditioners [3].

3.2. Difficulties in applying energy efficient standards in Kyo-machiya

3.2.1. Outline of Kyo-machiya [4]. A Kyo-machiya is a wooden detached or townhouse-type residential building built before 1950, having a construction called "traditional wooden post and beam construction" or "traditional construction." It has all or several features of the following: Tori-niwa (a thin and long earth floor running from the entrance facing the road to courtyard), Hibukuro (atrium part in the upper space of the Tori-niwa), Tsubo-niwa or Oku-niwa, Tori-bisashi (eaves installed along the roadside), and Koshi (e.g., Mushi-kago-mado or Kyo-goushi).

Figure 3. Outside appearance of Kyo-machiya (townhouse-type) [5]

3.2.2. Difficulty in satisfying the energy-saving standards in Kyo-machiya. We will face many difficulties when trying to apply the usual energy-saving strategies, such as insulation and airtightness, to Kyo-machiya.

With respect to the facade, external insulation cannot be applied to avoid changing the traditional external appearance (a white plaster wall), and the material that can be used as the external finishing is very limited from a fire-prevention point of view. Regarding the inner surfaces of the soil walls, inner insulation is not allowed to maintain tradition, particularly for the residents, who put great weight on the feeling of the soil wall. Although an improvement in the insulation performance of the soil wall is required, its poor airtightness is much more serious. Cracking occurs very often because of shrinkage of the soil.

As an open-type design of the inner spaces is the original style, the airtightness between the spaces and the connection with the outside (openings, verandah, corridor, passage to toilet, etc.) is very poor. Thus, specially designed and produced fittings are required for air tightening the wooden structure.

3.3 Examples of estimated energy consumption and countermeasures in Kyo-machiya [6]
In the following, the annual heating/cooling loads are estimated for a simplified Kyo-machiya model, and the possibility of satisfying the energy efficiency standard is examined.

A simple two-story detached-type residence with an attic space was analyzed (Figure. 4). The residents principally inhabit the 1st floor, and the 2nd floor comprises a storage space and a space used only to lodge temporary visitors for a few days. All walls were made of soil with a thickness of 50 mm, and the gable walls on the east and west were installed with 40 mm thick wooden insulation (Table 1). The air exchange rate was set as 10 [1/h] when ventilated and 0.5 [1/h] when there was no ventilation, and Automated Meteorological Data Acquisition System (AMeDAS) data were used as inputs.

Following the standard, the rooms were heated at a setpoint temperature of 18 °C when the daily mean outdoor temperature was lower than 15 °C, and otherwise cooled at a set-point of 27 °C. Based on these conditions, the annual heating and cooling loads were calculated. The heating/cooling load of the hypothetical standard residence in Kyoto is 460MJ/m^2.

- Influence of heated/cooled space (Figure. 5-1): By limiting the heated/cooled spaces from both 1st and 2nd floors (specified by the standard) to only the 1st floor, the heating/cooling load was reduced by 38% from 1079 MJ/m^2 to 667 MJ/m^2.
- Influence of heating hours (Figure. 5-2): By changing the heating hours from 24 h following the standard to 6:00–24:00 (waking hours), or 6:00–12:00 and 18:00–24:00, the heating/cooling load decreased by 18% and 26%, respectively. The effect of no heating during the nighttime, when the outdoor temperature is low, is significant.

- Influence of heating/cooling setpoint temperatures (Figure. 5-3): By changing the setpoint temperatures from heating 18 °C/cooling 27 °C to heating 17 °C/cooling 28 °C, the annual load decreased by 17%, and the annual heating/cooling load was 438 MJ/m^2, which satisfies the standard.

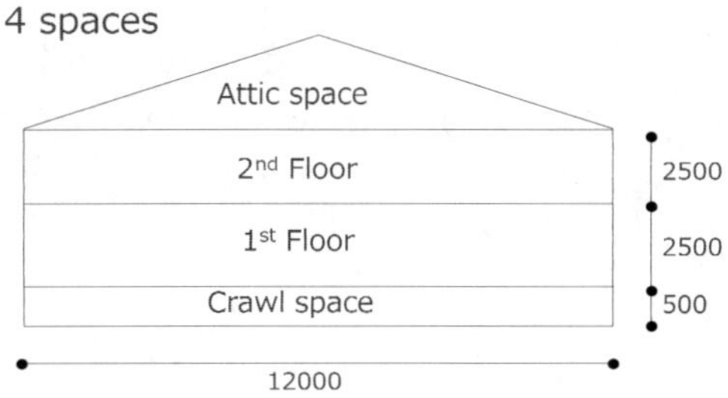

Table 1 Specification of wall [6]

N-S External wall	Soil layer 50mm
E-W External wall	Soil layer 50mm
	Wooden insulation 40mm
	Cedar board 10mm
Partition wall	Soil layer 50mm

Figure 4. Analyzed Kyo-machiya model [6]

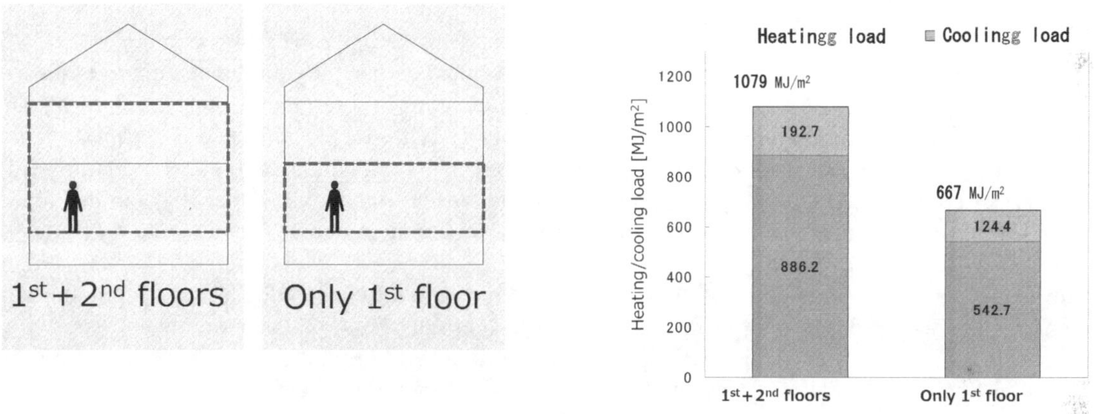

Figure 5-1. Heated/cooled spaces [6]

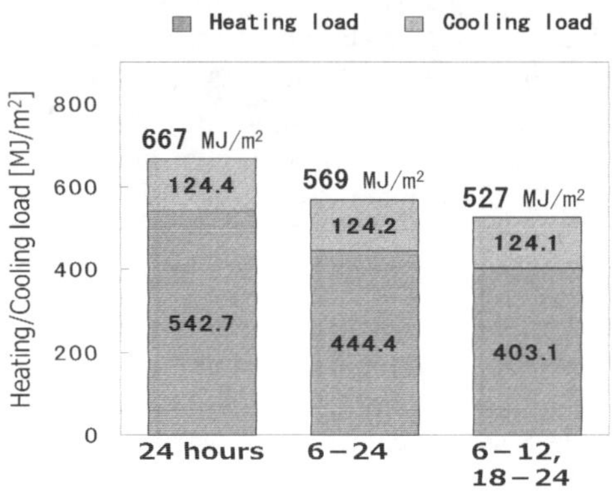

Figure 5-2. Heating/cooling hours [6]

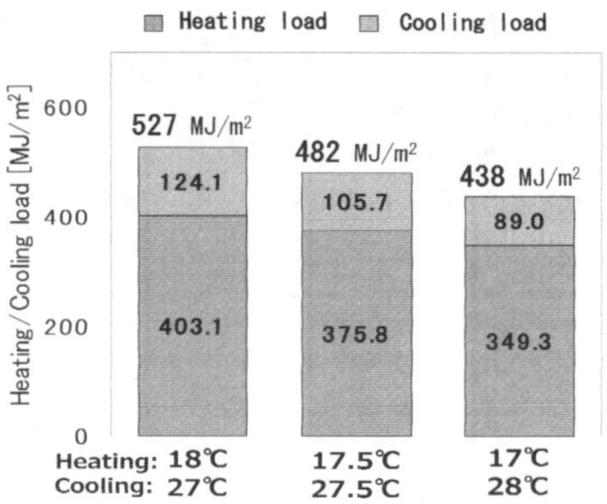

Figure 5-3. Setpoint temperatures [6]

As described above, the simulation conditions used here do not satisfy the specifications of the energy efficiency standard. However, the no use of the 2nd floor is not far from the present actual situation (aging society) and may be regarded as an energy-saving strategy in line with the reality of the present Kyo-machiya life. The changed heating/cooling hours and setpoints are also not far from the actual living conditions (very common to all areas in Japan except for Hokkaido) and seem reasonable changes.

Regarding the cooling hours and setpoint temperature in summer, their rationalization is discussed in Sections 4.3, 4.4, and 4.5 from the viewpoints of night ventilation (purge) and ventilation cooling (lowering of perceived temperature by increased indoor air velocity). Regarding heating in winter, the cold draft and infiltration are discussed in Section 4.2 because the perceived temperature and insulation performance are increased by reducing them.

4. Traditional townhouses in Kyoto and their indoor climate (and control)

4.1 Control of draft in winter, ventilation cooling and thermal storage during night

Although the reduction in the heating load has a large influence on satisfying the standard, as shown in Section 3, possible measures are very limited in Kyo-machiya. In a terrace house type Kyo-machiya, the boundary walls with neighboring houses can be ideally regarded as adiabatic if symmetry can be assumed, resulting in no heat loss. In addition, an Engawa (a verandah connected to the courtyard), a thermal buffering zone, can play an important role as an insulation layer by controlling the opening/closing of the doors on both sides of it, although these measures are not allowed by the present standard.

In winter, from the viewpoint of health, the indoor temperature should be kept high. In such a situation, the cold draft has a large influence on cold thermal sensation. In modern housing with an atrium, this influence is significant. Because cold sensation usually leads to an increase in heating load, reducing the cold draft is also important from an energy-saving point of view.

The indoor airflow is important to consider both thermal comfort (cooling in summer) and thermal discomfort (cold draft in winter). The indoor air flow, including the cold draft, is highly related to the building design, such as the atrium, and fittings such as sliding doors, Fusuma, and Shoji, used for partitioning off a house.

The cooling load fraction of all loads is presently small. However, considering that abnormally hot days and sultry nights often occur because of the heat island effect and global warming, symptoms such as heat stroke and insomnia have increased, and thus it is necessary to maintain a suitable indoor temperature in summer.

From these points of view, the cold draft and countermeasures in winter, night ventilation in summer, and the relationship between the indoor airflow and opening/closing of openings were measured and analyzed in this section.

4.2 Coldness and draft in winter: Issues related to atrium

4.2.1. Hibukuro in the Tori-niwa and atrium (Figure. 6). The upper part of the Tori-niwa is an atrium called Hibukuro. The main and original purpose of the Hibukuro is to discharge the smoke and soot generated by burning in cooking. In these days, a comfortable space with a high ceiling is strongly desired; thus, an atrium space is introduced in many residential buildings, but in a residence without sufficient insulation of the external walls, uncomfortable cold drafts are not uncommon in winter.

IOP Conf. Series: Earth and Environmental Science 1007 (2022) 012003 doi:10.1088/1755-1315/1007/1/012003

Figure 6. Hibukuro in Tori-niwa [5]

4.2.2. Measurement of air flow in a staircase of a model residence (Heisei-no-Kyo-machiya) [6]. The temperature and airflow distributions in a staircase were measured in winter, when the Zashiki on the 1st floor was air-conditioned by an air conditioner at 25 °C with the door (Fusuma) connected to the staircase kept open. The hot air flows into the staircase through the door, rises up to the 2nd floor, and then flows down along the stair as cold air. The influence of the opening or closing of the door on the 2nd floor was examined. (Figure. 7)

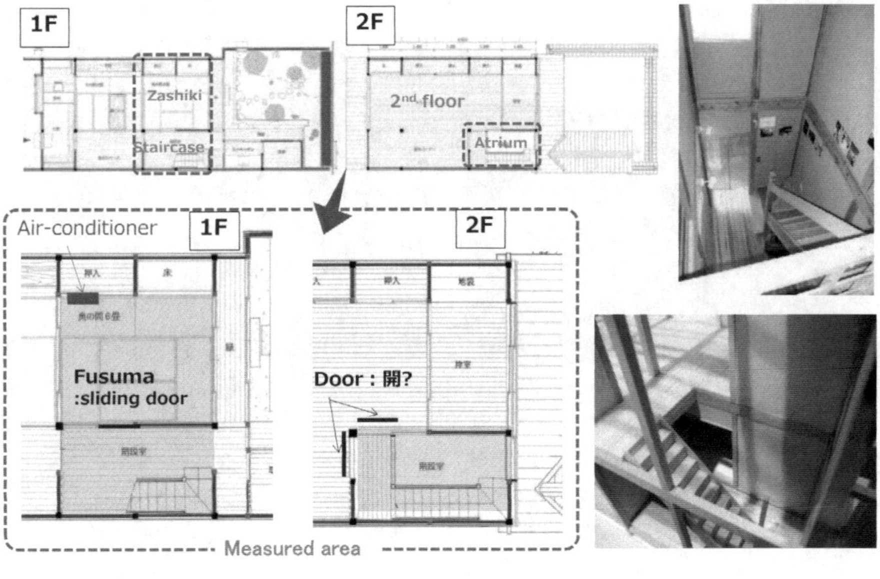

Figure 7. Plan and overview of measured staircase [6]

When the door on the 2nd floor was open, the hot air at 29°C that flowed out from Zashiki on the 1st floor at 0.3–0.4 m/s through the upper part of the door, rose up to the 2nd floor at 0.5 m/s. From the 2nd floor, cooler air at 19°C flowed down the stair at 0.2–0.35 m/s, then returned to Zashiki at 21°C through the lower part of the door on the 1st floor.

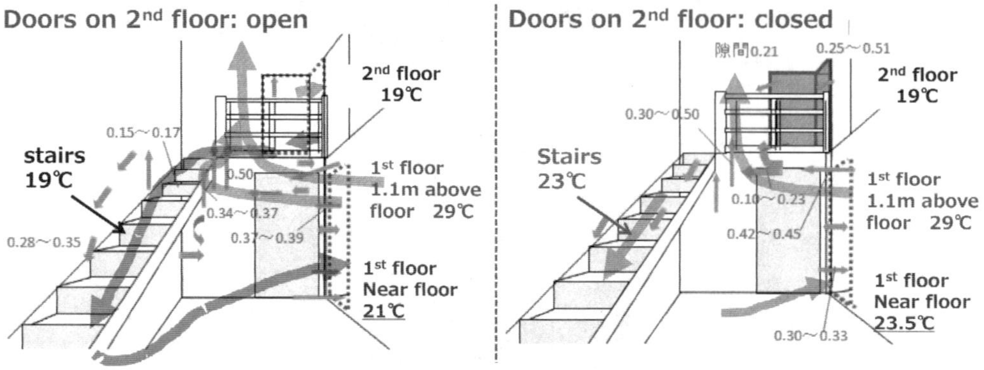

Figure 8. Air velocity and temperature in the staircase (left: open case, right: closed case) [6]

When the door on the 2nd floor was closed, the downward air velocity decreased and returned to Zashiki at 0.3–0.33 m/s and 23.5°C. The draft in Zashiki was weakened by closing the door on the 2nd floor.

From this measurement, it can be understood that the atrium (staircase) influences the temperature and air flow in other rooms, and the results are significantly influenced by opening or closing the door on the 2nd floor. Thus, the separation of the atrium from the neighboring spaces is important.

Figure. 9 shows examples of movable doors and windows installed in the atrium and staircase to prevent cold draft. The measured results of the temperature and air flow (not shown here) showed that the draft could be suppressed by suitably closing the openings.

Figure 9. Fittings isolating the atrium sliding doors and windows [6]

4.2.3. Atrium and staircases: The Hibukuro in the Tori-niwa (Figure. 6). Although there is a window connected to the room on the 2nd floor on the upper part of the Hibukuro, this window is usually closed (naturally, considering the role of the Hibukuro in discharging smoke), which positively acts to suppress the draft in winter.

4.2.4. Application to the Engawa. The improvement of the airtightness of openings has a more important effect on the Engawa, making it possible to use the Engawa more effectively. By suitably changing the degree of closure of the inner and outer doors of the Engawa depending on weather conditions, it can effectively utilize solar heat in winter (like a sunroom) and also work as an insulation layer on the outside (courtyard), thus effectively reducing the heating load and improving the indoor thermal environment that can be achieved.

4.3 Thermal (cooling) storage of the building structure by night ventilation [6]

Regarding night ventilation in summer, the cooling effect of thermal storage of the building structure on the following day was examined. The measurements were carried out in a model (mockup) residence

(Heisei-no-Kyo-machiya). On the first day of the measurement, the doors were opened during the daytime and closed during the night following the usual usage, while the opposite control was done on the second day (thermal storage). Figure 10 shows the flow of the measurement, including the opening and closing of the doors.

The measured room temperature during the night decreased by 2.5°C because of the night ventilation, and as a result, the increase in the room temperature on the following day was suppressed by 1.7 °C (Figure. 11).

Night ventilation was also shown to be useful in Indonesia [3], although special care such as mosquito nets is required for practical use.

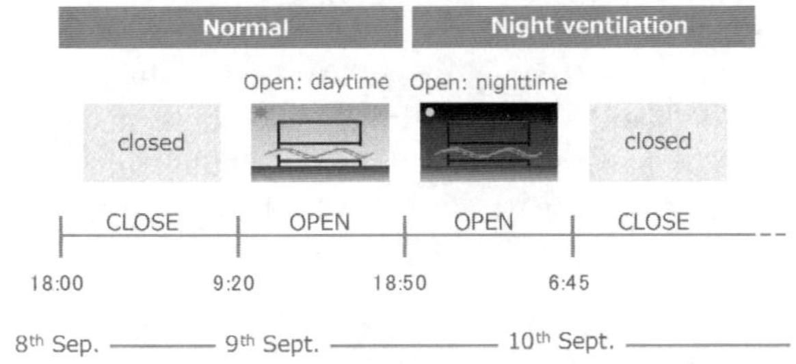

Figure 10. Flow of measurement and open/close of openings [6]

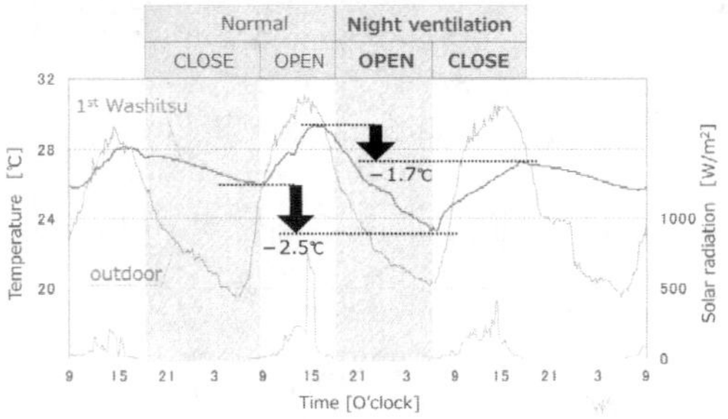

Figure 11. Measured results [6]

4.4 Ventilation cooling in summer [6]

The indoor air flow during ventilation in Kyo-machiya was measured with respect to opening or closing of the openings, with or without mosquito nets or bamboo screens, and based on the measured and calculated results of the indoor airflow in summer and autumn, a suitable opening or closing of the openings was examined.

The measured residential building is a traditional Kyo-machiya built about 80 years ago (Figure. 12), whose Tori-niwa was renovated to a living room by wooden-flooring the original earthen floor. The living room and Mise-no-ma on the 1st floor have an atrium space. The three Japanese-style rooms (Washitsu) on the 1st floor connect straight from the street to the courtyard. The air direction and velocity were measured by streamers and an anemometer, respectively.

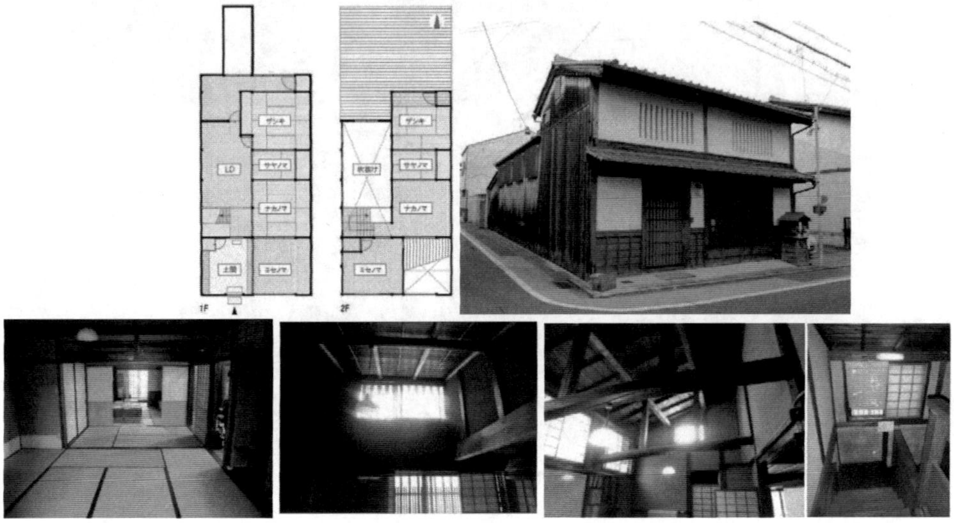

Figure 12. Measured Kyo-machiya [6]

Measured results (Figures.13-1–13-3)

- Figure. 13-1 Living-dining room (LD)+staircase: effect of west side windows. Weak airflow can be seen near the windows when they are opened, but there was no stable airflow in the LD. The air flowing through the windows diffused into the atrium space and the velocity became small. The outdoor air flowing through the west windows diffused into the upper space and weakened; thus, it was not effective for ventilation cooling in the living room. The utilization of ventilation in the rooms on the 2nd floor may be more effective.
- Figure. 13-2: south of Mise-no-ma.: Both sides of south windows were slightly open. The streamers swung strongly near the windows and velocity was 0.8m/s in strong wind, but airflow weakened with distance from the windows.
- Figure. 13-3: Window-mosquito net-bamboo screen of the Engawa. W-Open, M-Closed, S-Closed: weak flow in lower part. W-Open, M-Open, S-Closed: air velocity increased to 0.35–0.4 m/s. W-Open, M-Open, S-Open: air flow occurred in upper part (700mm from floor). Mosquito net and bamboo screen weakened ventilation.

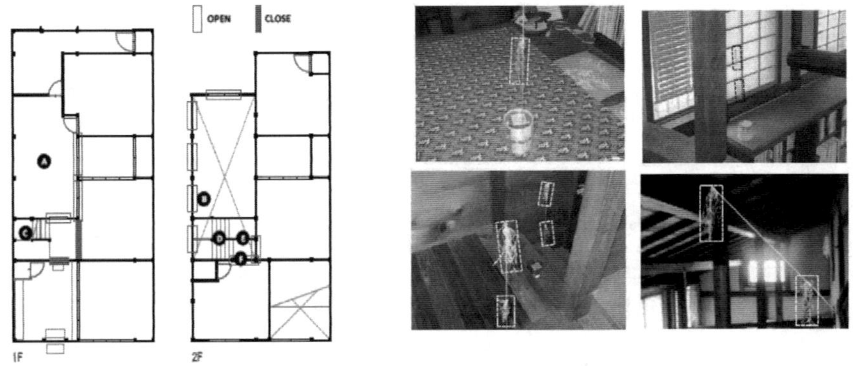

Figure 13-1. Living-dining room (□) + staircase: effect of west side windows [6]

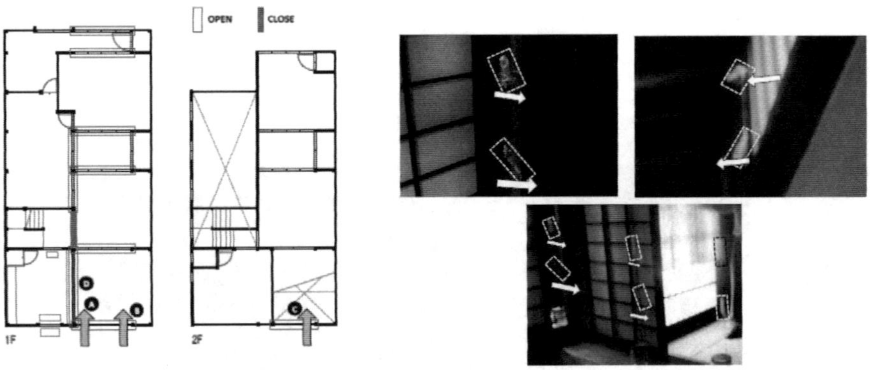

Figure 13-2. South windows of Mise-no-ma [6]

Figure 13-3. Window-mosquito net-bamboo screen of Engawa [6]

4.5. Influence of partition doors on ventilation cooling [6]

Because the position and closure of the partition doors influence the indoor air flow, they strongly influence the performance of the ventilation cooling in summer. Although the usual strategy is to open all the openings, the air may flow into a space where there are no occupants, thereby wasting natural energy. Therefore, the influence was examined by CFD analysis.

Figure 14. Heisei-no-Kyo-machiya [7]

A Heisei-no-Kyo-machiya, shown in Figure. 14, was simulated. The effect of opening/closing Fusuma (Japanese-style partition door) 1–3 was examined (Figure.15).

Simulation results (Figure. 16): By closing two doors along the corridor, the airflow route can be clearly specified, and significant air velocity can be obtained in the Washitsu. The air velocity in the corridor did not change significantly even if the wind velocity increased. The Washitsu air velocity can be increased by suitable closure of the partition doors.

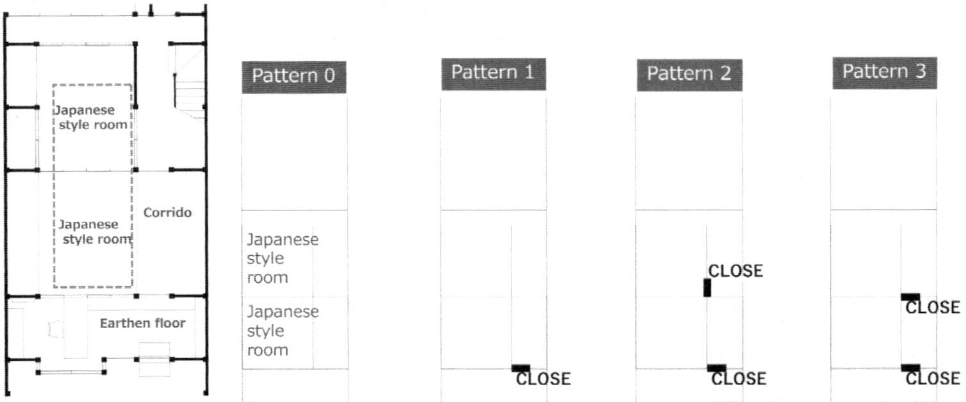

Figure 15. Closure of doors (Fusuma) in Heisei-no-Kyo-machiya [6]

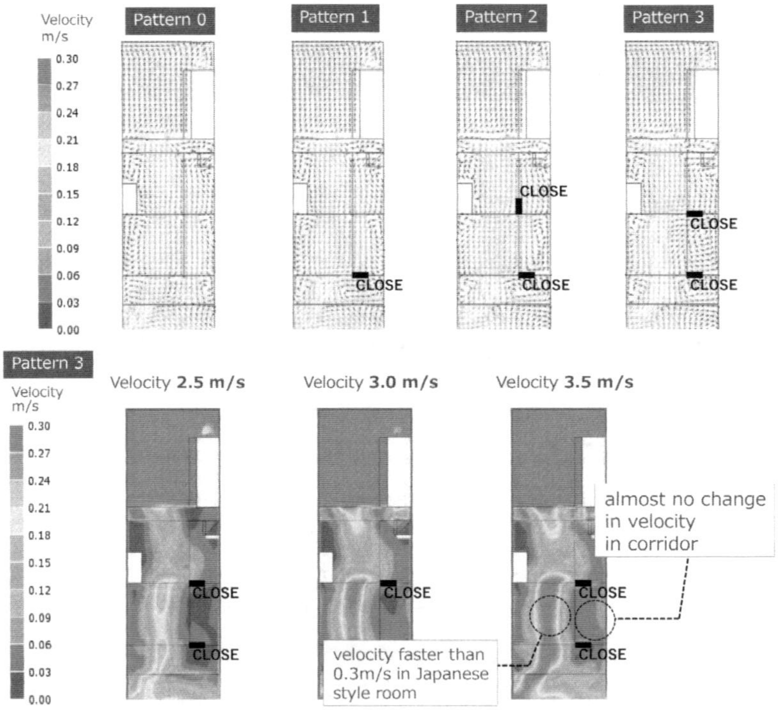

Figure 16. Simulated results of airflow pattern and velocity [6]

4.6 Control of indoor airflow as a countermeasure to suppress virus spread

Control of indoor airflow is also important to prevent the urgent issue of COVID-19 infection. Ventilation is strongly recommended for preventing COVID-19 infection. Dilution of the virus concentration is generally thought to suppress the disease. However, depending on the direction of airflow, the spread of the virus may be accelerated. From this point of view, it is important to specify the air flow route from the supply inlet to the discharging outlet and to separate the flow route from the occupied spaces to avoid air mixing. Displacement ventilation may be a good option in that regard. For the control of the horizontal airflow, suitable fittings for partitioning should be designed and placed as described above. The atrium also has to be designed with consideration of a well-specified air flow from a lower-occupied place to the air outlet in the upper space.

5. From the present to the future

5.1 What is the present issue from a viewpoint of eco-friendliness (SDGs)?

Technology in building construction and the utilization of natural energy has developed to a very high level, and zero-energy houses (ZEHs) are now realizable targets. In addition, recognition of the importance of our health is now a matter of course. What should we think about regarding our target "eco-friendliness" or SDGs?

Now, greater concern about regional culture, individual life, and human relationships may be suitable candidates, something more than energy and environment, something more than global issues, although these are important, of course.

In the following, it is discussed about locality, local culture, history, individuals, families, neighbors, and the public in an era of globalization, using the traditional residential building Kyo-machiya, and the modern version of it, "Heisei-no-Kyo-machiya" in Kyoto City, as an example.

5.2 Heisei-no-Kyo-machiya and an environmental buffer zone

5.2.1. Background. Kyoto city held a civic meeting in 2008 to realize "Kyoto, a city taking care of wood culture." The civic meeting adopted "Heisei-no-Kyo-machiya" as one of the main themes to be discussed. "Heisei-no-Kyo-machiya" is a new residence model inheriting tradition and related knowledge of the Kyo-machiya, a traditional residential building in Kyoto city, and integrating the most advanced environmental technology with it. Based on the report of the civic meeting in 2010, Kyoto city has been taking measures to disseminate and spread "Heisei-no-Kyo-machiya," in cooperation with the "Heisei-no-Kyo-machiya" consortium.

What is the problem with Kyo-machiya? Traditional residences disappear and are replaced by high-rise flats. Therefore, the preservation and inheritance of the wooden houses, Kyo-machiya, in which people live as cultural heritage and tourist resources, and the conservation of forests surrounding Kyoto city are areas of concern.

5.2.2. Concept of "Heisei-no-Kyo-machiya" (Figure. 17) [7]. In the "Kyo-machiya," weight is placed on the relationship between the inside and outside, and the connecting spaces such as the "Tori-niwa" and "Engawa" (veranda) are adjusted in the indoor environment. Through this relationship, a fruitful living culture has been cultivated, depending on the season. Appropriate design of this intermediate region, the "environment adjustment space," connecting the inside and outside constitutes the space concept of "Heisei-no-Kyo-machiya." Figure. 18 shows examples of "environment adjustment space". "Heisei-no-Kyo-machiya" emphasizes the relationship between an individual and its environment and the relationship between a residence and surrounding area, with the "environment adjusting space" intervening them.

In Malaysia, this kind of space is common with the use of pilotis in high-rise residential buildings as a semi-common space for community functions, such as children's playgrounds, gatherings of homemakers, and weddings [8].

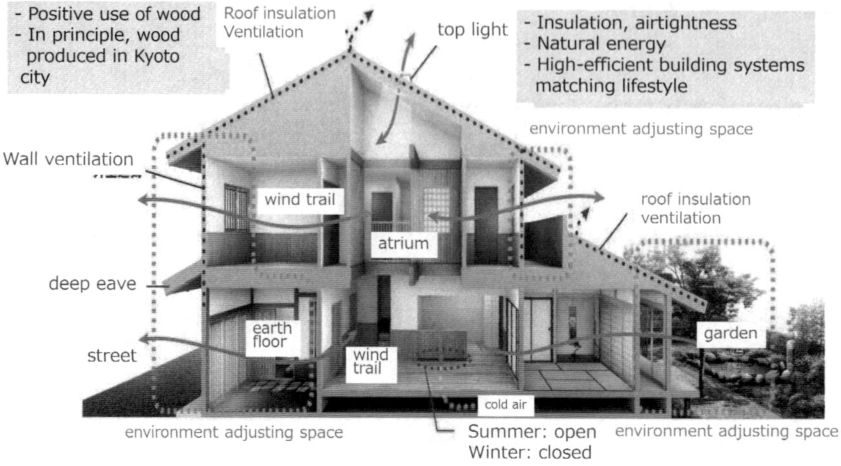

Figure 17. Concept of the Heisei-no-Kyo-machiya [7]

Figure 18. Examples of environment buffer spaces [7]

6. References

[1] Ministry of Land, Infrastructure, Transport and Tourism, Outline of amendment of energy efficiency standard (in Japanese) https://www.mlit.go.jp/common/001012880.pdf

[2] Noor Hanita A M, Takagi N, Hokoi S, Uno T and Ekasiwi S N N 2019 *Oman in the 21st Century: Issues and Challenges* (New York: Nova Science Publishers)

[3] Uno T, Hokoi S, Ekasiwi S N N and Noor Hanita A M 2012 J. Asian Archit. Build. Eng. **11**(2): 407-414.

[4] https://machiya-kyoto.jp/about

[5] https://www.city.kyoto.lg.jp/tokei/cmsfiles/contents/0000004/4241/kyomachiya-plan.pdf

[6] Arai K 2015 Effects of Opening/Closing of Windows and Doors on Improvement of Thermal Environment and Energy Conservation in Kyomachiya, Master thesis, 2015.

[7] https://www.city.kyoto.lg.jp/tokei/page/0000247796.html

[8] Sapian A R, Abdul Majid N H and Hokoi S 2011 *Proc. of 9th Nordic Symposium on Building Physics* (Tampere, Finland: Tampere University of Technology) p 1315-1322

Sustainability for whom? Cities and buildings through the lens of older people

V Soebarto[1,2]

[1]School of Architecture and Built Environment, The University of Adelaide, Adelaide SA 5005, Australia

Abstract. As we move towards larger and more complex urban developments and with increasing occurrences of extreme weather resulting in significant environmental, societal and economic impacts, it is no longer a question that cities around the world must aim for sustainability. The past few decades have seen efforts by many countries to tackle these critical issues. At the same time, many countries also face significant demographic changes, with almost a quarter of their population over 65 years will be ageing in place, in their own homes. Yet cities, particularly our inner-city built environments, are spaces thatare usually imagined, planned and structured for a younger, working-age demographic. This paper discusses the current gap in knowledge in addressing sustainability of our cities and buildings, and proposes a new way of engaging people whoare often forgotten, yet they have many years of experience and wisdom about how to live sustainably: the older people. Two examples of ongoing research with older 'citizen scientists' are presented. Both cases demonstrate that it is critical to consider cities and buildings from the eyes of older people in society.

1. Introduction

In less than three decades from today, two-thirds of the world's population will be living in cities (Ritchie and Rosser, 2018). As we move towards larger, more complex urban spaces with higher-density living, greater pressure is placed on creating community and enabling socialization without adversely impacting on the environment. However, as cities become more densely populated and new developments are inevitable, more resources are needed to support the livelihood of people. As a result, resource consumptions, and consequentlycarbon emissions, from cities will continue to increase. A recent estimate by the Intergovernmental Panel on Climate Change (IPCC) indicates that cities consume between 67% and 76% of global energy and generate about 75% of global carbon emissions (Seto, et al 2014) and the shares of cities in consuming energy and producing carbon emissions are expected to increase with increasing city populations (United NationsDepartment of Economic and Social Affairs, Population Division, 2012; Seto, Güneralp, Hutyra 2012).

Increasing global carbon emissions has been linked to the warming of our environment and climate change, resulting in various catastrophic events, such as heat waves, wildfires, droughts in some parts of the world yet heavy floods and storms in the other parts, as well as the disappearance of various species.

[2] Corresponding author: veronica.soebarto@adelaide.edu.au

Studies have also shown that people in poorer countries are at least four times more likely than those in rich countries to be severely impacted by extreme weather (Richard and Bradshaw, 2017). In other words, 'city people's activitiesand demands have significant impacts not only on the city environment itself but also and more critically on amuch larger scale beyond city boundaries.

The past few decades have seen many countries around the world trying to reduce the adverse impacts from buildings and the built environment in general through various policies, codes, strategies, and tools. The oil crisis in early 1970's that resulted in high fuel costs had triggered the first 'green building movements' particularly in developed countries such as the USA, with groups of architects recommending 'passive' design approaches (Cassidy et al, 2003). Whilst after the energy concern was over the momentum for green buildings slowed down, dedicated core groups of architects "continued to push their green building concept forward". Following the publication of *Our Common Futures,* also widely known as the Bruntland Report, in 1987 and the United Nations' landmark conference on Sustainable Development, known as the Rio Conference, in 1992,the International Union of Architects (UIA) and the American Institute of Architects (AIA) held a World Congress of Architects in 1993 and produced the Declaration of Interdependence for a Sustainable Future (Kilbert, 2004). The AIA subsequently formed its Committee on the Environment (COTE) and established the US Green Building Council at the end of 1993, which then produced the LEED (Leadership in Energy and Environmental Design) Standard in 1994. Just prior to this, the British Research Establishment (BRE) produced BREEAM (BRE Environmental Assessment Method) in 1992. While these green movements mainly focusedon the environmental performance of buildings instead of the whole notions of sustainability (which also include the social and economic dimension), these early movements started to have a major impact on buildingdesign, construction and operation, shifting the goal to simply design and build buildings, to designing and constructing high-performance buildings. Other countries were soon to follow suit, with the Japan Sustainable Building Consortium producing Comprehensive Assessment System for Built Environment Efficiency (CASBEE) in 2001, the Australian Green Building Council producing Green Star sustainable rating systems for buildings in 2003, the Building and Construction Authority (BCA) of Singapore producing Green Mark Scheme in 2005, the PAM (Pertubuhan Akitek Malaysia) Council producing Green Building Index in 2008 and Green Building Council Indonesia (GBCI) producing Greenship in 2009. The scope of these assessment systems or tools has also been broadened to assessing the environmental impacts of the entire development, such as addressed in LEED for Cities and Communities or Green Star – Communities (GBCA, 2016).

The pursuit towards green buildings is no longer seen as just a movement. It is now in the mainstream, at least in the above countries. It is also realized that only focusing on the 'greenness' of the buildings is not enough. With continuing increase of greenhouse gas emissions and global temperatures, many countries have pledgednot only to reduce their energy consumption and carbon emissions but to also achieve zero carbon. Likewise, achieving near or net-zero carbon buildings is now a common goal for many building authorities and organizations around the world (e.g. World Green Building Council, 2021). At the same time, it has been realized that the wider sustainability issues – social sustainability and occupant health – must also be addressed. The WELL Building Standard, Wellness Index and Health Building Index, are some of the tools available to evaluate indoor environment to ensure "more thoughtful and intentional spaces that enhance human health and well-being" (International WELL Building Institute, 2021). The current COVID-19 pandemic has further awaken building authorities, designers and building services engineers, that reducing energy consumption and greenhouse gas emissions must not be the sole priority of the construction industry. As, or even more, critical, is ensuring the indoor environment quality of buildings to be conducive to the health of the occupants.

2. Sustainability for whom?

But have we done enough? Alongside with the above environmental changes and challenges faced by the construction industry to ensure public health, the world also experiences significant demographic changes in the past few decades. In 1990, the proportion of older people, or those aged 65 years and over, was 6%. This proportion increased to 9.3% in 2020 and by 2050 the share of older people globally is expected to increase to16% (United Nations Department of Economic and Social Affairs, Population Division, 2020) with the largest proportion of older people will be in Eastern and South-Eastern Asia. In Indonesia, the percentage of older people is even expected to reach 20% in 2024 (TNP2K, 2020). As public health improving, which has resulted in an increase in global life expectancy, the proportion of older people in all parts of the world will continue to increase, with women accounted for more than half the population of older people (United Nations Departmentof Economic and Social Affairs, Population Division, 2020).

Yet cities, particularly our inner-city built environments, are spaces that are usually imagined, planned and structured for a younger, working-age demographic (United Nations Department of Economic and Social Affairs, Population Division, 2015). To date, the age-friendliness and usability of buildings and public spacesin cities by older people are not (yet) addressed in any of the aforementioned assessment schemes and tools, including those that are supposed to evaluate designs to address occupant wellbeing and health. Whilst Goal 11 of the Sustainable Development Goals (United Nations, n.d) clearly aims to "make cities and human settlements **inclusive**, safe, resilient and sustainable", and despite the fact that older people share a significantproportion of our population, older people are still not the focus on mainstream thinking, planning and designing our cities up to the present time. Among the many studies that have been conducted, Morellec, Anastova and Falzon (2013) show that many older people have difficulties in navigating inside public buildings. They do not feel safe in crossing the road due to poor design (Zito et al, 2015) and are discouragedfrom using public benches due to lack of shading and low seats that make it difficult for them to stand up without being helped (Coman, Caponecchia and Gopaldasani, 2021).

Figure 1. Are our "sustainable" cities and buildings age friendly?

This lack of focus on older people is however expected to change. The UN has now set the 2030 agenda to befocusing on sustainable development for ageing, older persons (UNDP, 2021). Indeed, it is important to addresssustainability not just for the health of our environment but also for the health of people – all people. The quality of built environment has been shown to support the physical, mental and social health of people (Sugiyama, 2010) while 'unfriendly' public spaces can deter people, particularly older people, from using them. These can result in lack of social engagement, isolation and loneliness, which can diminish the health and well-being of older people (WHO, 1946; Donovan and Blazer, 2020). The opposite of this is age-friendly cities, cities that encourage "active ageing by optimizing opportunities for health, participation and security in order to enhance quality of life as people age" (WHO, 2007).

Addressing the needs of older people when designing buildings, from housing to public buildings, is also critical because older people spend more than 70% of their waking hours, indoors (Spalt et al, 2016). So far, however, the most common issues relating to older people that have been addressed in city planning, buildingdesigns and standards are accessibility and safety (for example LAHC, 2014). These issues are addressed through principles for universal design and access, included in building codes and education, though mostly these only occur in the developed countries. In the home environment, these issues often lead to home modifications, including through architectural solutions and smart home technologies, to accommodate the occupants as they age and particularly as they become frail (Peek et al, 2016).

Notwithstanding the progress made in addressing some needs of older people in buildings, considerations andassessments of building designs to ensure thermal comfort needs of older people are still absent in any designguidelines, building codes and standards, environmental assessment schemes, and design evaluation for wellbeing. Addressing the thermal environment is a critical step in planning our cities and building designs tobe age-friendly, because, as we age, we experience physiological changes, such as reduced vascular reactivity, lower metabolic rate, and reduced muscle strength (Blatteis, 2012). These impact on older people's thermal sensation and preference due to changes in their thermoregulation (van Hoof et al, 2017). The decline in the efficiency of cold and warm defense mechanisms in older people means that they become vulnerable to heat- related mortality and morbidity (Klein et al, 2014). Many medical conditions, such as cardiovascular and respiratory diseases, can be exacerbated by being in hot (or cold) environments (Hansen and Soebarto, 2019). It is therefore not surprising that excess deaths during heat waves occur predominantly in older people (Kenney, Craighead and Alexander, 2014). The decline in health among older people due to living in unsuitable environments for their health and wellbeing can put a significant burden not only to their families but also to the public health system.

With cities around the world experiencing unprecedented rapid growth and increasing ageing populations, ensuring that cities and buildings continue to allow older people to be engaged and have quality of life has now become more critical than ever. It is critical for the built environment to continue to minimize its adverse impacts on the environment, yet it is as critical to plan and design our cities and buildings to ensure that the needs of citizens – users of building spaces and occupants of buildings, particularly those who are often forgotten: the older people – are addressed.

This paper discusses some work that places older people not just as passive recipients of city planning and building designs, but instead as active players, termed as 'citizen scientists'. In the first study, with the older citizen scientists we developed an audit tool to be used by older people to gather information about, and reflecttheir experience in using, public and green spaces around their neighbourhoods. In the second study, we tried to understand the living environment of older people – through their eyes and experience – particularly the thermal qualities of their homes and how they responded to them, while we also conducted monitoring / measurements to gather quantitative data about the indoor environment. Results of the study have led to the development of design and operational guidelines to achieve thermal comfort at home while minimizing energy use (thus subsequently reducing greenhouse gas emissions), with the older people playing an active role in developing the guidelines. Through the eyes of older people, we have attempted to address the sustainability of our cities and buildings.

3. Addressing age-friendliness of public and green spaces

Through *Global age-friendly cities: a guide*, the World Health Organization (WHO, 2007) sets up eight topics to be addressed. They are: public spaces and buildings, transportation, housing, social participation, respect and social inclusion, civic participation and employment, communication and information, and community support and health services. In the topic of public spaces and buildings, eleven areas are to be addressed including: pleasant and clean environments, the importance of green spaces, somewhere to rest, age-friendly pavements, safe pedestrian crossings, accessibility, secure environment, barrier-free buildings, and adequate public toilets. Based on this guideline, many local governments have developed and published their own age-friendly built environment guidelines. For example, in Australia, the Government of South Australia publishes *Age-friendly Living: Guidelines for Residential Development* and *Age-friendly Neighbourhoods: Guidelines and Toolkit for Local Government* (Gov. of South Australia, 2012a; 2012b), *Design Guidelines for SustainableHousing & Liveable Neighbourhoods* (Gov. of South Australia, 2017). The Planning Institute of Australia (2013) also publishes *Healthy Spaces and Places. A national guide to designing places for healthy living.*

Notwithstanding the plethora of multiple guidelines and documents such as above, we do not necessarily knowhow older people actually engage with public spaces and the neighbourhoods around them. While the "Age- friendly Cities" movement of the WHO, followed by the involvements of more than 1000 cities from over 40countries, have attempted to address the growing trend of urbanized ageing population since 2007, there is little evidence, based on the voice of older people themselves, to confirm that our cities have been developed and operated to be age-friendly (Scharlach, 2016).

To fill this gap in knowledge, a team of researchers at The University of Adelaide in South Australia conducted a pilot study in 2018-2019 to explore actual utilization of public spaces by older people in the Greater Adelaide Metropolitan area. However, unlike most previous studies that employed methods such as observations of, and focus group discussions and interviews with older people to gather the information, this research employed a'citizen science' approach. This approach is a partnership between professional researchers and 'citizens', whereby the citizens themselves undertake the tasks traditionally done by the professional researchers (Ankeny and Barrie, 2019; Bonney et al, 2016). This cooperation with the citizens can take three different forms: contributive, collaborative and co-created (Bonney et al, 2009). In this pilot study, not only did the citizens contribute to the research by collecting data and collaborate in data analysis and interpretation, they also co-created the research questions and helped developing the research design.

3.1 Older Citizen Scientists

The older citizen scientists were recruited from advertisements in an online magazine produced by the Office for Ageing Well South Australia, newsletter of the City of Adelaide and a community web site of the Councilon the Ageing (COTA) South Australia. Whilst thirty-two people expressed their interests to be involved, only fifteen finally participated, twelve were female and three were male, with age ranging from 60 to 84 years old. The majority lived in the metropolitan area. All except three rated their health to be good to very good.

3.2 Citizen Science Audit Tool

Together with the older citizen scientists a neighbourhood audit tool was developed, based on the principles set out by the WHO's guide for age-friendly cities (WHO, 2007; Barrie et al, 2019). The tool had two formats: online to be used on a smartphone and as a paper booklet. The online version was developed and hosted in anESRI platform called Survey123 (ESRI, n.d). The tool allowed the older citizens to record their experiences in using public and green spaces during their normal daily activities, by answering Likert scale questions relating to a number of key areas including: (1) their state of general health and wellbeing at the time; (2) overall perceptions such as visual/aesthetic perception, aural perception, thermal comfort, lighting quality, feeling of safety and security, state of cleanliness; (3) nature of the visit including reasons for and frequency of the visit,and mode of travel (e.g. walking, by public transport or by private vehicle), as well as (4) any public facilitiesavailable to them around the space (e.g. public toilets, seating). For those using the tool on their smartphone, they were also asked to take photographs of each space they were auditing and enter short comments about their perception of that space. Each time a citizen scientist hit the 'submit' button on their phone, the data were sent to the Cloud, which was then accessed by the university research team. For those using the paper booklet, a disposable was provided, and on the booklet, they were asked to note the exact location of the space being photographed as well as any comments they wanted to add. As the photographs taken by a smartphone were automatically geocoded, and with the citizens using the paper booklet indicating the location of the space they were visiting, the research team was able develop a clear picture about places often visited by older people, places that they preferred or liked, or places they would want to avoid, as well as the characteristics, important aspects and problems that were present in those spaces. Examples of the online version of the audit tool are shown in figure 2.

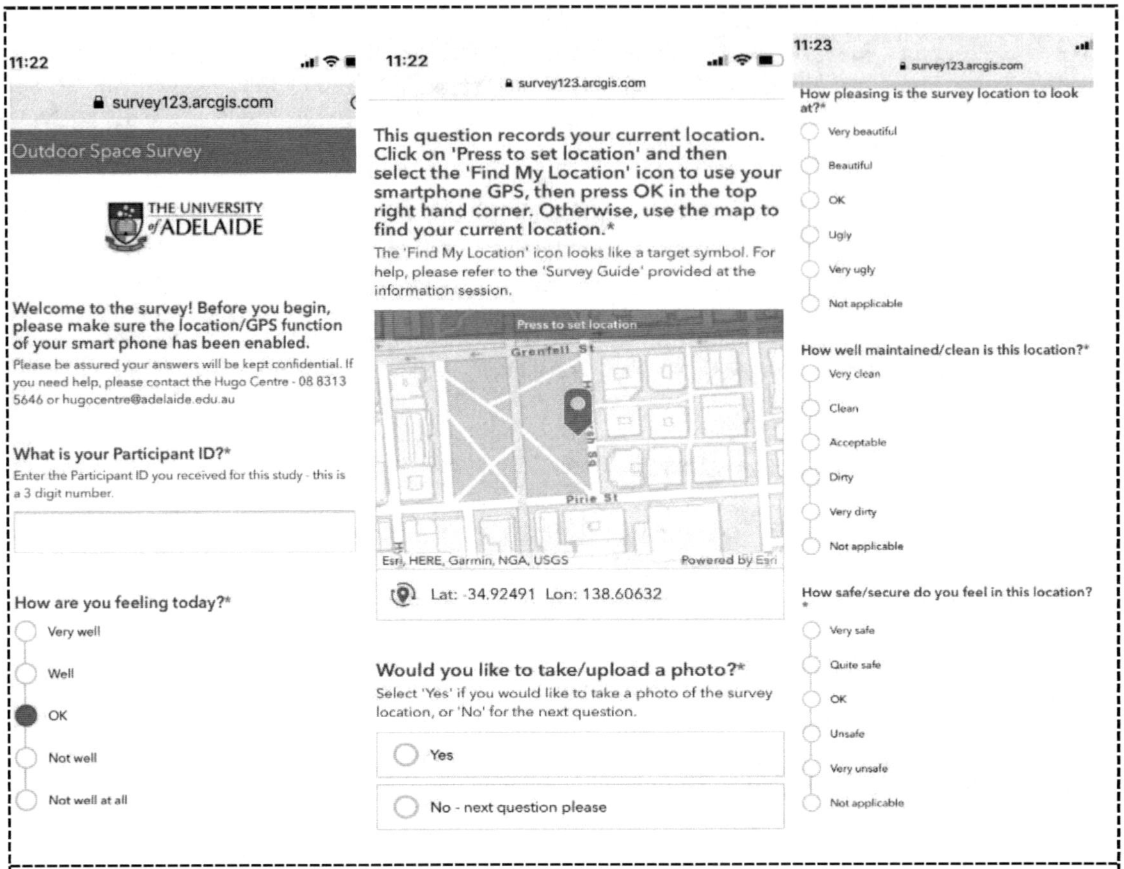

Figure 2. Some screen shots of the smartphone-based citizen science audit tool

3.3 Training and deployment

Each older citizen scientist received an hour training on how to use the audit tool, whether it was by using their smartphone or on paper booklet. Telephone and email assistance was also available to them during data collection. It is interesting to note that although it was initially thought older people would be reluctant to usetechnology (such as a smartphone) and preferred to use the paper booklets, all but two citizen scientists choseto use the smartphone-based audit tool. One person who preferred to use the paper booklet did so simply because this person preferred to use their own camera (as this person was a keen photographer). The other person could not use their mobile phone as it was quite old and did not have enough data allowance. The other13 citizen scientists did not encounter any major issues in learning and using the smartphone-base audit tool. The tendency to use smartphones over the more traditional paper-based material among these older citizen scientists aligned with findings in other studies, for example by Rosales and Fernandez-Ardevol (2016) that demonstrated how older people embraced this relatively new technology.

3.4 Results and lessons learned

The older citizen scientists used the audit tool for 3 months, recording their experiences in using public and greenspaces that they came across during those months at different times of day. In total, 264 audit entries were received with an average 17.6 audits per person. Almost all audits were undertaken within 1.6 km of their home, indicating the radial of the area considered to be their neighbourhood. Spaces audited included cafes, theatres, library, shopping precincts, and public green spaces.

After all audits had been completed, the university researcher presented each citizen scientist with a summary of their audits and each of them was asked to explain their findings. This way the interpretation of the results was done not just by the university researchers but in collaboration with the citizen scientists. Based on all the audits, key design elements in public spaces were identified. They were: seating or benches, street trees, native bushland, park trees, and water feature. A 'good' design of either outdoor spaces or buildings was considered by the citizen scientists as important and having impact on their wellbeing. The presence and quality of green spaces were also critical to them. Many citizen scientists chose to go through a certain route every day instead of taking other routes, due to the presence of trees, grass and birds along the one they chose.

Providing an opportunity for older people to be the 'scientists' to collect data and directly reflect on their experience in using public spaces means that we will have first-hand information about how inclusive and age-friendly our cities are. Out of the eleven areas of the WHO Age-friendly cities guide (2007) that need to be addressed, the older citizen scientists placed importance on, first, "somewhere to rest", and second, "the importance of green spaces" over the other nine areas. This is not to say that the other areas of the guide should not be addressed in planning or developing our cities, but this pilot study has indicated that there is no point of creating public spaces that were safe and secure, accessible, and pleasant, if there is nowhere for the older people to sit down or rest or have access to green space. In other words, it is critical to understand the characteristics of spaces considered to be sustainable, livable and enjoyable, not just for the general population but also from the eyes of our older citizens. Further details about this study can be found in a previous publication by Barrie et al. (2019).

4. Addressing thermal comfort of housing to improve wellbeing

As mentioned earlier, most guidelines, building codes and standards that address the needs of older people onlyfocus on the issues around accessibility and safety. Studies that explore the relationship between building design, indoor environment, thermal comfort and wellbeing of older people have only emerged relatively recently in the early 21st century. The WHO published its first *Housing and Health Guidelines* only in 2018, to provide "provide evidence-based recommendations for healthy housing conditions and interventions" (WHO, 2018), placing the importance of addressing indoor temperatures suitable for older people to minimize health risks. The guideline recommends that for countries with temperate or colder climates, indoor temperatures need to be maintained no lower than 18 °C. Below this temperature adverse health effects, such as cardiorespiratory diseases, are likely to occur particularly among older people, children and those with chronic illnesses. On the other end of the spectrum, 24 °C seems to be the recommended cut-off point based on a previous WHO finding that "there is no demonstrable risk to human health of healthy sedentary people living in air temperature of between 18 and 24 °C" (WHO, 1987). Nevertheless, it is important to note that while high temperatures have been linked to increased mortality, heat stroke, hyperthermia, dehydration and hospital admissions, there is no evidence that temperatures above 24 °C will increase health risks dramatically as people are acclimatized to different temperatures (WHO, 2018). In Boston, USA, for example, the acceptable maximum indoor temperature is 25 °C, whereas in people in Thailand can accept indoor temperature of 32 °C.There is currently no standard of acceptable indoor temperatures for older people in Australia that will support the wellbeing of older people, despite the fact that there are abundant sustainable design guidelines and building code to address thermal comfort in housing in Australia.

4.1 Survey of older people

To fill this gap in knowledge, we have recently completed a 3-years study, funded by the Australian ResearchCouncil, to investigate the living environment of older South Australians who lived at home, termed as 'ageingin place', to understand the relationships between house design, indoor environment quality particularly the thermal quality, thermal comfort, occupant behavior and wellbeing, as well as energy use and cost to achieve thermal comfort.

The study was conducted in three stages, involving more than 300 older people who lived in three climate zones in South Australia: warm arid, warm temperate and mild temperate. The first stage involved a telephone survey of 250 older people, aimed to gather information about the types of housing in which older people lived,their heating and cooling behaviors, and their perceived health and wellbeing status. Results from the study show that while the majority (over 70%) indicated that they had good health and wellbeing, they often had to rely on using heating and cooling to be comfortable in their house, mainly because the house was not properlyinsulated (resulting in the need to heat the house in winter) and not properly shaded (resulting in the need to cool the house in summer). The reliance on heating and cooling meant that it was challenging to reduce energy cost, and the majority of the participants did express concern over this issue. The survey results highlight the importance of good design to provide 'affordable' thermal comfort (Soebarto et al, 2019).

4.2 Conversations with older people

The second stage of the study involved a series of focus group discussions with older people in the three climatezones to gain a better understanding of the survey results. The discussions aimed to understand the thermal qualities of the existing living environment of the participants, their strategies to keep cool in hot weather andwarm in cold weather, their use of heating and cooling, and to identify existing problems related to the planning and design of their homes. Through the discussions, we found that whilst most of our participants had issues with the planning and design of their houses, they were able to take actions to achieve thermal comfort while reducing their energy cost. More importantly, we found that their thermal behaviours were influenced or determined by three factors, namely (1) personal factors, such as age, sex, living arrangement (alone or with other), financial situation and health/wellbeing); (3) knowing, including the knowledge about how to achieve thermal comfort, beliefs and attitudes (particularly towards the environment); and (3) the feeling they would like to have in their home, including their thermal sensation (whether they feel hot, warm, neutral, cool or cold), thermal comfort (whether they feel comfortable or not comfortable) and thermal delight (which is not necessarily about hot or cold, but about the delight of having the sun inside, or smelling the cool breeze). Thesemain factors affected their 'doing', which could be in the form of personal adaptations (such as wearing clothing appropriate for the weather, opening windows when feeling hot), having home modifications (such as doing renovation), or adopting technological solutions (such as installing air-conditioners). Further details about the second stage of this study can be found in van Hoof et al. (2019).

4.3 Actual experience of older people

The third stage of the study adopted the 'citizen science' approach, whereby the older people (who were interested to participate) reflected their thermal experience, daily, or at least twice a week, using online digital survey tablets, for 9 months, covering a period of summer, autumn and winter. During this 9 months-period, we also monitored and recorded the indoor environment parameters (dry bulb temperature, globe temperature, relative humidity, air velocity, and CO_2) every 30 minutes using sensors connected to a central data logger that sent the data to the Cloud, accessible by the researchers, every week. However, as we aimed to compare the results of our study with that of other studies, we had to use an international standard for thermal comfort survey, ASHRAE 55 (ASHRAE 2017), which meant that we could only adopt the first form of the citizen science approach, i.e. the contributive form.

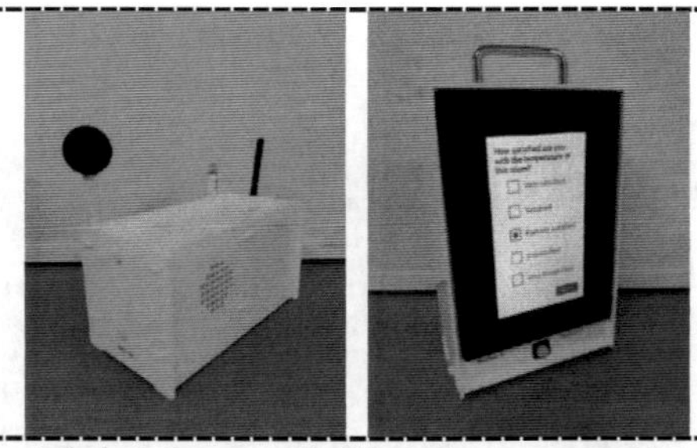

Figure 3. Indoor environmental data logger (left) and survey tablet (right) used in the housing study of older South Australians

In total 71 older people from 57 homes participated in this long-term, with age ranging from 65 to 87 years old and comprising of 48 female and 23 male. The majority had income between $30,000 and 50,000 per year anda similar percentage of participants was in the lower bracket of income as they were in the higher income bracket. We also asked them to indicate their perceived health and wellbeing through questions that could determine their quality of life score using the EQ-5D-5L questionnaire (covering mobility, ability for self-care, ability to do usual activities, experience of pains or discomfort, and experience of anxiety or depression (van Reenen and Janssen, 2015)). We also asked questions that could indicate their frailty level using the Edmonton Frailty scale (Rolfson et al, 2016).

Most of participants' houses had the typical construction types for South Australia: either had brick-veneer walls or cavity bricks with insulated roofs of corrugated metal or concrete tiles. Few of the older houses had uninsulated stone walls or lightweight timber framed construction. Except in the mild temperate climate, most of the houses did not have wall insulation. The older houses had uninsulated timber floors whereas the newer houses had uninsulated concrete slab on ground. On average the floor area was around 156 m^2, which was smaller than the average floor area of houses in Australia (186.3 m^2), indicating the tendency of older people to 'downsize'.

During those 9 months period, the participants or 'citizen scientists' recorded various aspects through the survey tablets, including which room they were in (e.g. living room or bedroom), their clothing layers, their activity, whether or not windows or doors were open, whether or not the heater, cooler or fan was running, andtheir thermal sensation, thermal preference, thermal satisfaction, perception of indoor air quality and air movement, as well as their perceived health and wellbeing at the time. More than 10,000 responses were collected after 9 months together with more than 1,500,000 records of the indoor environmental conditions.

As details of the study have been reported elsewhere (Soebarto et al, 2019; Williamson et al, 2020; Williamson et al, 2022) only highlights of the results are mentioned here. First, while all houses had heaters and all but two had coolers, most participants did not use any air-conditioning to be cool or warm, with less than 40% using the heaters in winter and no more than 15% using the coolers in summer. So instead of relying only on heating and cooling a vast majority applied personal adaptive actions to stay thermally comfortable, for example by wearing lighter clothing and reducing activity during hot weather or wearing heavier clothing and staying active during cold weather.

It is important to point out that while on average the indoor temperature throughout the year in all homes wasaround 22 °C, we found that in some houses the indoor temperature could go down to as low as 6.8 °C and ashigh as 38.1 °C. This is quite a concerning finding, considering that health risks have been found to increase in very low or very high temperatures (WHO, 2018). Concerns over running cost were once again stated as the main reason for those who rarely used heating or cooling.

It is also important to mention that, while the majority of the participants perceived their wellbeing to be 'reasonable' to 'good', our analysis shows that the perception of 'good' wellbeing occurred when the indoor operative temperatures were between 15 and 28 °C. This also coincided with the condition in which they felt the most satisfied with their thermal environment.

The study and findings once again indicate the importance of collecting evidence about actual experience of older people in the built environment, or in this case in their own home. While from the point of view of energyuse and environmental impact, the houses of these older people could be considered as 'energy efficient' due to the low usage of heating and cooling, resulting in relatively low overall energy bills, from the point of view of potential health risks these houses are not as ideal as they may seem. The main reason for the older people to not using much energy for heating and cooling was not because they had a 'sustainable' building that ticked all the boxes of environmental assessment rating tools, but because of the concerns over the running cost. Forpeople who no longer earn regular incomes, reducing utility bills (by not turning on heating or cooling when they actually need it) is often necessary even if that would mean doing something (or not doing something) that would have a negative impact on their health and wellbeing. Our further investigation through building simulation indicated that the thermal performance of many of the houses could be improved, such as by addinginsulation, reducing air leakage by applying drought-proofing (e.g. sealant around window frames), or by adding external shading devices; however, the financial implication of such improvement needs to also be carefully considered (Arakawa Martins et al 2021).

4.4 Improving housing with older people

The final stage of the research was to develop a guideline to achieve thermal comfort in housing of older people. This guideline is based on the main factors identified in the focus group discussions and supported with quantitative data collected through the indoor environment monitoring and the participants' reflections of their daily experiences. Before the actual guideline was written, we discussed the ideas for the structure and content of the guideline with the participants. In other words, we adopted the third form of citizen science – co-create or co-design – in developing the guideline.

Figure 4. Focus group discussions with older 'citizen scientists'

The guideline recommends four main steps that can be undertaken by older people to achieve thermal comfortat home: personal actions, doing something around the house (such as knowing when to open or close the windows and blinds), using equipment and technologies (by understanding different types of heating, coolingand fans, the pros and cons), and doing house renovation, downsizing or moving to another and better house. Some cases based on real house scenarios (but with hypothetical occupants) are presented at the end of the guideline to provide examples for implementing the guideline.

The draft version of the guideline was then presented to and consulted with the participants, and these older 'citizen scientists' had an opportunity to provide very comprehensive feedback to us. Based on the feedback by the citizen scientists, the final guideline has been refined and is currently underway to be made available to general public, community groups, public libraries, and any relevant organizations associated with older people. Since the material that informed the guideline as well as the guideline document itself was sourced from and co-created with the citizens – the older people, we believe the guideline will be a very useful resource to older people. We also expect that policy makers, planners, architects and designers will have a better idea about practices and strategies that are more relevant to older people to achieve thermal comfort in their homes while still maintaining or improving their wellbeing. We do not believe that it is sufficient nor appropriate to simply recommend strategies that are more relevant to the younger, working-age groups, as normally addressed in many existing design guidelines, building code and standards.

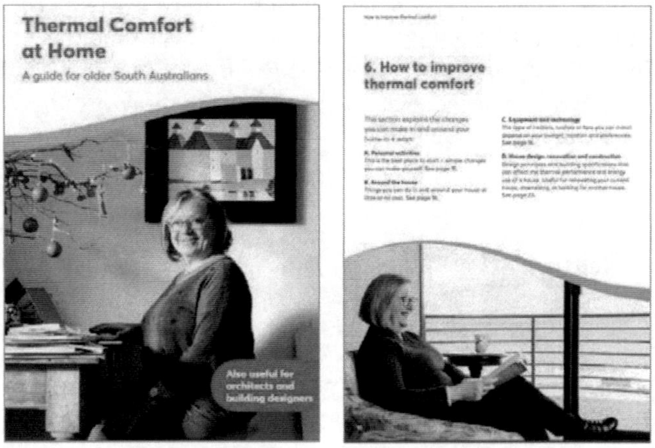

Figure 5. Example pages from Thermal Comfort At Home Guideline

5. Conclusion: Sustainability for all

This paper started with a brief discussion about the pressing issues our cities are facing, from increasing urban density and increasing demand of resources that has resulted in adverse impacts on the environment, to the adverse impacts of our cities and buildings on the environment. Some actions taken by the construction industry in trying to reduce the environmental impacts from buildings have been discussed, particularly through the development and implementation of various environmental assessment scheme and tools. The wider issues ofsustainability to be addressed in cities and buildings – the social and wellbeing of people – are then discussed.

The paper has pointed out that, despite realizing the importance of addressing social issues and wellbeing of people, our cities and buildings tend to be developed with the younger, working-age demographic in mind. Older people, which actually accounts for a significant proportion of the population and will continue to increase its share in the future, are usually forgotten.

On the positive note, recent years have seen some progress being made to ensure that our cities and buildings become more age friendly. The United Nations through the Sustainable Development Goals, as well as the World Health Organization, have set the pathway to achieve this. However, there is still little evidence showing that our cities, public spaces and buildings are indeed relevant to and responding to the need of older people.

Two studies have been put forward to demonstrate how we can improve our cities and buildings by actively engaging with older people and placing them at the centre of such effort. Instead of developing guidelines, standards and policies *for* older people, these studies *gathered evidence, learned from,* and *worked with* older people to develop guidelines and strategies that will be relevant to improving the wellbeing of older people. In the first study, with the older people acting as 'citizen scientists', the quality of public spaces was assessed to identify problems but also elements in the neighbourhoods that were considered important by older people. Such activity is critical if we really want to achieve cities that are truly sustainable, environmentally and socially. The second study focuses on one aspect of sustainable buildings: the thermal quality of the indoor environment. Likewise, the older people played an active role in sharing, reflecting and recording their daily thermal-related experiences in their homes, while the related parameters of the indoor environment were being measured and recorded by the university researchers. The study has identified not only the thermal conditionsthat are considered conducive to the wellbeing of the occupants, but also resulted in a guideline to achieve thermal comfort in housing, co-created with the older people.

While the two studies only addressed specific issues in achieving sustainable cities and buildings, they respond to the Sustainable Development Goal 11 to achieve cities that are inclusive, safe, resilient and sustainable (UN, 2007) by including the older citizens to have a say about their living environment. Sustainability and sustainable cities, communities and buildings cannot be achieved by simply applying formulas and focusing only on the actively working population in the society. With the changing demographics around the world and the ever-increasing proportion of the older population, it is time to look at how to achieve a truly inclusive, safe, resilient and sustainable world through the lens of older people.

6. References

[1] Ankeny, R. and Barrie, H. (2019). Learning Not Just from but with Citizens: The Importance of Co-Design in Health-Related Social Research. American Journal of Bioethics 19: 54–56.

[2] Arakawa Martins, L., Williamson, T., Bennetts, V., and Soebarto, V. (2021). The use of building performance simulation in developing thermal comfort guidelines for the living environment of older South Australians. Journal of Building Performance Simulation. Under review.

[3] ASHRAE (2017). ANSI/ASHRAE Standard 55-2017, Thermal Environmental Conditions for Human Occupancy. American Society of Heating, Refrigerating and Air-Conditioning Engineers, Inc., Atlanta, 2017.

[4] Barrie, H., Soebarto, V. I., Lange, J., Breen, F., & Walker, L. (2019). Using citizen science to explore neighbourhood influences on ageing well: pilot project. Healthcare, 7(4), 126-1-126-13.

[5] Blatteis, C.M. (2012). Age-dependent changes in temperature regulation- a mini review. Gerontology 58(4): 289-295.

[6] Bonney, R., Phillips, T.B., Ballard, H.L., and Enck, J.W. (2016). Can citizen science enhance public understanding of science? Public Understanding of Science 25: 2–16.

[7] Cassidy, R. et al. (2003). White paper on Sustainability. A report of Green Building Movement. Building Design & Construction. Supplement, 2-47, November.

[8] Coman, R.L., Caponecchia, C.D. and Gopaldasani, V. (2021). Impact of public seating design on mobility and independence of older adults. Experimental Aging Research, 47(3): 262-272, DOI: 10.1080/0361073X.2021.1884445.

[9] Donovan, N.J. and Blazer, D. (2020). Social Isolation and Loneliness in Older Adults: Review and Commentary of a National Academies Report. The American Journal of Geriatric Psychiatry 28(12): 1233-1244.

[10] ESRI (n.d). ArcGIS Survey 123. Available at: https://www.esri.com/en-us/arcgis/products/arcgis-survey123/overview. Accessed 26 October 2021.

[11] GBCA (2021). Communities. Green Star Rating system. Green Building Council Australia. Available at https://new.gbca.org.au/green- star/rating-system/communities/. Accessed 26 October 2021.

[12] Gov. of South Australia (2017). Design Guidelines for Sustainable Housing & Liveable Neighbourhoods. Adelaide, South Australia.

[13] Gov. of South Australia (2012a). Age-friendly Living: Guidelines for Residential Development. Adelaide, South Australia.

[14] Gov. of South Australia (2012b). Age-friendly Neighbourhoods: Guidelines and Toolkit for Local Government. Adelaide, South Australia.

[15] Hansen, A., & Soebarto, V. (2019). The Epidemiology of Health and Mortality at Extremes. In S. Roaf, & W. Finlayson (Eds.), Proceedings of the 1st International Conference on: Comfort at the Extremes: Energy, Economy and Climate, pp. 10-22. Cambridge: Ecohouse Initiative Ltd.

[16] International WELL Building Institute (2021). WELL. Available at

https://www.wellcertified.com/. Accessed 26 October 2021.
[17] Kenney, W.L., Craighead, D.H. and Alexander, L.M. Heat waves, aging, and human cardiovascular health. Medicine & Science in Sports Exercise 46(10): 1891–1899.
[18] Kibert, C.J. (2018). Green Buildings: An Overview of Progress. Florida State University Journal of Land Use and Environmental Law. 19: No. 2 , Article 11. Available at: https://ir.law.fsu.edu/jluel/vol19/iss2/11. Accessed 26 October 2021.
[19] Klein, R.J., Kinney, P.L., et al. (2014). Intra-urban vulnerability to heat-related mortality in New York City. Health Place 30: 45-60.
[20] Morellec, F.L., Anastassova, M. and Falzon, P. (2013). The environment: a source of capabilities for older adults? ECCE '13. Proceedings of the 31st European Conference on Cognitive Ergonomics, Aug, Toulouse, France. DOI 10.1145/2501907.2501945.
[21] Peek, S.T.M., Luijkx, K.G., Rijnaard, M.D., … van Hoof J, et al. (2016) Older adults' reasons for using technology while aging in place. Gerontology 62(2):226-237.
[22] Planning Institute of Australia (2013). Healthy Spaces and Places. A national guide to designing places for healthy living. Heart Foundation, Planning Institute of Australia, Australian Local Government Association: Kingston, ACT.
[23] Richards, J.A. and Bradshaw, S. (2017). Uprooted by Climate Change: Responding to the growing risk of displacement. https://www- cdn.oxfam.org/s3fs-public/file_attachments/bp-uprooted-climate-change-displacement-021117-en.pdf. Accessed 26 October 2021.
[24] Ritchie, H. and Roser, M. (2018). Urbanization. OurWorldInData.org. Accessible at https://ourworldindata.org/urbanization. Accessed 26 October 2021.
[25] Rolfson, D.B., Majumdar, S.R., Tsuyuki, R.T., Tahir, A. and Rockwood, K. (2006). Validity and reliability of the Edmonton frail scale, Age Ageing 35: 526–529.
[26] Rosales, A. and Fernandez-Ardevol, M. (2016). Smartphones, apps and older people's interests: from a generational perspective. MobileHCI '16: Proceedings of the 18th International Conference on Human-Computer Interaction with Mobile Devices and Services, September, pp. 491–503. Available at https://doi.org/10.1145/2935334.2935363. Accessed 26 October 2021.
[27] Scharlach, A. E. Age-Friendly Cities: For Whom? By Whom? For What Purpose? in (Ed.) Moulaert, T. & Garon, S. Age-Friendly Cities and Communities in International Comparison, Vol 14, Chapter 17: 305-329, Springer.
[28] Seto, K.C. et al. (2014), Human settlements, infrastructure, and spatial planning. Climate Change 2014: Mitigation of Climate Change: Contribution of Working Group III to the Fifth Assessment Report of the Intergovernmental Panel on Climate Change (IPCC, Geneva, Switzerland), Chap 12, pp 923–1000.
[29] Seto, K.C., Güneralp, B., Hutyra, L.R. (2012) Global forecasts of urban expansion to 2030 and direct impacts on biodiversity and carbon pools. Proceedings of National Academy of Science USA, 109(40):16083–16088.
[30] Soebarto, V., Bennetts, H., Hansen, A., Zuo, J., Williamson, T., Pisaniello, D., . . . Visvanathan, R. (2019). Living environment, heating-cooling behaviours and well-being: Survey of older South Australians. Building and Environment 157: 215-226.
[31] Soebarto, V., Williamson, T., Carre, A., & Arakawa Martins, L. (2019). Understanding indoor environmental conditions and occupant's responses in houses of older people. In IOP Conference Series: Materials Science and Engineering Vol. 609: 1-7. online: IOP.
[32] Spalt, E.W., Curl, C.L., Allen, R.W., …, Kaufman, J.D. (2016). Time-Location Patterns of a Diverse Population of Older Adults: The Multi-Ethnic Study of Atherosclerosis and Air Pollution (MESA Air). Journal of Exposure Science & Environmental Epidemiology 26(4): 349–355.
[33] TNP2K (2020). The Situation of the elderly in Indonesia and access to social protection programs: Secondary Data Analysis. The National Team for the Acceleration of Poverty Reduction. Jakarta, Indonesia.

[34] United Nations (n.d). The 17 Goals. Sustainable Development Goals. Department of Economic and Social Affairs. Accessible at https://sdgs.un.org/goals. Accessed 26 October 2021.
[35] United Nations Department of Economic and Social Affairs, Population Division (2020). World Population Ageing 2020 Highlights: Living arrangements of older persons (ST/ESA/SER.A/451). United Nations, New York. Available at https://www.un.org/development/desa/pd/sites/www.un.org.development.desa.pd/files/undesa_pd- 2020_world_population_ageing_highlights.pdf Accessed 26 October 2021.
[36] United Nations Department of Economic and Social Affairs, Population Division (2015). Population 2030: Demographic challenges and opportunities for sustainable development planning (ST/ESA/SER.A/389). United Nations, New York. Available at https://www.un.org/en/development/desa/population/publications/pdf/trends/Population2030.pdf
[37] United Nations Department of Economic and Social Affairs, Population Division (2012). World Urbanization Prospects, the 2011 Revision: Highlights. United Nations, New York. Available at www.un.org/en/development/desa/publications/world- urbanization-prospects-the-2011-revision.html. Accessed 26 October 2021.
[38] United Nations Development Programme (2021). Ageing, Older Persons and the 2030 Agenda for Sustainable Development. United Nations, New York. Available at https://www.undp.org/publications/ageing-older-persons-and-2030-agenda-sustainable-development. Accessible 26 October 2021.
[39] van Hoof, J., Bennetts, H., Hansen, A., Kazak, J., & Soebarto, V. I. (2019). The living environment and thermal behaviours of older South Australians: a multi-focus group study. International Journal of Environmental Research and Public Health 16(6): 935-1-935-19.
[40] van Hoof, J., Schellen, L., Soebarto, V., Wong, J., & Kazak, J. (2017). Ten questions concerning thermal comfort and ageing. Buildingand Environment 120: 123-133.
[41] van Reenen, M, and Janssen, B. (2015). EQ-5D-5L User Guide Basic Information on How to Use the EQ-5D-5L Instrument, EuroQol Research Foundation.
[42] WHO (2018). WHO housing and health guidelines. World Health Organization, Switzerland.
[43] WHO (2007). Global Age-friendly Cities: A Guide. World Health Organization: Geneva.
[44] WHO (1987). Health impact of low indoor temperatures. Report on a WHO meeting. Copenhagen, 11–14 November 1985. World Health Organization, Switzerland.
[45] WHO (1946). Preamble to the Constitution of the WHO as Adopted by the International Health Conference, New York, 19–22 June 1946. World Health Organization: Geneva.
[46] Williamson, T., Soebarto, V., Bennetts, H., Arakawa Martins, L., & Pisaniello, D. (2020). Thermal comfort, well-being and health of older residents in South Australia. In S. Roaf, F. Nicol, & W. Finlayson (Eds.), Proceedings of the 11th Windsor Conference: Resilient comfort, pp. 171-186. online: Windsor, UK.
[47] Williamson T, Soebarto V, Bennetts H, Arakawa Martins L, Pisaniello D, Hansen A, Visvanathan R, Carre A, van Hoof J. (2022) Assessing human resilience: A study of thermal comfort, wellbeing and health of older people (Chapter 7). In Nicol, Roaf, Rijal (Eds) The Handbook of Resilient Thermal Comfort, Routledge, UK.
[48] World Green Building Council (2021). Advancing net zero. Available at: https://www.worldgbc.org/thecommitment. Accessed 27 October 2021.
[49] Zito, G.A., Cazzoli, D., Scheffler1, L., Jäger1, M., Müri, R.M., Mosimann, U.P., Nyffeler, T., Mast, F.W. and Nef, T. (2015). Streetcrossing behavior in younger and older pedestrians: an eye- and head-tracking study. BMG Geriatrics. (2015) 15:176. DOI 10.1186/s12877-015-0175-0.

Acknowledgments
I would like to acknowledge Dr Helen Barrie of The University of South Australia (was at The University ofAdelaide) who led the citizen science project to investigate the use of public spaces by

older people. I wouldalso like to acknowledge my University of Adelaide collaborators in the study to investigate and improve thethermal environment of housing of older people: Professor Dino Pisaniello, Dr Alana Hansen, Professor Terence Williamson, Dr Helen Bennetts, Professor Jian Zuo, and Larissa Arakawa Martins, as well as the Partner Investigator, Professor Joost van Hoof from The Hague University of Applied Sciences. The contributions of all research participants, who cannot be named for privacy reasons, are highly valued and greatly acknowledged. The first study was funded by The Office of Ageing Well, South Australia. The second study was funded by the Australian Research Council through the Discovery Project scheme, ARC DP180102019.

IOP Conf. Series: Earth and Environmental Science 1007 (2022) 012005 doi:10.1088/1755-1315/1007/1/012005

Passive cooling performance on Indonesia contemporary tropical facade in producing the present comfortable space

A M Nugroho[12]

[1] Department of Architecture and Center of Environmental Studies, University of Brawijaya, Veteran Malang Street, Indonesia

Abstract. This paper investigates passive cooling solutions on an Indonesia Contemporary House in response to housing challenges in dense urban areas. The research approach involves detecting temperature and humidity in current settings, then using ventilated and bioclimatic facade to reduce the room's temperature and compare it to its neutral temperature value. The bioclimatic facade has the best performance, lowering the temperature by 2.5°C when compared to the ventilated wall, which reduced the temperature by 2.2°C. Passive cooling techniques, such as bioclimatic and ventilated facade applications, can create a comfortable space all day long.

1. Introduction

Indonesia architecture empirically can be interpreted as an art, knowledge and technology of built environment, that focuses on form, space, place and culture in tropical islands between two continents and two oceans (Southeast Asia) containing natural and human values. The science of Indonesia architecture is inseparable from the system of belief-knowledge and its architectural dimensions. Indonesia architectural dimensions begin with the first dimension which is the observation point, every human needs an observation point with the result that Indonesia architecture always contains focus and detail elements. The second dimension is the plane, which embodies ornaments diversity as the face of nature in the built environment. The third dimension is the space as Pangarsa states [1] that vernacular space characteristics are used by many activities (producing space) and not used individually (consuming space). The fourth-dimension concerns aspects of place that related to changes in the meaning of space over time (morning, afternoon, evening and night) and users. The fifth dimension is the knowledge aspect in the present world [2]. These five dimensions are also the development tendency of scientific architecture in the world, that appears with the term: sustainable architecture, green architecture, climatic responsive architecture, participatory design and smart building [3].

The tradition of tropical adaptation is a long process and becomes a form of creation and innovation according to the era. Tropical Architecture as part of Indonesia architecture science or knowledge is a cross-scientific study through deepening on tropical architecture theoretically and empirically in the context of Indonesia architecture sustainability. Research and design steps to actualize Indonesia tropical architecture sustainability, among others: first, identifying dimensions of the vessel, users and built environment. Second, developing in two directions, to the outside as a universal scientific contribution and to the inside as a root of national identity. Third, efficient use of Indonesia tropical architecture on dimensions of points, planes, spaces, places, and cultural knowledge [4].

[2] Corresponding author: sasimurti@yahoo.co.id

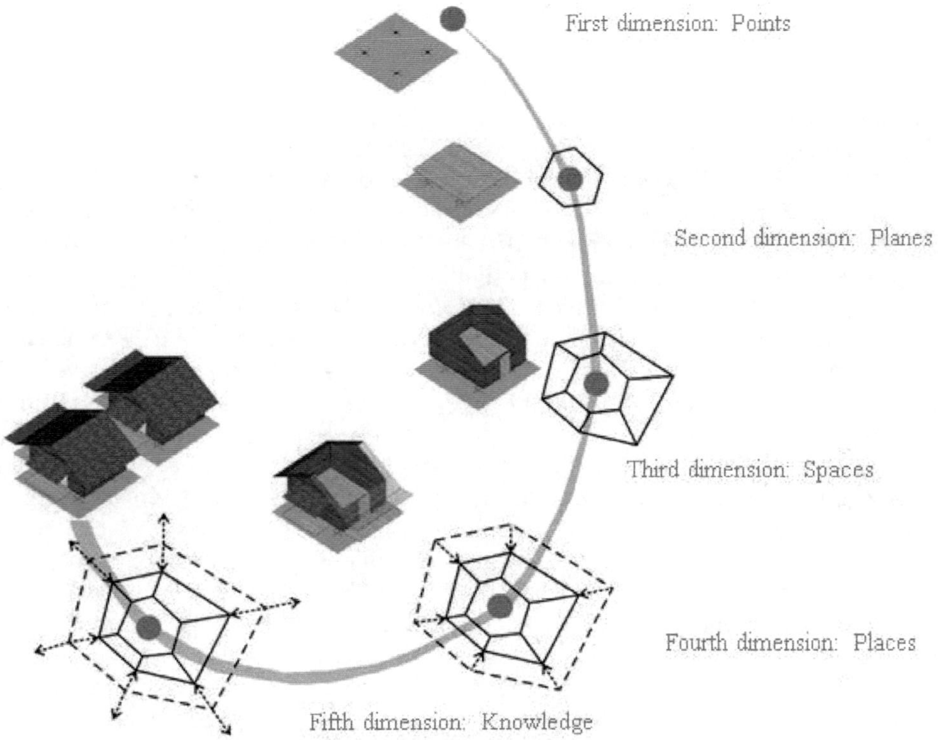

Figure 1. Indonesia architectural dimensions [4]

Karyono [5] defines Tropical Architecture as an architectural work that provides solutions to climate problems in its environment. Nugroho [4] defines Indonesia tropical architecture as an art, knowledge and technology of built environment located in tropical islands of Southeast Asia; focuses on nodes, forms, spaces, places and science through techniques of identification, growth and development for the renewal of nature and human values. The discussion regarding tropical buildings must be approached from climate aspects, particularly by the field of building science or architectural science, in order to provide more precise and measurable answers. The rapid growth of the city has changed the urban environment along with the loss of green areas that have a significant influence on comfort level in buildings. Comfort conditions in a building increased significantly due to excessive esposure to solar radiation outside the house [6]. In tropical climate, particularly windows element in building envelope has a big impact in heat conditions during the day [7]. Thermal comfort conditions of a building that are suitable with building user needs can be achieved through strategies of temperature reduction during the day, either through passive or active methods.

This temperature reduction technique can be done by modifying climate factors, human body heat transfer and building components. Climate factors include solar radiation, temperature, wind and humidity. The aspect of human heat transfer can be done by calculating the temperature physiologically to achieve human thermal comfort [8]. The thermal comfort index is determined by environmental parameters, they are temperature, solar radiation, humidity and wind speed. These environmental parameters can be changed by building components that are climate responsive [9]. Buildings that have enough shade indicate the condition of indoor temperature is lower than outdoor temperature. Several studies [10], [11], [12], [13] show the relationship between climate factors, building users and building components towards the creation of better thermal comfort conditions in the tropics.

Passive cooling is a design strategy for achieving temperature at human comfort threshold, with the following stages: formulating a neutral temperature of a place based on climate data (locus); recognizing dominant building elements related to function (focus); performing a series of measurement techniques to reach thermal comfort level (mode) and achieving optimal results for thermal environment (novus). Neutral temperature is temperature at a comfortable level in particular area based on the average temperature per month in a year by using climate data of 5 years minimum, according to the formula developed by Szokolay [8].

In vernacular buildings, ventilated facade placement, such as windows and doors at certain points, is for creating airflow. Ventilated facade that can be opened is able to reduce heat and moisture, but it can increase the room temperature due to the solar heat infiltration if it is unprotected. Ventilated facade on the east and west sides are the main parts that need to be protected, because solar heat is higher at those orientations [4]. Concerning ventilation size is the most efficient way to create successful cross ventilation. Comparison of indoor temperature shows a difference of more than 1.5°C between cross-ventilated and non-ventilated houses [14]. Prianto and Depecker [15] examined various types of window lattices to increase comfort level through increasing wind speed in the room. The lattice window with a 45 degree angle is able to reach comfortable condition for light and moderate activities.

Planting trees around the building is one way to control temperature in the building [17]. Research studies related to vertical garden as part of tropical houses nowadays are based on aesthetic and functional values. Vertical garden presents a number of advantages to cool spaces as well as productive plants. Based on results of the research by Nugroho [18], vertical garden with red spinach species produced a reduction in temperature maximum of 5°C while for green mustard produced a reduction in temperature maximum of 2°C. Red spinach that have darker color has more significant impact on reducing temperature. Types of plants with darker leaf colors influence the increasing solar heat absorption, so that they are able to reduce the surrounding temperature. Based on the Indonesia tropical architecture concept in the past, it can be formulated that there are several strategies in producing a comfortable space, through applications of ventilated facade and integration with plants or better known as bioclimatic.

2. Method

According to Malama and Sharples [19], there is a general view that traditional architecture is more climate responsive than contemporary architecture. This is interesting to prove through a series of measurable researches to determine passive cooling performance on contemporary architecture in tropical climate. Research studies conducted in dry season at the house object representing contemporary tropical house in Malang City, Indonesia. The passive cooling performance of the house is assessed by measuring temperature and humidity inside and outside the house. Measurements were made in June using hobo data logger that was placed in the centre of each room and measurements data recorded every hour at a height of 90 cm with parameters and equipments specification according to table 1. Measurements were made in rooms that represent existing conditions, a space with applications of ventilated facade and bioclimatic facade, simultaneously in June 2018.

Table 1. Measuring equipments parameters and specification on contemporary house.

Equipments	Equipments	Accuracy (◦C) (%)	Note
Indoor and outdoor temperatures	Hobo data logger	$\pm$ 0.1	Recorded every 60 minutes
Indoor and outdoor humidities	Hobo data logger	$\pm$ 1	

(a) bioclimatic facade

(b) ventilated facade

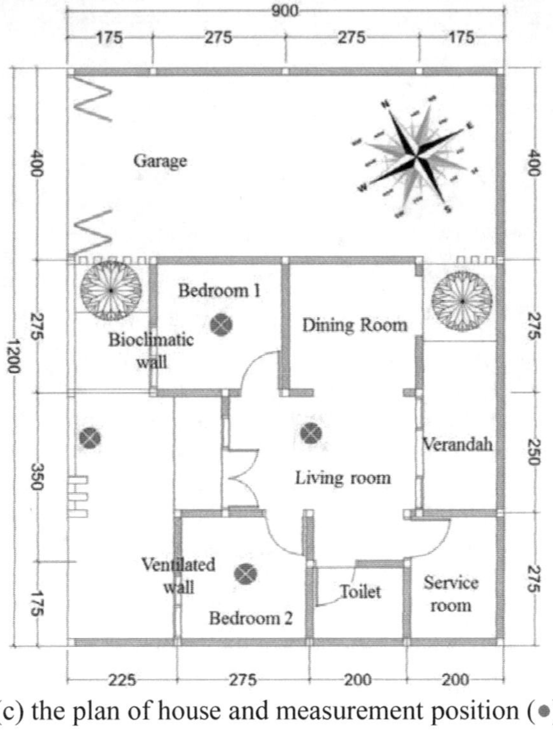

(c) the plan of house and measurement position (•)

Figure 2. Research object of contemporary Indonesia tropical house

3. Result and discussion

The results of passive cooling performance on contemporary Indonesia tropical house design according to research stages are including neutral temperature condition of the City; passive cooling performance on the existing house; temperature reduction performance on ventilated facade; temperature reduction performance on breathing facade; temperature reduction performance on bioclimatic facade; and comparison of passive cooling performances.

3.1. Neutral temperature in Malang City

The comfort zone is formulated based on Szokolay equation [8] by entering annual temperature data per month according to BMG weather data in Malang City. The range of comfortable zone is 5°C with the detail of 2.5°C above and below neutral temperature. The calculation result shows the neutral temperature of Malang City is 24.9°C, so the comfort zone upper limit is 27.4°C and lower limit is 22.4°C. This neutral temperature applies to conditions without airflow. The most comfortable period occurs between July and August, related to the northern turning point of the sun. The most uncomfortable period occurs between October and November, which is because it is a period of months between equinox and the southern turning point of the sun, in the same time also coincides with the change of dry season to rainy season and position of the sun is just above Malang City.

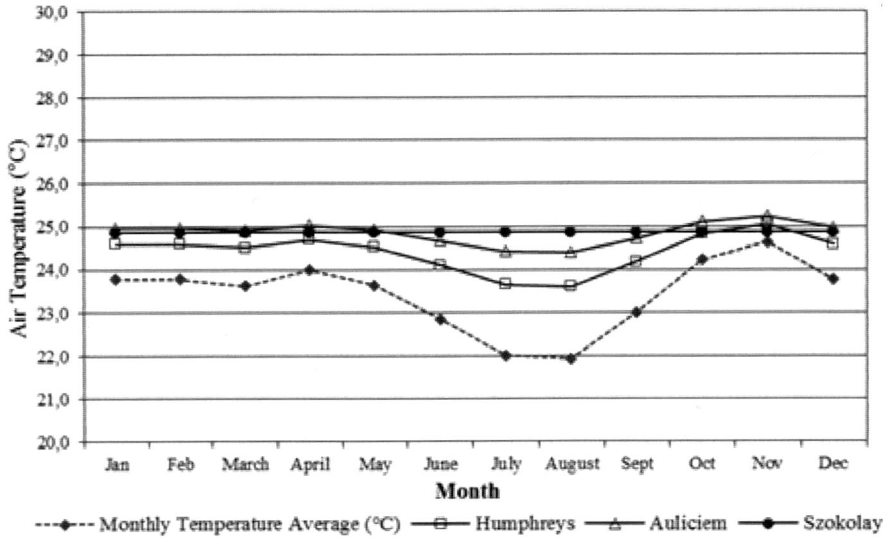

Figure 3. Neutral temperature in Malang City

3.2. Passive cooling performance on existing conditions of contemporary Indonesia tropical house

Hourly data on temperature in the existing room can be shown in Figure 4. The average outdoor temperature during March-August is above the neutral temperature in Malang City at 11:00 until 15:00. Based on measurements in the existing house, it can be seen that indoor temperature is below outdoor temperature at 08:00 until 14:00. The highest indoor temperature is around 28°C occurs at 14:00. The indoor temperature looks higher than outdoor temperature in the afternoon and night. The temperature starts to increase at 15:00 and reaches the peak at midnight until 07:00. The measurement also shows the existing house can reduce indoor temperature but is still above thermal comfort neutral temperature in Malang area (24.9°C) at 13:00 until 15:00. Based on measurements, it can be seen that indoor temperature is below outdoor temperature at 09:00 until 15:00. The highest temperature reduction is 1.5°C at 12:00 until 13:00. Efforts should be made to improve Passive cooling performance, so that indoor temperature is in neutral temperature area (22.4°C until 27.4°C), particularly at 13:00 - 15:00.

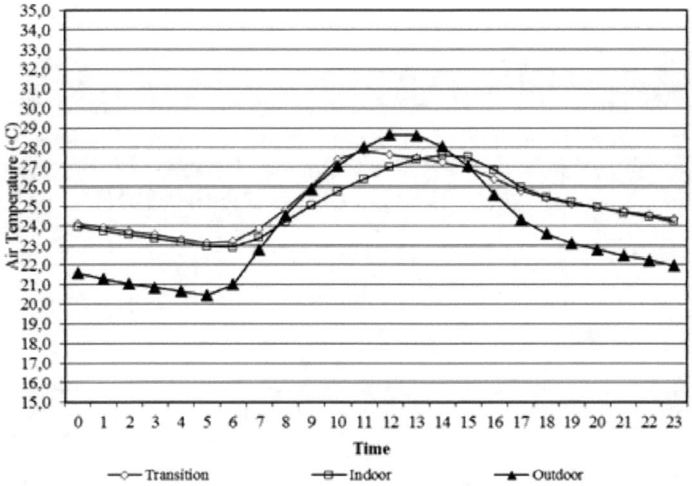

Figure 4. Comparison between indoor and outdoor temperatures on the object of contemporary Indonesia tropical house

3.3. Temperature reduction performance on ventilated facade application in contemporary Indonesia tropical house

The first strategy is ventilated facade application that cover the lower, middle, and upper parts of room in the existing house. Measurements were made for a month, in June 2018 with data logger equipment for temperature and humidity. The measurement results indicate that indoor temperature is below indoor temperature without ventilation and outdoor temperature at 10:00 until 15:00. The average indoor temperature with ventilated facade application is 25.3°C or within the range of neutral/comfortable temperature of Malang City. Average temperature reduction performance during the daytime from 09:00 until 15:00 with the ventilated facade application (2.2°C) is higher than the existing (0.9°C) which is equal to 1.3°C. The indoor temperature has been successfully within the area of neutral temperature (22.4°C until 27.4°C) throughout the day.

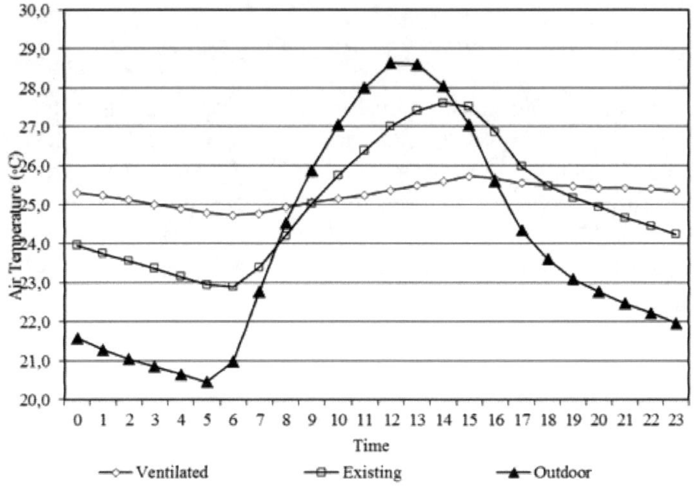

Figure 5. Comparison between the room with ventilated facade and the existing room on contemporary Indonesia tropical house

3.4. Temperature reduction performance on bioclimatic facade application in contemporary Indonesia tropical house

The second strategy is bioclimatic facade application that includes the double skin facade existence that is 1.5 meters from the wall and the presence of a pool in a room inside the existing house. Measurement results indicate that indoor temperature without bioclimatic double skin facade is below outdoor temperature at 09:00 until 18:00. The average indoor temperature with bioclimatic facade application is 25°C or within the range of neutral/comfortable temperature in Malang City. The average temperature reduction performance throughout the daytime from 09:00 until 15:00 on bioclimatic facade application (2.5°C) is higher than the existing (0.9°C) which is equal to 1.6°C.

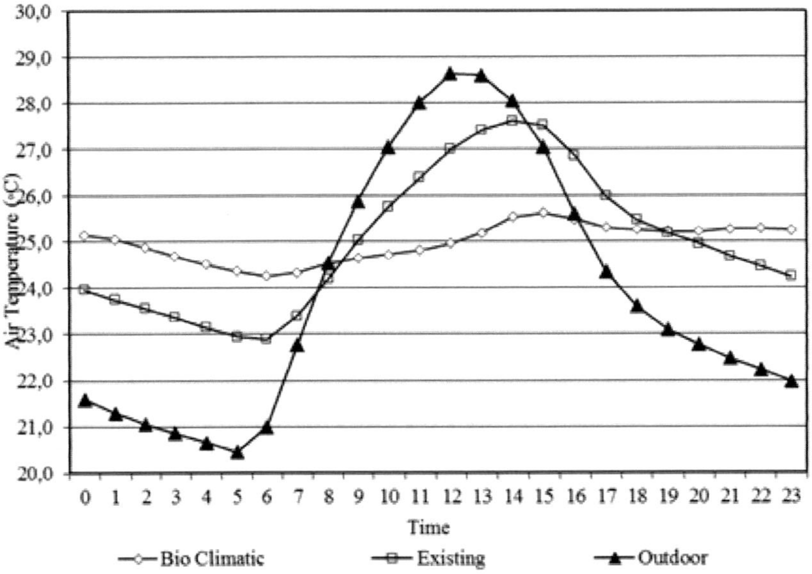

Figure 6. Comparison between the room with bioclimatic facade and the existing room on contemporary Indonesia tropical house

3.5. Comparison of Passive Cooling Performance on Facade Applications in Contemporary Indonesia Tropical House

Comparison of passive cooling performance is done by measuring the temperature reduction in a space simultaneously at four rooms, each room applies a variety of tropical facades. The average indoor temperature with each application strategy indicates that bioclimatic facade has the best performance in creating indoor temperature according to neutral temperature in Malang City at 25.2°C during the day and 25°C for 24 hours. The presence of a pool and vegetation is part of the bioclimatic wall concept, which also includes a double-skin facade that is 1.5 meters from the wall.

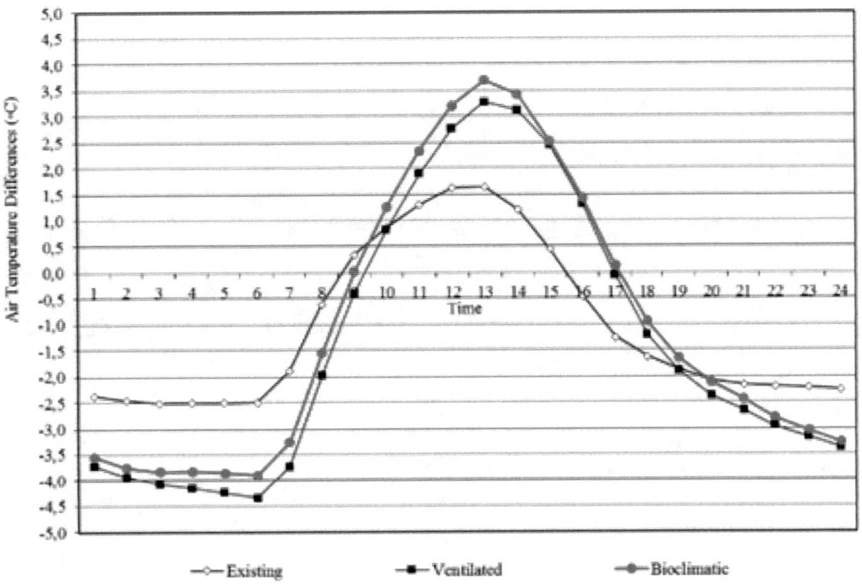

Figure 7. Comparison of temperature reduction performance between ventilated facade and bioclimatic facade on Contemporary Indonesia Tropical House

4. Conclusion

Based on the calculation of neutral temperature in Malang City, it can be seen that the comfortable room temperature is 24.9°C where the comfort level of temperature is between 22.4°C until 27.4°C. Thermal conditions in the existing house show that during the day the indoor temperature is higher than neutral temperature, so that passive cooling strategies based on Indonesia tropical facades are needed. Tropical facade applications are ventilated facade and bioclimatic facade. The results indicate that bioclimatic facade has the higher passive cooling performance, which can reduce temperature by 2.5°C, compared with ventilated facade at 2.2°C. Passive cooling strategies through applications of bioclimatic facade and ventilated facade on Indonesia tropical architecture concept are able to produce a comfortable space throughout the day, therefore artificial ventilation systems that waste energy are no longer needed.

5. References

[1] Pangarsa G W 2010 *Arsitektur Kemanusiaan, Teropong Visual Eko Prawoto* (Surabaya: Andi Offset)
[2] Nugroho A M 2013 *Proc. Seminar Temu Ilmiah Ikatan Peneliti Lingkungan Binaan Indonesia*
[3] Nugroho A M 2014 *Proc. Seminar Arsitektur Merah Putih*
[4] Nugroho A M 2018 *Arsitektur Tropis Nusantara Rumah Tropis Nusantara Kontemporer* (UB Press)
[5] Karyono T K 2000 *Desain Arsitektur.* **1** 7
[6] Ahmed A 2004 *Proc. Thermophysics Conference*
[7] Givoni B 1998 *Energy Build.* **28** 1
[8] Szokolay S V 1998 *Proc. Passive and Low Energy Architecture International*
[9] Hamdan M and Nugroho A M 2005 *Proc. World Renewable Energy Regional Congress and Exhibition*
[10] Rajeh M 1989 *Natural Ventilation in Terrace Housing of Malaysia : Effect of Air Well on Air Flow and Air Velocity* (University of Queensland)
[11] Abdul Malek A S and Young A 1993 *Proc. International Energy Conference*
[12] Abdulrahman S and Kannan K S 1997 *Proc. The Asia – Pasific Conference on the Built Environment: Integrating Technology with Environment*

[13] Sapian A R 2004 *Possibilities of using Void to Improve Natural Cross Ventilation in High-rise Low Cost Residential Building* (Doctoral dissertation, Universiti Teknologi Malaysia)
[14] Garde F, Adelard L, Boyer H, Rat C 2004 *Energy Build.* **36** 4
[15] Prianto E, Houpert P and Depecker J P 2001 *International Journal on Architecture Science* **1** 4
[16] Gut P, Ackerknecht D 1993 *Climate Responsive Building Appropriate Building Construction in Tropical and Subtropical Regions* (SKAT)
[17] Nugroho A M 2014 *Proc. Green Architecture for Better Future*
[18] Malama A and Sharples S 1996 *Archit. Sci. Rev.* **39** 95

Acknowledgments
The author thanks Universitas Brawijaya Indonesia for the support through the Research Grant Scheme: "Hibah Penguatan Kelompok Kajian 2021".

Influence of landscape material properties on microclimate change in a tropical area with case in Manado City

S Sangkertadi[1,2], **R Syafriny**[1], **C E V Wuisang**[1], **and Z Zahra**[1]

[1]Dept.of Architecture, Sam Ratulangi University, Manado 95115, Indonesia

Abstract. The objective of this research is to show the influence of thermal properties of landscape materials on microclimate change of tropical environment in the city of Manado. The microclimatic variables of this study include surface temperature, air temperature, and solar radiation. The landscape materials as object of the research were bricks, hard-soil, concrete-block, trees, and garden-pond water. The research used descriptive quantitative method. Measurement of thermo-physical characteristic of brick and paving stone were conducted by using calorimeter and quick-thermal-conductivity meter. Measurements of solar radiation, and temperatures were also performed in a commercial area in Manado city. Calculations of heat exchanges were realized by using HEAT2 computer program in order to evaluate magnitude of surface temperature of various landscape materials. The results show that a dense leaves tree can reduce solar radiation in its shade by up to 80% at a mid-daytime. The landscape materials of red brick, light concrete-block and hard soil can cause a surface temperature by up to 50 ^{0}C under a clear open sky. Water surface temperature of garden pond can also reach 50 ^{0}C, while at a depth 50 cm can decrease to 25 ^{0}C.

1. Introduction

In general, people who carry out activities in outdoor spaces feel physiologically comfortable when they are in an environment with temperatures that are not too hot and not too cold, especially with the addition of cool breezes. However, the heat of the sun which can be very hot during the day, causes thermal discomfort, so humans try to find shelter under the shade, for example under a tree or under some other shade, and also look for water for cooling. In this case, apart from atmospheric climatic factors, the morphology and physical anatomy of the outdoor landscape architecture also play a role in causing variations in the level of thermal comfort for humans who carry out activities, both at night and during the day. Landscape morphology that composed of various types of material can affect changes in microclimate, especially environmental temperature [7]. However, the thermophysical properties of landscape materials are an integral part of the morphological characteristics. Even the surface of water may also affect the microclimate [3]. The thermophysical properties of the material will affect its surface temperature and the surface take role as heat radiator to the environment when the material receives a heat penetration of solar radiation at outdoor space [9]. Radiant temperature of outdoor landscape environment mostly affected by the heat capacity and emissivity of surface material [12]. The physical morphology of the greenery also has a significant effect on the filtration of solar thermal radiation to the points under shade [5]. Since the thermophysical characteristics of the material are known, it will then be designing a landscape architecture with optimize to serve a comfortable microclimate to the user [1]. The microclimate variables that play the most role on the perception of thermal comfort in outdoor spaces are radiant temperature, air temperature and wind [2], [4]. Previous studies mostly used reference

[2] Corresponding author: sangkertadi@unsrat.ac.id

materials where the physical properties were be taken from literatures published in developed countries like in Japan, US, Europe, and Australia. Meanwhile in this study it used local materials, where the thermophysical properties are not mentioned in the literatures. Therefore, it should be measured in order to know the values that may be difference from the literatures. Specifically, this study aims to obtain information of the fluctuation of microclimate variables due to the application of various types of landscape materials, which include type of pavement in the location (brick, paving, soil), water and trees. The microclimate variables were the surface temperature of the material, air temperature, and penetration of solar radiation. As case study, the site is in the coastal area of the Mega Mas commercial area in Manado City, Indonesia. In this area, the landscape formation consists of pavement material, trees, and there is also a garden water pond.

2. Methodology

This research used a quantitative description method that include steps of measurements, practical calculations and computer simulations. Location of field measurements was in the Mega Mas beach area, in Manado city.

2.1. Measurement of thermal properties of materials

The selected local materials which are to be measured in this study were local red brick, and concrete-block. Measurements were to determine the value of specific heat capacity (c) in Joule/kg^{0}C, thermal conductivity in W/m^{0}C, and density in kg/m^3. The measurement instruments were Joule Calorimeter, digital balance, thermocouple, and quick-thermal-conductivity meter.

2.2. Microclimate measurement

Outdoor microclimate measurements that include air temperature, surface temperature and solar radiation, were performed in a clear day when the sun was not covered by clouds. The equipment consists of infrared thermometer (to measure the surface temperature of the material), solar power meter (to measure solar radiation) and digital thermometer (to measure air temperature). Measurements of surface temperature and radiation were carried out at a point under the shade of a dense leaves tree, and at another point that face to a clear open sky or exposed to the direct sunlight. In addition, measurements of the body of water in a garden pond were also carried out at the surface and at the bottom of 50 cm deep. Measurement of air temperature was conducted at the height of human activity, 1.5 m from the surface of the land. The activity of measurements was performed in a coastal area of Manado City (1.5 LU) on 10 January 2021.

2.3. Calculation simulation

The calculation and simulation of conduction heat transfer of the materials were performed by using the HEAT2 computer program. The objective is to determine the surface temperature of the material due to radiation penetration from the sky or by sun light. The geometrical modelling to be applied to the program is explained schematically in the Figures 1 and 2. The simulation by HEAT2, applied transient calculation mathematical model. The boundary at the top of the material was radiant temperature that varies every hour. This represents the penetration or product of sun radiation to the surface. A daily solar radiation in the tropics was estimated by using equations that refer to some references ([9],[11]) as shown in Figure.4. By approximation, a conversion calculation of solar and sky radiation to radiant temperature (*tr*) is applied (Equations (1) to (4)). This *tr* would be then as boundary on the upward of the material (Figure 2). The bottom boundary is characterized by soil at deep 10 m, where the temperature was set constant 25 ^{0}C. A research on soil heat flux in the tropics, show that in a depth of 10 m the temperature is relatively constant at 20 ^{0}C [14], while another study shows that the soil temperature is constant 28.95 ^{0}C also in a depth of 10 m [13]. Therefore, in this study the median value of 25 ^{0}C was taken.

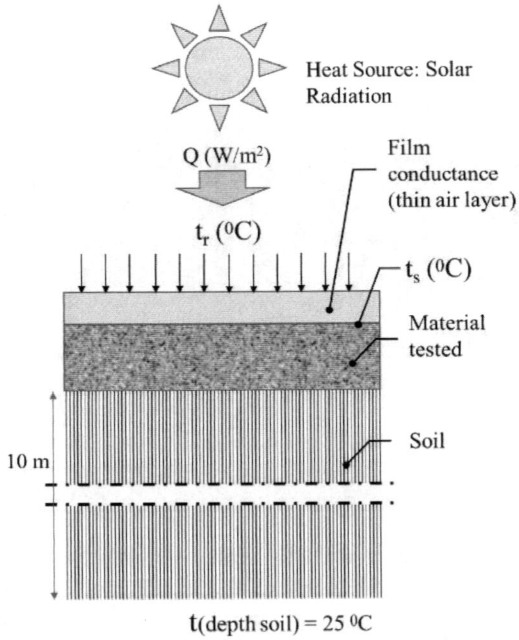

Figure 1. Material system modeling for computational simulation process

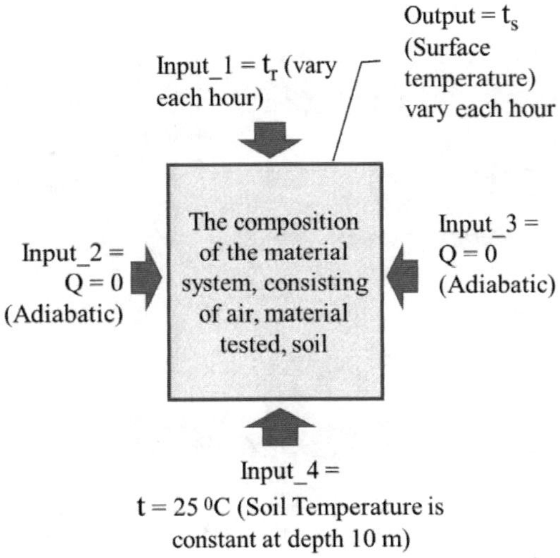

Figure 2. Input-boundary layer and output modeling for 2-dimensional heat transfer simulation using HEAT2

Figure 3 shows the modelling of the conversion from solar radiation (short wave radiation) and celestial radiation (long wave radiation) to radiant temperature, by introducing an imaginary black body that absorbs all radiation from above and transmits all heat energy downward. The radiation temperature conversion calculation process can be set by applying the general radiation equations as follows:

$$Q_r = \varepsilon\sigma T_r^4 \tag{1}$$

$$I_{dir} + I_{dif} + L_w = \sigma T_r^4 \tag{2}$$

$$L_w = h_r(t_{sky} - t_a) \tag{3}$$

$$T_{sky} = 0.0553\, T^{1.5} \tag{4}$$

Where I_{dir} is direct sun radiation, I_{dif} is diffuse radiation, and L_w is longwave radiation. Equation (4) is known as swinbank equation [10], where T is in Kelvin. t_r (in °C) can be then estimated by those equations. In a situation where t_a is bigger than t_r, and impact of wind speed is neglected, tr can be then approximatively taken as same as t_a. Figure. 4 shows the variation of sun radiation (I_{dir} and I_{dif}) and radiant temperature that be applied in the simulation.

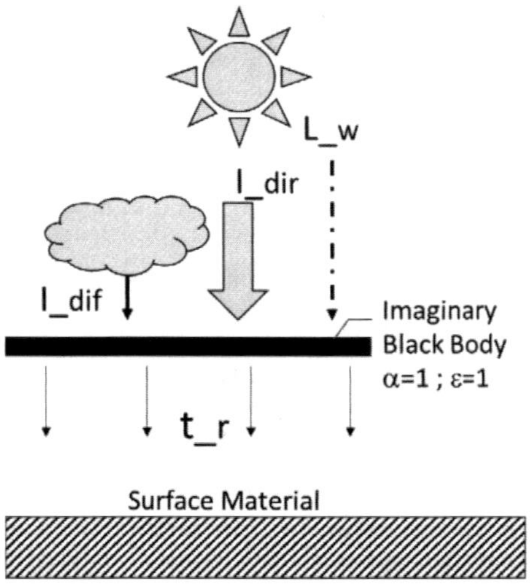

Figure 3. Heat penetration on surface

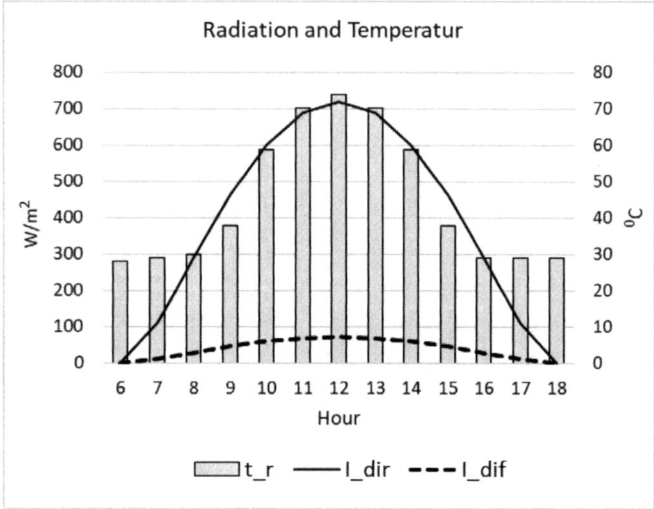

Figure 4. Solar radiation and radiant temperature

3. Results and discussion

3.1. Microclimate in the coastal area of Manado City

The measurement showed that due to hot solar direct radiation, the surface temperature of the hard and dry soil was almost reach 50 ^{0}C at the peak hottest during the day (Figure 5). Meanwhile, when under the shade of a tree (Tanjung tree species), the highest surface temperature of the material was close to 35 ^{0}C at 13.00 or about 60% compared to other point that face directly to skylight. The surface temperature of the soil from those two situations begins to differ after 08.00, when the influence of the sun has begun to make the material hotter. The pattern of heat changes, showing a curve that peaks at midday, and decreases in the morning and evening. The surface temperature pattern of the material shows the same tendency as solar radiation fluctuation that measured at the same time. The amount of solar radiation was almost 1050 W/m^2, at 12 noon at a point that face to open sky. On the other hand, at another point were shaded by the leaves, the highest value was close to 150 W/m^2 at 12 noon or reduced drastically to around 80%. Other researcher has also found the similar result. A study in tropical area, in Penang Malaysia, a dense leafy tree could reduce the soil temperature by 17% and could filter solar radiation by about 80% [8].

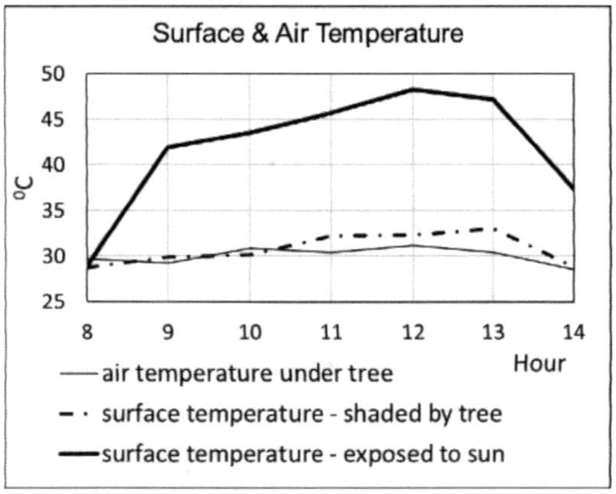

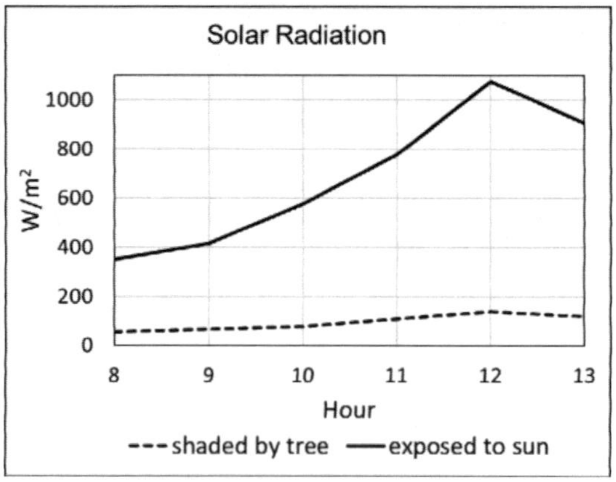

Figure 5. Results of measurement of surface temperature (left) and sun radiation (right)

3.2. Thermophysical properties of local materials

Measurement of heat capacity and thermal conductivity of concrete block and red brick materials showed that different results from those that are generally found in several literatures. Specific heat capacity of local concrete block reaches 1423 J/kg^{0}C, with thermal conductivity of 1.06 W/m^{0}C. While the local red brick has specific heat capacity of 1090 J/kg^{0}C and thermal conductivity 0.43 W/m^{0}C (Table.1)

Table 1. Thermal properties of local materials

Properties	Unit	Concrete tile block	Local Red brick
Density[a]	kg/m3	1947	1662
Specific Heat Capacity[a]	J/kg ^{0}C	1423	1090
Thermal conductivity[a]	W/m^{0}C	1.06	0.43
Absorption coefficient from solar radiation [b]	%	60	90
Emissivity[b]	-	0.9	0.9

[a] By measurement
[b]Data from the SNI (National Indonesia Standard)

3.3. Hard surface temperature

Calculations of surface temperature of the local materials i.e. light concrete block and red brick, and dry soil were conducted by using HEAT 2. The results show (Figure 6) that at midday, the highest surface temperatures can reach in range between 45 ^{0}C to 50 ^{0}C, that occurs from 13.00 to 15.00 (Figure 6). In this simulation, the highest air temperature applied is around 32 ^{0}C at 12.00, according to the results of microclimate measurements. Peak hour of air temperature is generally following the peak heat radiation of the sun at 12.00. There are delay in peak heat flux of surface temperature due to the inertia property of materials. This is also known as time-lag of the material against the heat load. In the morning, the surface temperature can be lower than the air temperature (Figure 6). This can happen, because the air warms up faster due to the influence of human activities in the outdoor space, and also because of the influence of the sun rising since 06.00. It appears that the surface temperature of the concrete block is lower than the surface of the red brick at midday. The concrete block material has a light colour, so there is enough heat from the sun that is reflected, and the rest is absorbed which causes the heat of the material and its surface. While the red bricks are darker in colour, so more heat is absorbed by the material which also causes the high surface temperature of the material. Colour can affect to a higher or lower surface temperature.

A study in a tropical climate, in Malaysia, also showed that a white-coated asphalt can reduce temperatures by up to 60% compared to dark-coloured asphalt [16]. In addition, the conductivity of concrete is higher than that of red brick, which causes the heat flow from the concrete into the soil to be greater than that of red brick, so that less heat is retained in the concrete. The higher temperature of surface material will consequently cause a radiant heat increase of the outdoor microclimate. The results of the calculation also show that the surface temperature of the material starts to be higher than the air temperature from 08.00 until the evening. It reaches a peak close to 50 ^{0}C at 13.00. In Japan, as comparison, the surface radiation temperature of paving-concrete material can reach a peak of 40 ^{0}C even though the climate is cooler than in the tropics, as stated by Takebayashi and Kyogoku [6].

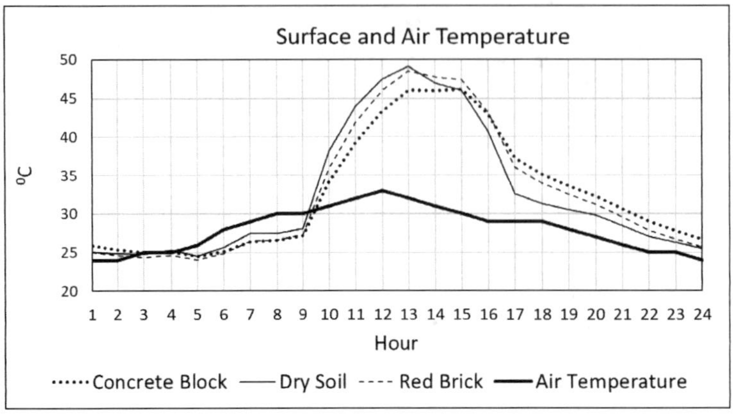

Figure 6. Simulation of surface temperature by HEAT2

3.4. Water temperature

A simulation calculation of water temperature using HEAT2 can show changes in temperature according to depth. At the surface, the surface water temperature can be almost the same as the surface temperature of other hard pavement materials, which can reach peak peaks of around 50 °C (Figure 7). However, at a depth of 20 cm, the peak temperature has decreased considerably to around 25 °C. Even at a depth of 50 cm, the water temperature can relatively constant around 25 °C as same as ground soil temperature, in period from 7.00 to 17.00. This simulation is only limited to the case of water that is not moving, there is no mixing of temperatures between the bottom and the surface, due to convective movement. To lower the surface temperature of the water, it may be necessary to mix it with the bottom, or there is a movement of flowing water. Even the surface temperature of a flowing river water in tropical climates can still be quite high and can reach almost 40 °C [15].

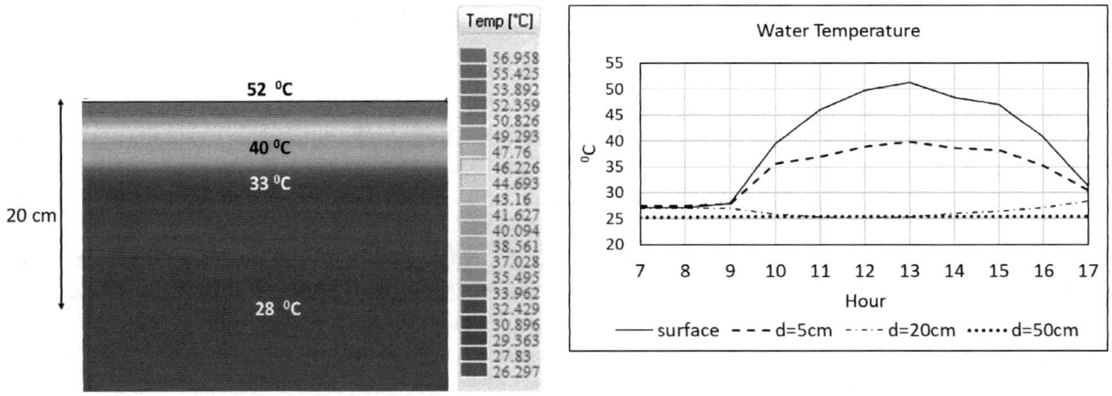

Figure 7. Water temperature

4. Conclusion

The thermophysical properties of landscape materials affect the change of microclimate, in particular the mean radiant temperature that be caused by the surface temperature of the material. The vegetation factor in the form of dense leafy trees has the potential to reduce or filter solar radiation that can up to 80% at midday. In addition, it can also reduce the surface temperature of the material under their shade by up to 30%, when the sun is vertically above it.

The surface temperature of the pavement materials, namely paving stone, brick and hard soil, during the day can reach almost 50 °C at midday. The surface temperature of the garden pond water under open sky can quite hot at midday, therefore, to cool it, it is needed an intervention to move the water in order to gain a temperature mixture with the other side which is cooler.

5. References

[1] G Y Choi, H S Kim, H Kim, J S Lee, 2021, Int.J of Clim. Ch.Str and Mgmt, **13,**2 97-119 (2021)
[2] B Givoni, Climate Considerations in Building and Urban Design, (J Wiley & Sons, New York, 1998).
[3] M Lesi, Iklim Mikro di Sekitar Permukaan Air. (Project Research Report, Pascasarjana IPB, 2017)
[4] Sangkertadi & R Syafriny, J. of Urb. and Env.Eng, **10**, 2, 177-185 (2016)
[5] FM Shahidan, E Salleh, KMS Mustafa, 2007. Effects of tree canopies on solar radiation filtration in a tropical microclimatic environment. In Proceeding of The 24th Conf. of PLEA, Singapore 22-24 Nov, 2007.
[6] H.Takebayashi and S Kyogoku, MDPI Sust. J. **10**, 337, 1-11 (2018)
[7] SKSO Thani & S M S Abdullah, Int. J of Behav. Std,. 2, **7**, 31-41 (2017).
[8] J Md Tukiran, J Ariffin J, AN Abdul Ghani, Int. J. of Geomate, **11**, 24 2275-2283 (2016)
[9] SV Szokolay S V. Introduction to Architectural Science: The Basis of Sustainable Design. Oxford: (Architectural Press , 2010)
[10] Y Qin, Jacob E. Hiller, 2011, Const & Build Mat, **25**, 3753-3761 (2011).
[11] J Widen & J Munkhammar. Solar Radiation Theory. (Uppsala University, Swedia, 2019)
[12] J Huang, J G Cedeno-Laurent, J D Spenglen, Build & Env, **80,** 84-95 (2014)
[13] A Mukhtar, MZ Yusoff, NK Ching. IOP Conf. Series: J.of Phys.**949,** 012011 (2017)
[14] W Cui, Q Liao, G Chang, Q Peng, T C Jen, Measurement and prediction of undisturbed underground temperature distribution. In Proceedings of the ASME 2 international mechanical engineering congress & exposition, IMECE2011, Denver, CO, USA, January 1 2011.
[15] G Bwire, D A Sack, A Kagrita, T Obala, A K Debes, M Ram, H Komakechi, CM George, and CG Orac, 2020. BMC Pub. Health.**20,**1128.
[16] J Md Tukiran, J Arifin, A N Abdul Ghani, Jurnal Teknologi (Sci. & Eng), **78**,5, 63-70 (2016)

Acknowledgements

Acknowledgement is addressed to the Ministry of Education, Culture, Research & Technology, Republic of Indonesia for financing the research in fiscal year 2021.

IOP Conf. Series: Earth and Environmental Science 1007 (2022) 012007 doi:10.1088/1755-1315/1007/1/012007

Measurement and calculation of indoor climate of swallow's nests house in Minahasa

L J Lambey[1], Sangkertadi [2,3], M Regar[1], R Syafriny [2] and S J K Umboh[1]

[1]Sam Ratulangi University, Dept. of Animal Husbandry, Manado, Indonesia
[2]Sam Ratulangi University, Dept. of Architecture, Manado, Indonesia

Abstract. Swallow birds (*Collocalia fuciphaga*) are found in many islands in Indonesia. In general, swallows like sites which are warm and humid for their house and making nests. Swallow's nest is a product of high economic value. Many people are interested in building houses for swallow to gain profit from the nest. The aim of the study is to describe the thermophysical characteristics of building materials of the house, and its indoor air temperature. The methodology used is descriptive analysis. The study includes measurements of outdoor and indoor climate variables which are air temperature, surface temperature and humidity. Identification of thermophysical material include heat specific capacity, thermal conductivity and density of wall and roof material. The results of climatic measurements carried out on 7 and 8 May 2021 in Minahasa showed that the indoor temperature was in the range of 25 to 27 ^{0}C, and relative humidity of around 84 to 85 %. Geometric and thermophysical modelling were made for simulation calculations by using HEAT2 software. The calculation results show that the distribution of indoor air temperature is in range of 25.6 to 25.8 ^{0}C, where in average there is a difference of about 0.5 ^{0}C lower than the measurement results.

1. Introduction

Swallow's nest is a product of high economic value, among others, as a highly nutritious food. Indonesia's export volume of swallow nests reaches almost 1300 tons with a value of US$ 540.4 million in 2020 according to the National Statistics Agency of Indonesia. 80 percent of the world's swallow's nest requirement is supplied from Indonesia in 2020, as mentioned by Indonesia Minister of Trade (https://www.suarasurabaya.net/ekonomibisnis/2021). Since about 2 decades, many people and businessmen are interested in building houses for swallow nests and enjoying the products that have economic value in trade and exportation. In this case, knowledge in building sciences can contribute to the development of swallow's nest houses, especially regarding heat exchange performance in order to ensure the indoor climate favoured by swallow. The swallow's macro habitat is around the coast and areas overgrown with plants or forests [1]. The preferred habitat types are forests, rice fields, settlements, rivers, lakes and vacant land/grasslands [3]. The ideal swallow microhabitat is an area that has air conditions with a temperature of 27-29 ^{0}C and humidity above 70% [4], quiet, safe, hidden and not much disturbed by predators, therefore swallow can be easy to attach their nests and easy to get out and enter the room. The same condition was also expressed by Arif Budiman in Setiawan [5], that the preferred air temperature in the swallow's house is between 26 to 29 ^{0}C. Temperatures that are too low can reduce nest productivity, while humidity that is too high can cause fungi growth in nesting sites and mosquito growth in puddles in the swallow house [2].

[3] Corresponding author: sangkertadi@unsrat.ac.id

The purpose of this study is to reveal the thermophysical characteristics of the material and indoor space of the building favoured by the swallow bird. This is an application of methods in building science to the field of biology or animal sciences. A Case study and observation were carried out in a swallow house building that have existed for a long time ago and is still favoured by swallows from several generations. A case in this study is a swallow house in Lemoh Barat Village, Minahasa Regency, North Sulawesi (Figures 2 and 3). In addition, heat transfer simulation calculations were also carried out from a similar design model house to obtain information of the effect of building envelope materials to indoor climate.

2. Methodology

This research applied a quantitative description method that includes measurement, and computer simulations. Figure 1 show the flowchart of methodology. A sample of house of swallow's nest was taken as object of the study (Figures 2 and 3). The dimension of the house is 15.4 (l) x 8.2 (w) x 14.2 (h). Number of floors is 5. Main structure is made from concrete, zinc roof, light concrete hollow brick for walls, and light concrete for floor.

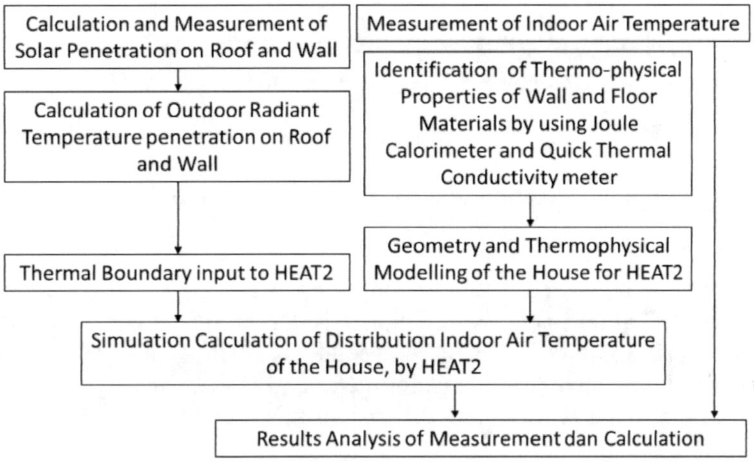

Figure 1. Flowchart of methodology

Back & Side Elevation Side Elevation

Figure 2. Elevations of the sample house

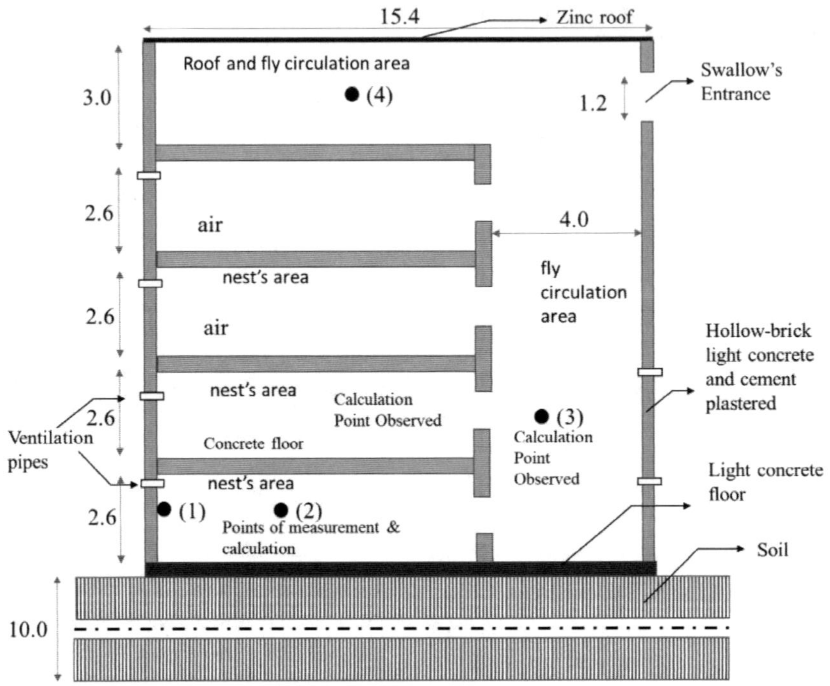

Figure 3. Section of the sample house (without scale)

2.1. Measurement of thermal properties of materials

The material as object to be measured the thermophysical properties were light concrete hollow-brick. The material is a product of a local small-scale industry. Measurements were to determine the value of specific heat capacity (c_p) in Joule/kg^{0}C and thermal conductivity in W/m^{0}C, as well as density in kg/m^3. The measurement instruments include Joule Calorimeter and Quick Thermal Conductivity-meter. The thermal properties of other materials that are also part of this research, such as zinc and light concrete of floor, are taken from the literatures.

2.2. Microclimate measurement

Indoor and outdoor climate measurements include air temperature, surface temperature, solar radiation and humidity. The equipment consists of infrared thermometer (to measure the surface temperature of the material), and digital thermo-hygrometer (to measure air temperature and humidity). Locations of measurement of indoor air temperature were in the centre of room and near the inner surface of wall in 1st floor.

2.3. Calculation simulation

Calculation of transient heat transfer simulation uses the HEAT2 computer program, which was to determine the distribution of indoor air temperature due to radiation penetration from sun to the walls and due to a very little air ventilation. Geometrical modelling that to be applied to the program is explained graphically in Figure 3. Penetration of solar radiation resulting radiant temperature on surfaces of the building. These temperatures which is also known as sol-air temperature (Equation 1)were then introduced as thermal input boundary at outer side of walls and roof in the calculation process by HEAT2. The value of sol air temperature (Equation 1) depend on global solar radiation (G in W/m^2), outdoor air temperature (t_0), surface absorption to solar radiation (α), and heat exchange coefficients (h). It was used value of h=22.73 W/m^{2}K, refer to Indonesia Standard on Energy Efficiency of Building Envelope: SNI-6389:2011. The value of G was estimated by using equations of common solar radiation theory [6], as shown in Figure 4.

$$t_{sa} = t_o + \frac{G\alpha}{h} \quad (1)$$

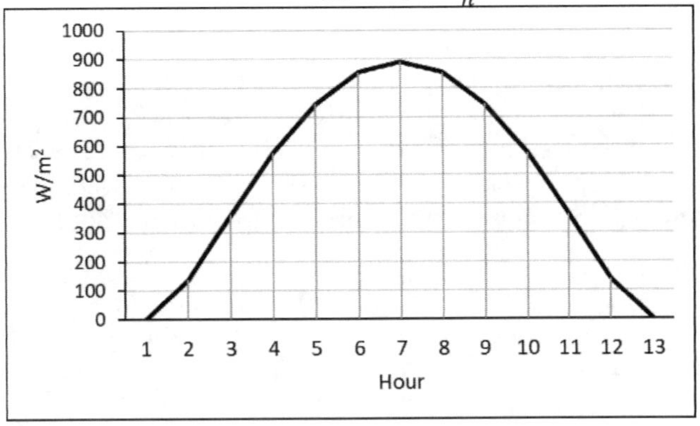

Figure 4. Solar radiation on horizontal surface for simulation

3. Results and discussion

The results of measurements of specific thermal capacity, density and thermal conductivity of the hollow brick light-medium concrete are shown in the Table.1.

Table.1. Measurement results of thermal properties of hollow brick concrete

Properties	Unit	Value from Measurement	Value from Reference
Density	kg/m3	1688	1300 – 1700[a]
Specific Heat Capacity	J/kg ^{0}C	1408	960[a]
Thermal conductivity	W/m^{0}C	0.29	0.1 – 0.3[a]

[a]www.engineeringtoolbox.com

Measurement of indoor air temperature were carried out at two points: near wall and at the centre of room. Air temperature near the wall is greater 0.5 to 1.0 ^{0}C than those in the middle of the room in average. The mean air temperature in the centre of the room were 26.4 ^{0}C, while those near the walls were 26.9 to 27.4 ^{0}C. The distribution of air temperatures is still within the range of the swallow's preference [4]. The average inner side wall surface temperature ranges from 27 to 29 ^{0}C, while the outer side of wall surface temperature can reach greater than 40 ^{0}C (Figure 5). Outside air temperature varied from about 25 ^{0}C to about 34 ^{0}C (Figure 6). Measurement of indoor air humidity show that relatively constant in range of 84.3 to 85%, while at outdoor was 60% to 87% (Figure 6).

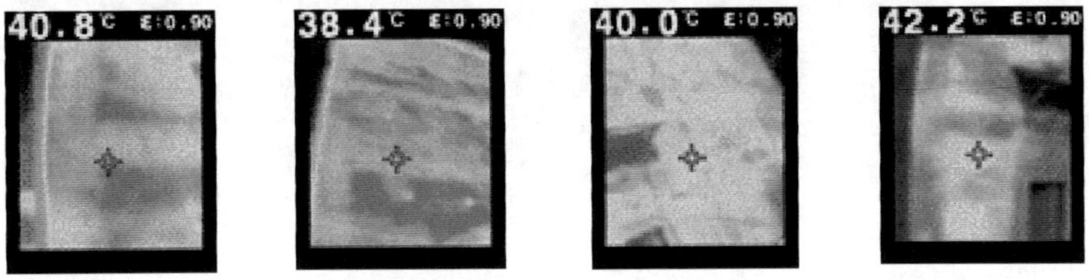

Figure 5. Outside vertical wall surface temperature at some positions at 15.00

Through simulation calculation by using HEAT2, it can obtain many values of result, with possibility to visualize air temperature distribution in the rooms. Figure 7 shows visualization of temperature pattern of the simulation at 14.00, where it appears that the hotter air temperature is just under the zinc roof and around the opening or around the edge of inner wall. The hot inner surfaces are due to a significantly solar heat penetration from outdoor by means of heat conduction process through building

IOP Conf. Series: Earth and Environmental Science 1007 (2022) 012007 doi:10.1088/1755-1315/1007/1/012007

envelopes. The magnitude of air temperature in the centre of the room was varied between 25 to 27 ^{0}C, which means it is close to the measurement results (Figure 8). The outdoor temperature varies from about 25 ^{0}C to about 34 ^{0}C (Figure 6). The air temperature in the centre of room at first floor is relatively constant around 25 to 27 ^{0}C although air temperature outside the room varies with a maximum-minimum difference of up to 9 ^{0}C. This condition can occur due to two things, firstly it is supported by the thermophysical properties of the wall material, which is able to store heat energy, and restrain the rate of heat from the outside, secondly, is due to the ventilation openings which are very small to introduce a large amount of air flow from the outside, therefore the hot air from outdoor does not affect the heat changes in the room.

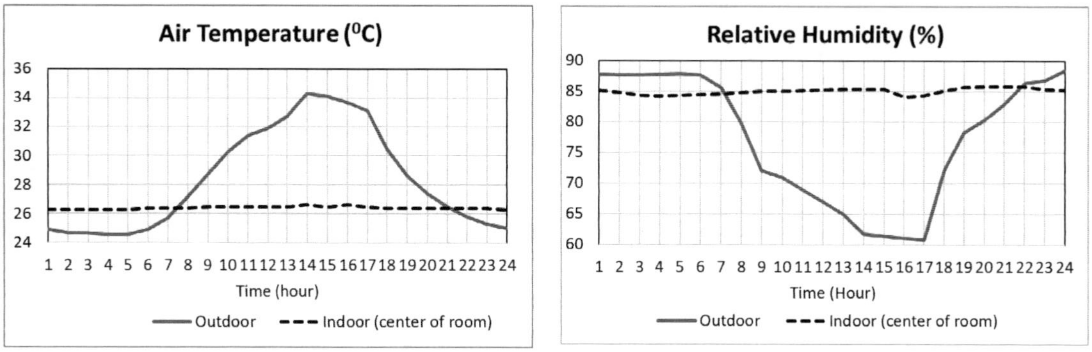

Figure 6. Indoor and outdoor air temperature (left) and humidity (right) from measurement

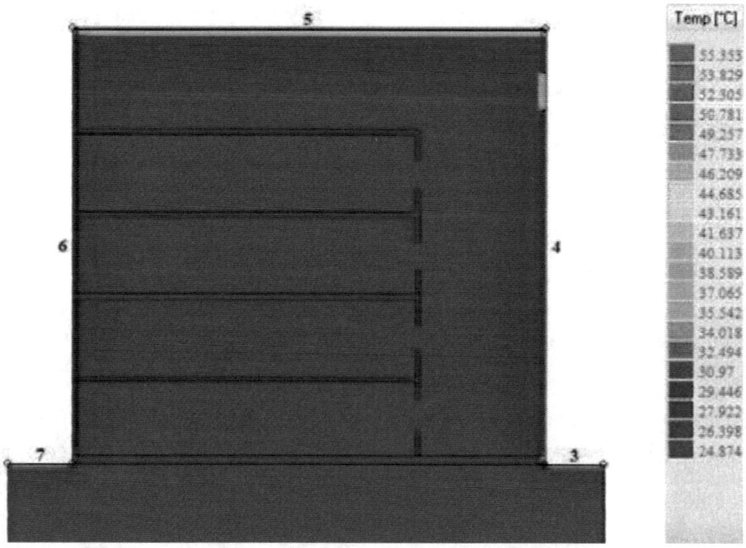

Figure 7. Result of HEAT2: indoor air temperature

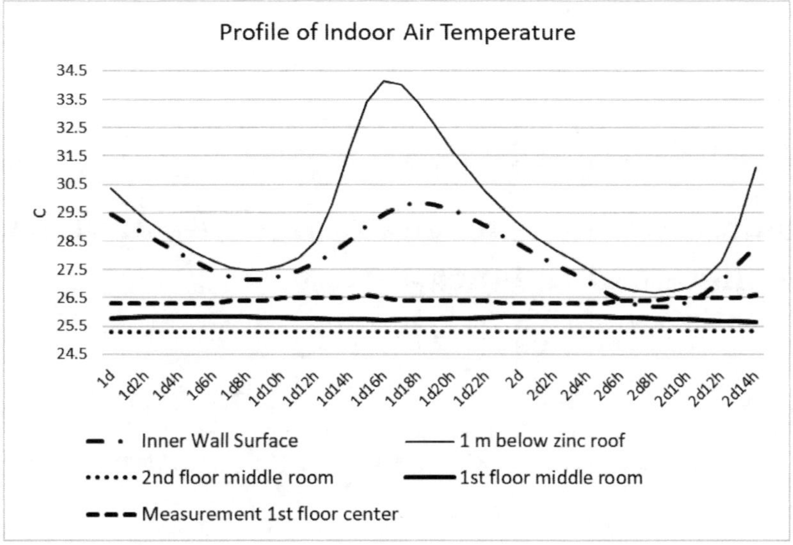

Figure 8. Indoor air temperature by simulation and measurement

4. Conclusion

Swallow's nest has a high economic value, even becoming an Indonesian export product that sells well in the international market. Therefore, the cultivation of swallow's nest in a building attracts the attention of the public and entrepreneurs. Swallows like to live in an environment where the air temperature is relatively constant between 26 to 29 ^{0}C, with the humidity that more than 70% but not exceed 90%. Buildings for swallow nests are generally multi-storey, where on the top floor, not for nests, because the air temperature is too hot due to being near the roof and consequently the minimum maximum fluctuation of air temperature is too large. Knowledge in building science can solve swallow house construction problems, among others, by applying materials that have certain thermophysical properties in order to achieve an indoor climate preferable by the swallow.

The thermophysical properties of envelope materials and microclimate surrounds the building may affect the performance of indoor air profile and humidity of a swallow house. A sample of swallow building house showed that indoor air temperature was relatively constant about 26.4 ^{0}C in average, with deviation about 0.5 to 1^{0}C. The air temperature difference between centre of room and those near inner side wall is about 0.5 to 1 ^{0}C.

5. References

[1] Gosler A 2007 *Birds of The World: A Photographic Guide* (Richmond Hill, Ont: Firefly Books)
[2] Ibrahim S H, Teo W C and Bahrun A 2009 *UNIMAS E-Journal of Civ. Eng.* **1** 1
[3] Margareta R, Abdullah M 2010 *J. Sains dan Tek.* **8** 2
[4] Sofwan A and Winarso P 2005 Rancang bangun sistem pengendali suhu dan kelembaban udara pada rumah burung walet berbasis mikrokontroler AT89C51. *Proc. Seminar Nasional Aplikasi Teknologi Informasi* (Yogyakarta)
[5] Setiawan TH 2013 *J.Tek.Sipil* **12** 2
[6] Widen J and Munkhammar J 2019 *Solar Radiation Theory* (Uppsala: Uppsala University)

IOP Conf. Series: Earth and Environmental Science 1007 (2022) 012008 doi:10.1088/1755-1315/1007/1/012008

Development of a wind catcher for high-rise apartments in the hot-humid climate of Indonesia

T Kubota[1,6], T Takahashi[2], A R Trihamdani[3], H Mori[4], and T Asawa[5]

[1] Graduate School of Advanced Science and Engineering, Hiroshima University, 1-5-1 Kagamiyama, Higashi-Hiroshima, 739-8529 Hiroshima, Japan
[2] Chiyoda TechnoAce Co., Ltd.
[3] YKK AP R&D Center Indonesia, PT. YKK AP Indonesia, Tangerang, Indonesia
[4] Central Research Laboratory, YKK AP Inc., Japan
[5] School of Environment and Society, Tokyo Institute of Technology, Japan

Abstract. This study aims to develop a new device, "cross-wind catcher", for mid-to-high-rise affordable apartments for improving the ventilation performance in the tropics, where the natural ventilation is considered essential. We conducted a primary simulation study to find optimum wind catchers, followed by a field experiment using a full-scale experimental building to confirm the effects of the proposed device. The results showed that the proposed cross-wind catcher was able to increase the ACH by approximately 1.2 to 1.4 times compared with the control unit.

1. Introduction

Natural ventilation is essential for residential buildings in the hot and humid climates for indoor air quality, thermal comfort and therefore energy-saving [1]. Outdoor winds at the ground level are often weak especially in urban areas in the topics. Hence, it is important to utilize relatively stronger winds at the higher levels for efficient natural ventilation [2]. To improve the performance of wind-driven ventilation, various devices such as wind walls [3], exhaust cowls [4], wind tower or wind catcher [5] were adopted in the previous studies. Meanwhile, a comprehensive review on wind-driven ventilation devices has been conducted in [6]. The details of various devices, flow rates, features of wind ventilation were reported. This study aims to develop a new device, "cross-wind catcher", for mid-to-high-rise affordable apartments for improving the ventilation performance in the tropics. First, we conduct a series of CFD analysis to find optimum wind catchers for the apartment buildings. Second, a field measurement was conducted in a full-scale experimental unit located in Tangerang, Indonesia to confirm the effects of the proposed wind catchers on the ventilation rates.

2. Methodology

2.1. CFD simulations

As a primary simulation study, a commercial CFD software, scSTREAM ver. 14, was used to solve the RANS equation to investigate wind flows in and around the typical apartment buildings. The computational settings are summarised in table 1. The CFD models were validated by comparing with the results of a field measurement conducted in the above-mentioned full-scale experimental house [7-

---

[6] Corresponding author: tetsu@hiroshima-u.ac.jp

8]. It was found that the goodness to fit was higher when the modified LK k-ε model was used as a turbulence model. Thus, the same turbulence model was adopted for this study.

Table 1. Computational settings

Analysis type	Steady state analysis with isothermal condition	
Turbulence model	Modified LK k-ε model	
Discretization scheme	First order upwind scheme	
Algorithm	SIMPLE	
Domain volume (m)	Root: 132 × 152.6 × 68.7	
	Block: 12.4 × 9.7 × 8.7	
Standard length (mm)	Root: 225	
	Block: 60	
Mesh elements	Root: 71 × 68 × 44	
	Block: 233 × 168 × 159	
	Total: 6,436,328 elements	
Boundary conditions	Inlet	Velocity: 3.0 m/s
		Ref. height: 10 m (α=0.2)
	Outlet	Fixed static pressure
	Walls	Wall: No slip
		Symmetry: Free slip

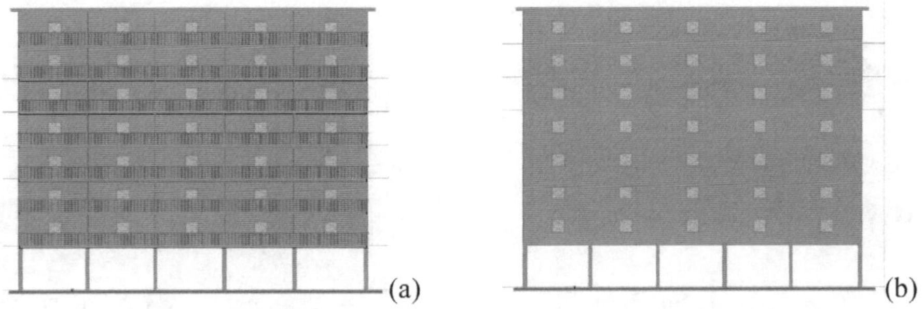

Figure 1. Simplified models for CFD simulations; (a) apartment with verandas and (b) apartment without verandas

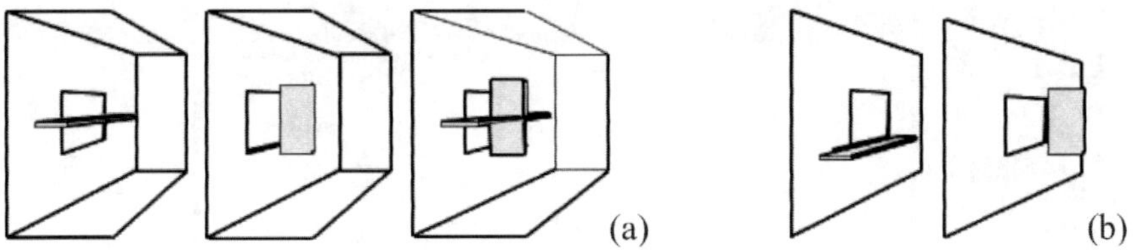

Figure 2. Wind catchers for (a) apartment with verandas and (b) apartment without verandas

The simulation models for the primary study were seven-storey simplified apartment buildings, representing typical affordable apartments (figure 1). Each unit has simple openings of 1.2 m by 1.2 m on both sides of the units. As shown, since wind flows in and around the units can be significantly influenced by the existence of veranda, we analysed the wind flows in two models (with or without verandas) to find the optimum wind catchers.

A simple board-shaped device was used as the wind catcher in this study (figure 2). We analysed the wind flows with changes of positions and sizes of the wind catchers. We examined horizontal wind

IOP Conf. Series: Earth and Environmental Science 1007 (2022) 012008 doi:10.1088/1755-1315/1007/1/012008

catchers and vertical wind catchers respectively, and we further investigated the combined wind catchers for the two apartment buildings (with/without verandas), respectively.

2.2. Field experiments

The field experiments were conducted in Tangerang, Indonesia (6.22°S, 106.57°E) from October 2020 to January 2021, experiencing not only the dry season (October) but also the rainy season (November to January). The experimental building has two identical rooms (Room A and Room B), and both rooms have the same size of 4.98 m × 5.15 m, with a floor to ceiling height of 3 m (figure 3). The imitated veranda with a depth of 1.5 m was attached to both units. As shown, simple openings of 1.14 m × 1.14 m were located on the front windward side of the units (north-facing) and rectangular openings of 1.7 m × 0.85 m were designed on the rear leeward side (south-facing).

The experimental cases are presented in figure 4. Case 1 is the default conditions, in which no device was attached for both units. The proposed wind catcher (cross-fin) was installed for Room A in Case 2. The size of the wind catcher was increased in Case 3. Detailed vertical and horizontal spatial distributions of wind speed were measured by non-directional anemometers (6332D with probe 9065-3, KANOMAX, Japan) with a measurement interval time of 5 s in the two units simultaneously (figure 5). Ultra-sonic 2D anemometers (CYG-81000, Young Company, USA) were installed at the immediate outside of the front openings (central positions). The proposed wind catcher was installed only for Room A and compared the resultant ventilation rates with Room B (control unit). Outdoor weather conditions were also recorded at a weather station located approximately 10 m away from the experimental building.

Figure 3. Experimental building in Tangerang, Indonesia.

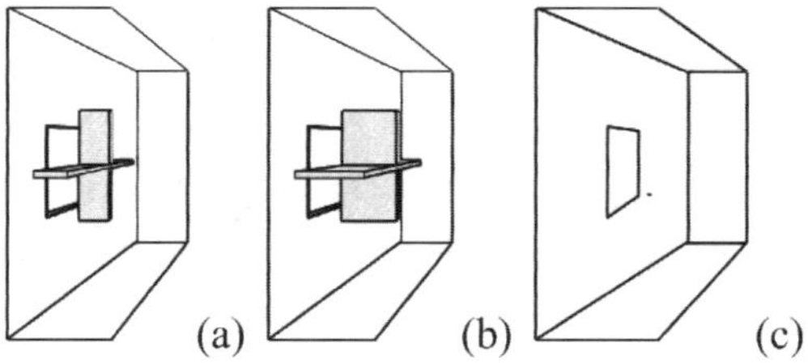

Figure 4. Experimental cases. Wind catchers for Room A in (a) Case 1 (no device), (b) Case 2 and (c) Case 3.

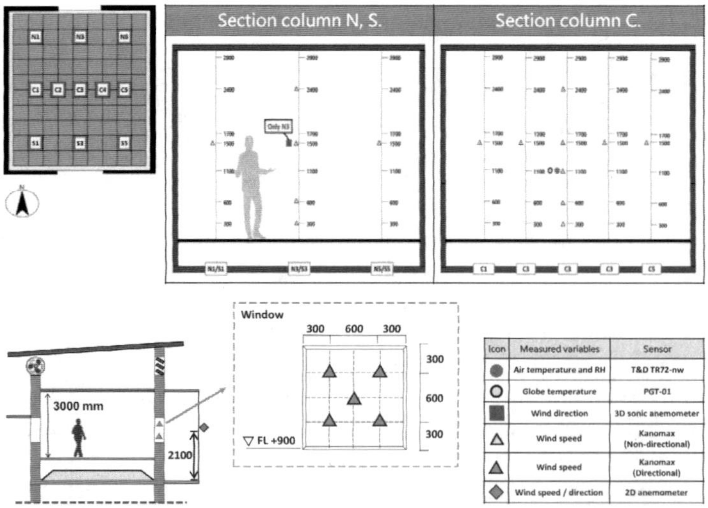

Figure 5. Measurement points and measured parameters in the field experiment [7-8].

Temporal average wind speeds were calculated in three different ranges of wind direction, which is N (±22.5°), NW (31.5°-337.5°) and W (247.5°-270°), based on the results of field experiments. The wind velocity data of 1 m/s or more at the weather station at 10 m height above ground were used for the analyses. The spatial average wind speeds on the inlet openings were calculated based on the results of multiple regression analyses using the results of CFD simulations with substituting the data from the field experiment. Then, ventilation rates (ACH) were calculated for each unit, respectively.

3. Results and discussion

3.1. Effects of verandas on wind flows

Figure 6 depicts the wind velocity at the external wall plane (section) of the two types of apartment building under the wind direction of 0° (perpendicular to the external wall). As shown, a stagnation point can be seen at the centre of the 5th floor units in the case of apartment without verandas. Winds spread radiantly from the stagnation point. The winds are slightly stronger at the edges of the building compared to those around the centre. As shown in figure 6, verandas strongly influence the wind flows near the external wall. Veranda plays a role to divide the wind flows and distribute similar wind flows to most units. Omrani et al. [9], for example, reported that natural ventilation in a 36-storey building was improved by adding a balcony on the building.

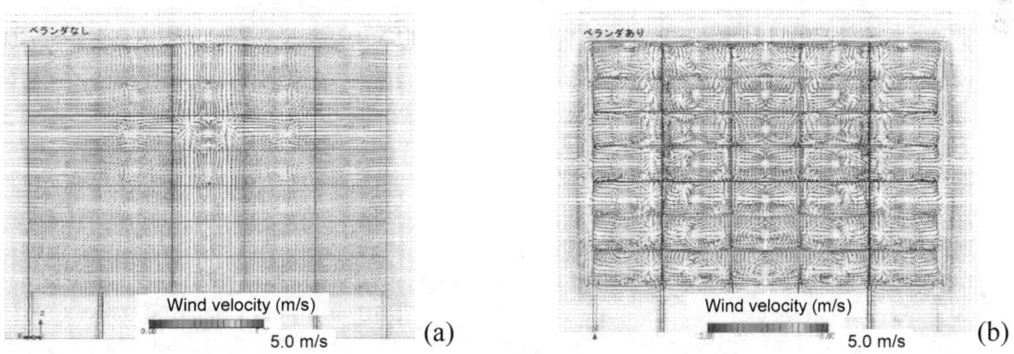

Figure 6. Wind velocities at the external wall plane of (a) apartment without verandas and (b) apartment with verandas (wind direction: 0°).

3.2. Optimum wind catchers for the apartment building with verandas

First, a horizontal wind catcher was optimized through a series of CFD simulation under the wind direction of 0° (perpendicular to the external wall). The main findings are summarised as follows:

- We compared the ventilation effects in different two sizes of wind catcher: 0.5 m × 1.2 m and 0.75 m × 1.8 m for each. The larger wind catcher gave better ventilation rates for all units.
- The wind catchers should be placed insides of verandas rather than outsides for all the units. When the wind catchers were placed immediate outside of the openings, the winds flowed straightforwardly in the units and increased the indoor wind speeds at the occupant levels.
- The wind catchers should be placed at the middle of the openings regardless of the floor levels.
- The maximum effect was observed in the centre of 7th floor units. The average wind speeds in the unit were increased by approximately 1.2 times than those without the device.

As indicated above, optimum settings for the horizontal wind catchers are the same regardless of floor levels. This is primarily because inflow features were almost uniformed due to the existence of verandas as discussed before.

Second, a vertical wind catcher was similarly examined under the wind direction of 0°. The main findings are as follows:

- Just like the horizontal wind catchers, the larger wind catcher showed better ventilation rates for all units.
- Vertical wind catchers should also be placed insides of verandas.
- Vertical wind catchers should also be placed at the middle of the openings regardless of floor levels as well as horizontal positions.

As same as the horizontal wind catchers, optimum settings for the vertical wind catchers are also the same regardless of floor levels as well as horizontal positions because of the existence of verandas.

Based on the above results, we proposed a new device, "cross-wind catcher", by combining horizontal and vertical fins (see figure 2a). The results showed that the horizontal wind catchers did not increase indoor wind flows especially in the corner units very much. The vertical wind catchers, on the other hand, contributed to increase the indoor wind speeds particularly for the upper units. Interestingly, the combined cross-wind catchers effectively increased the wind speeds in the units for every floor regardless of the horizontal positions. The proposed cross-wind catchers were able to increase the wind velocities for all the units. The average wind speeds with cross-wind catchers were found to be 1.15 times higher than those without the devices with the maximum difference of 1.23 times.

Furthermore, similar CFD analyses were conducted under the wind direction of 45°. The results showed that the average wind speeds with cross-wind catchers were found to be 1.1-1.18 times higher than those without the devices, which is almost the same as those of 0°. This implies that the effects of the cross-wind catchers are expected even under a diagonal wind direction.

3.3. Optimum wind catchers for the apartment building without verandas

First, a horizontal wind *catcher* was examined through a series of CFD simulation under the wind direction of 0° (perpendicular to the external wall). The main findings are summarized as follows:

- The larger wind catchers gave better ventilation rates for all units, particularly for the upper and lower floor levels.
- Unlike the apartment with verandas, optimum vertical positions of the horizontal wind catcher are different, depending on the floor levels. For the lower floors such as 1st and 2nd floors, the horizontal wind catchers should be placed at the bottom edges of the openings. For the upper floor such as 7th floor (top floor), the horizontal wind catchers should be attached at the upper edge of the openings instead, whereas middle positions were found to be better for the medium floor levels. The optimum positions depend on the stagnation point and the building height.
- A horizontal wind catcher for a unit did not affect the wind flows in the surrounding units very much. This means that a wind catcher can be designed for each unit individually without considering the effects on the surroundings.

Second, a vertical wind catcher was similarly examined under the wind direction of 0°. The main findings are as follows:

- Just like the horizontal wind catchers, the larger wind catcher showed better ventilation rates for all units, particularly for the corner units.
- Optimum horizontal positions of the vertical wind catcher are also different, depending on the horizontal positions of units. On both sides of the corner units, the optimum positions of the vertical wind catchers were found to be outsides of the building.
- Similarly, a vertical wind catcher for a unit did not affect the wind flows in the surrounding units very much. This means that a wind catcher can be designed for each unit individually without considering the effects on the surroundings.

Similar CFD analyses were conducted under the wind directions of 15°, 30° and 45°, respectively. As for the vertical wind catchers, the ventilation rates were increased especially in the corner units with the increase in wind direction. Optimum positions of the vertical wind catchers should be different, depending on the wind directions.

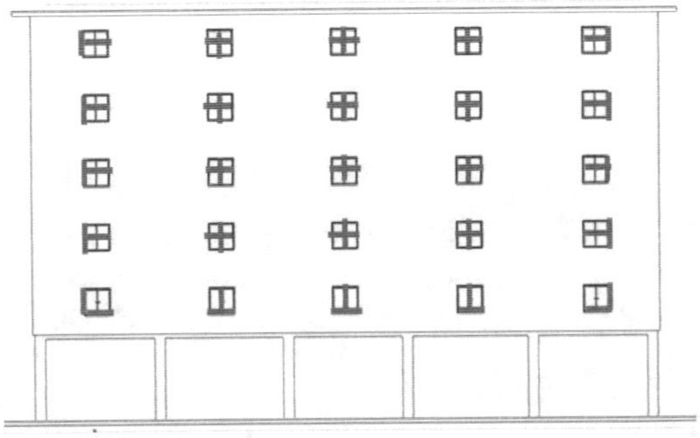

Figure 7. Proposed wind catchers for the apartment without verandas.

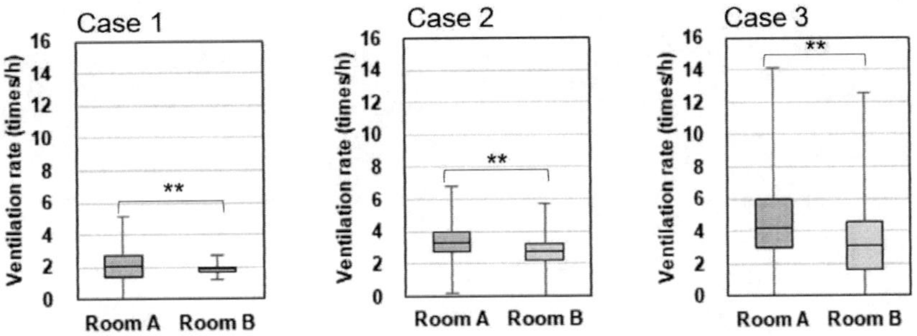

Figure 8. Ventilation rates of the two units in Cases 1-3 (wind direction: north). ** 1% significant level.

Based on the above results, we proposed the optimum positions of the proposed cross-wind catchers in the case of an apartment without verandas (figure 7). The results showed that the proposed wind catchers were able to increase the ventilation rates for all the units. For example, in the 1st floor and 5th floor units, the ventilation rates with the proposed wind catchers are approximately 1.5 times larger than those of default units without the devices with the average increase of about 1.27 times.

3.4. Field experiments in the full-scale experimental building

In Case 1, we compared the vertical distributions of wind velocities between the two units (Rooms A and B), where the wind catchers were not installed for both units. As a result, significant differences between the two units were not seen in most measurement points. As shown in figure 8, the resultant ventilation rate was averaged at 2.1 times/h in Room A while the corresponding rate was 1.9 times/h in Room B.

The medium size of cross-wind catcher (0.5 m × 1.8 m) was attached to Room A in Case 2. Figure 9 shows the vertical distributions of wind velocities measured at several points between the two units under the wind direction of north (perpendicular to the external wall). As shown, particularly at the 1.5 m above floor, average wind speeds in Room A (2.2 m/s) were clearly higher than those of Room B (1.82 m/s). The increase in wind velocity can be seen until the leeward side of the unit. Instead, in upper and lower parts of the units, wind velocities in Room B were slightly higher than those in Room A. This means that the proposed cross-wind catcher made the indoor wind flows straightforward until the leeward side of the unit. The resultant ventilation rate in Room A was averaged at 3.4 times/h, which was about 1.23 times larger than that of Room B (figure 8).

IOP Conf. Series: Earth and Environmental Science 1007 (2022) 012008 doi:10.1088/1755-1315/1007/1/012008

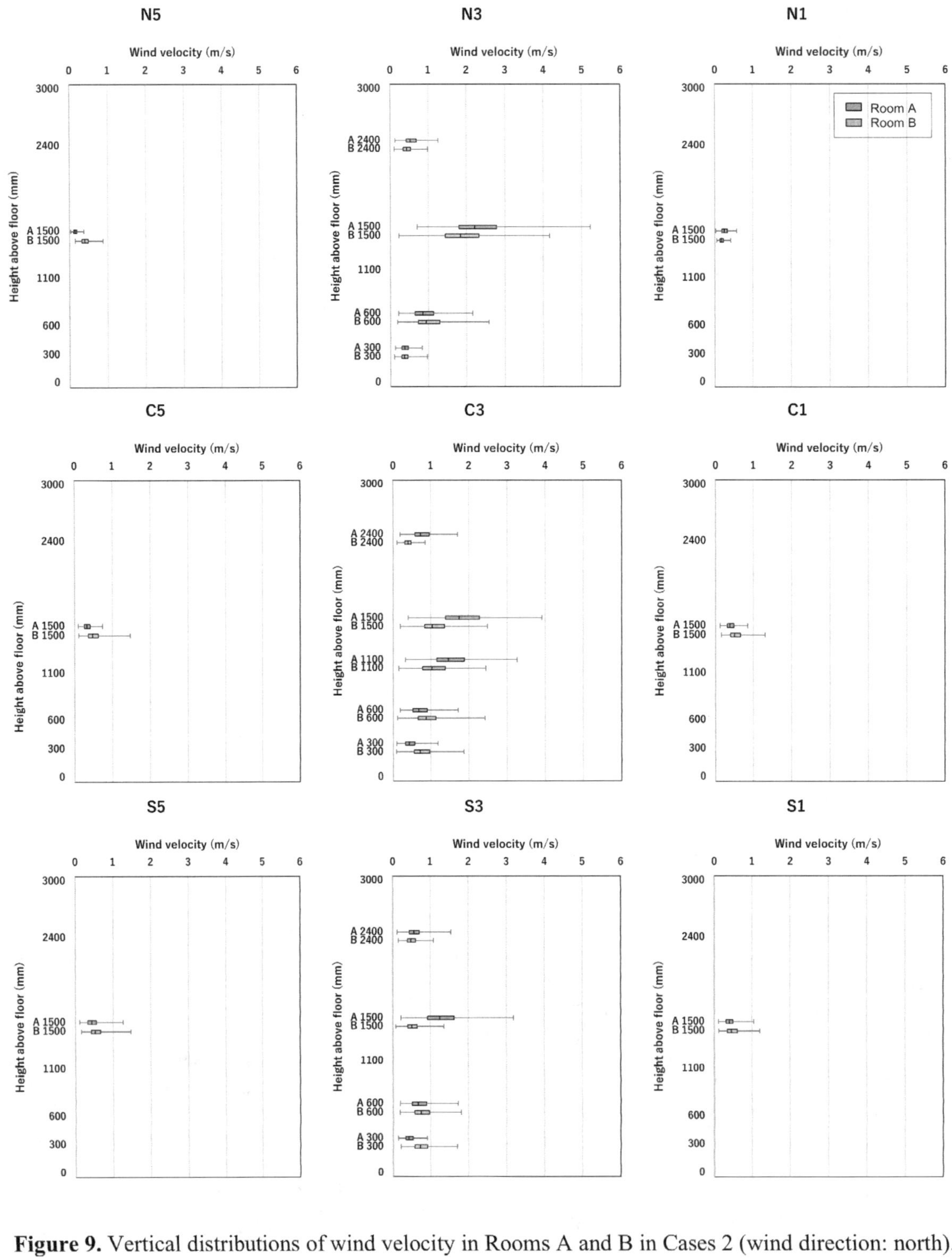

Figure 9. Vertical distributions of wind velocity in Rooms A and B in Cases 2 (wind direction: north).

When the wind direction is northwest (45° west), the average indoor wind speed at 1.5 m height in Room A was about 1.19 times higher than that of Room B, which is almost the same as that of the above perpendicular wind direction (0°). As expected, the increase in wind velocity was not observed when the wind direction was 90° west.

The size of the cross-wind catcher was increased to 1.5 m ×1.8 m in Case 3. The results showed that wind velocities were increased especially at 1.5 m above floor in Room A just like Case 2. The average wind speed at 1.5 m height was 2.5 m/s in Room A whereas that of Room B was 2.2 m/s. The former is about 1.13 times higher than the latter, showing a similar effect with Case 2. As shown in figure 8, the resultant average ventilation rate in Room A was 1.35 times larger than that of Room B. This means that the increase in wind catcher size did not increase the ventilation rate very much.

4. Conclusions

- The results of CFD simulations showed that wind flows near the external wall of a typical mid-rise apartment were dramatically influenced by the existence of verandas.
- We proposed new cross-wind catchers for a typical apartment building with verandas and that without verandas, respectively.
- The results of the field experiment confirmed that the proposed cross-wind catcher is able to increase the indoor wind speeds particularly at the occupant level (approximately at 1.5 m above floor) and thus increase the ventilation rates significantly. The resultant ventilation rates were increased by approximately 1.2 to 1.4 times compared with those without the device.

5. References

[1] Kubota T, Rijal H B, Takaguchi H (Eds.) 2018 *Sustainable Houses and Living in the Hot-Humid Climates of Asia* (Springer)
[2] Kubota T and Supian A 2006 *J.Asian Archit. Build. Eng.* **5(1)** 199-206
[3] Mak C M, Niu J L, Lee C T, Chan K F 2007 *Energy Build.* **39** 995-1002
[4] Preiffer A, Dorer V, Weber A 2008 *Build Environ* **43** 1361-1372
[5] Sherman M H, Chan R 2004 *Building Airtightness: Research and Practice*
[6] Khan N, Su Y, Riffat S B 2008 *Energy Build.* **40** 1586-1604
[7] Kitagawa H, Asawa T, Kubota T, Trihamdani A R, Mori H 2022 *Build Environ* **207 B** 108442
[8] Kitagawa H, Asawa T, Kubota T, Trihamdani A R, Sakurada K and Mori H 2021 *Build Environ* **188** 107483
[9] Omrani S et al. 2017 *Build Environ* **123** 504-516

Acknowledgements
The authors would like to thank the Science and Technology Research Partnership for Sustainable Development (SATREPS), in collaboration with Japan Science and Technology Agency (JST, JPMJSA1904) and Japan International Cooperation Agency (JICA) for their supports. This research was also strongly supported by YKK AP Inc. and the Asahi Glass Foundation.

Experimental study on CO_2 concentration and indoor thermal comfort pf different classrooms during new normal in the University of Kitakyushu

D Sekartaji[1,3], N Sakai[1], D Novianto[2], and Y Ryu[1]

[1]Department of Architecture, The University of Kitakyushu, Hibikino Campus 808-0135, Japan
[2]Architecture Department, Faculty of Civil, Planning, and Geo-Engineering, Institut Teknologi Sepuluh Nopember. Kampus ITS, Sukolilo, Surabaya 60111, Indonesia

Abstract. Indoor thermal comfort and indoor air quality in educational buildings are essential for students' health and learning performance. In this pandemic situation, regulation of door and ventilation opening in the classroom for offline class become important for infection spread prevention. Based on Japan Ministry of Health, Labour and Welfare standard, CO_2 concentration level shall not exceed 1,000 ppm to prevent infection spreading. Yet significant amount of door and ventilation opening may reduce the level of thermal comfort because the Air-conditioning leakage. This study aims to evaluate CO_2 concentration and indoor thermal comfort in two classrooms with door and ventilation opening regulation. In this study, the experimental measurements conducted in spring and summer season in The University of Kitakyushu. The method used in this study is by measuring air temperature, relative humidity, and CO_2 concentration. The research results are expected to be a reference for the optimal application of regulations for thermal comfort and infection spread prevention.

1. Introduction

COVID-19 which was first discovered in Wuhan, China in 2019 has brought changes in various sectors including in the education field. Due to COVID-19 cases outbreak that entered to Japan between January 15 and February 10, 2020, Japan`s central government has declared state of emergency with new restrictions imposed in several prefectures. Based on Article 25 Paragraph 1 of the University Establishments Standards of Correspondence order of MEXT, Japan (updated: July 27, 2020)[1], after taking into account the implementation status of 2020 year classes and students` condition and wishes, comprehensively considering the status of local infections, the size of classrooms, the number of students, educational effects, it is recommended (depends to local infection situation) to conduct face to face class implementation method if it is deemed appropriate to conduct an face-to face class after taking infection control measures.

The University of Kitakyushu, located in Kitakyushu City of Fukuoka which selected as an environmental model city in 2008 [2], adjusts to the possibility of a new normal when the COVID-19 state of emergency declared in Kitakyushu or when the COVID-19 cases declined. The University of Kitakyushu made regulation based on the level of COVID-19 status that is in principle, when the state of emergency is declared, theory classes must be conducted online, yet experimental and practical classes can be conducted face to face. When the state of emergency revoked, in principle every class is

[3]Corresponding author: d-sekartaji@kitakyu-u.ac.jp

conducted face to face [3]. Based on The University of Kitakyushu first semester 2021 regulation [4], the classroom entrance / exit doors must be fully open or slightly open and open windows to ventilate once every 30 minutes if the weather is good. This regulation is related to building environmental hygiene management standards of Japan Ministry of Health, Labour and Welfare standard that stated indoor CO_2 concentration standard level shall not pass 1000 ppm [5]. Even the opening window regulation is once every 30 minutes, yet the regulations are not implemented correspondingly by this time.

The indoor air quality is believed become an effective way for decreasing spread of infectious diseases, including COVID-19. Excess CO_2 level increase the risk of spreading infections, hence keeping CO_2 as low as possible in rooms optimize the protection of infection spreading, which can be provided by optimal use of ventilation [6]. From the author previous research, window and door opening in long period during the day may has substantial relation with indoor thermal comfort [7]. Yet significant amount of door and ventilation opening may reduce the level of thermal comfort because the Air-conditioning leakage.

Based on the results of student questionnaires who run online classes, although time at home increases, students feel that their productivity and learning performance has decreased [8],[17]. Thus, the face-to-face classes in the new normal with optimal indoor air quality with indoor thermal comfort observing must become consideration. This study aims to evaluate CO2 concentration level and indoor thermal comfort in two classrooms with door and ventilation opening regulation. In this study, the experimental measurements conducted in spring and summer season in The University of Kitakyushu. The method used in this study is by measuring CO2 concentration, air temperature, and relative humidity. The research results are expected to be a reference for the optimal application of regulations for thermal comfort and infection spread prevention.

2. Outline of investigated building

The target university, The University of Kitakyushu Hibikino Campus (Figure 1) has two main buildings: north side building and south side building. Each building is a four-storey and double loaded corridor with a long axis in the east-west direction. North building has several courtyards in the corridor. The target measurement building is North side building. The measurement was only conducted in North building because all classrooms are located in North building while South building are laboratories. The actual measurement classrooms were located at the north side of building on the first floor near the courtyards. Both classrooms have one double door, one single door, and several smoke exhausts windows.

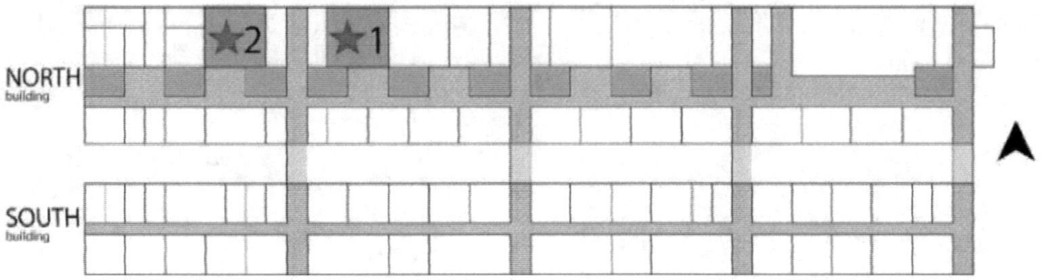

Figure 1. Target building floor plan (☆1: Measurement target N113 classroom, ☆2: Measurement target N107 classroom)

3. Formatting the text

3.1. Field measurement

Measurement of temperature, humidity and CO2 was conducted with different day and time in two classrooms, which have same total area, N113 and N107. N113 measurement was conducted in spring season, May 27th, 2021 (emergency state was declared) where air conditioner was turned off, only mechanical ventilation was activated. N107 measurement was conducted in summer season, July 30th, 2021 (emergency state was revoked) where both air conditioner and mechanical ventilation were turned on.

Class type in both classes was different when measurements were conducted, class in N113 was experimental class which discussion was allowed while class in N107 was theoretical class while discussion was not allowed. Both classrooms` entrance doors were opened when the measurement conducted. The measuring instrument used in this research was CO2 Recorder TR-76Ui (Figure 4). The smoke exhaust windows were opened when the CO2 concentration level in the classroom were increased. One smoke exhaust window size is 0.5 x 1.8 m. Total smoke exhaust windows (Figure 8) is 6, so that total area of smoke exhaust windows opening is 5.4 m2 in each classroom. AC outlets type in both classrooms are shown in Figure 7.

Figure 2. Measurement in N113.

Figure 3. Measurement in N107.

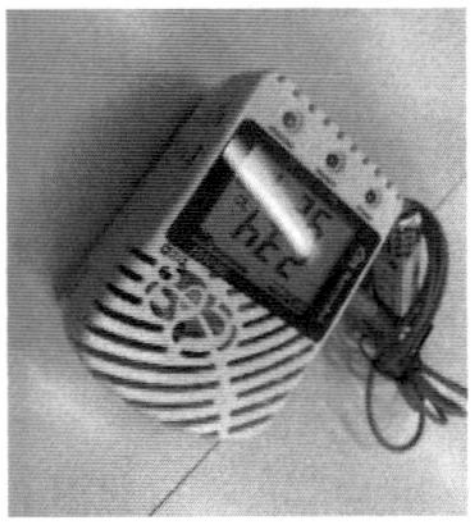

Figure 4. Measuring instrument.

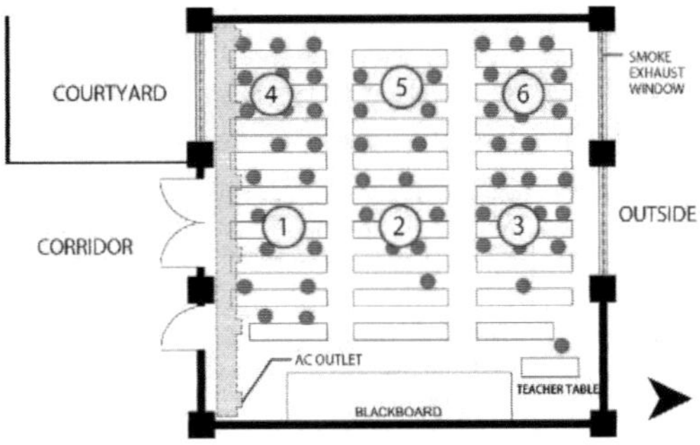

Figure 5. N113 seating positions and measurement point.

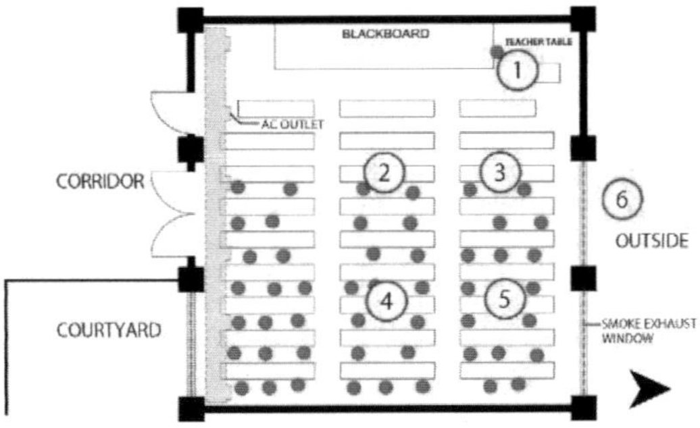

Figure 6. N107 seating positions and measurement point.

Figure 7. N113 seating positions and measurement point.

Figure 8. N107 seating positions and measurement point.

Table 1. Measurement item.

Room	Volume (m^3)	Total people	Class type	AC	Measurement point	Measurement day	Measurement time
N113	419.9	55	Experimental class and discussion	off	① - ⑥	May 27th, 2021 (Spring)	12:50-15:40 (2 h 50 m)
N107	419.9	51	Theory	On (25.5℃)	① - ⑥	July 30th, 2021 (Summer)	08:50-10:30 (1 h 30 m)

The measurement points area shown in figure 5 and figure 6. The installation height of the measuring instrument was 75 cm (table height). The measurement interval for each data logger was 1 minute. N113 measurement time was 2 hours 50 m in the afternoon while N107 measurement time was 1 hour 30 m in the morning (Table 1). Smoke exhaust windows in both measurements were set opened before the class began. When class began, N113 at 13:00 and N107 at 09:00, smoke exhaust windows were closed but the entrance doors remained open.

3.2. Meteorological conditions during the actual measurement period
The graphs of air temperature and wind velocity during the actual measurement period from AMeDAS meteorological information in Yahata [9], Kitakyushu are shown below (Figure 9). On May 27th, 2021, the air temperature around measurement time was around 20°C to 22°C. On July 30th, 2021, the air temperature was coming up steadily from 28.2°C to 31.5°C. Although wind velocity on May 27th is faster than wind velocity on July 30th, the wind direction in both days were generally same, which is from west side.

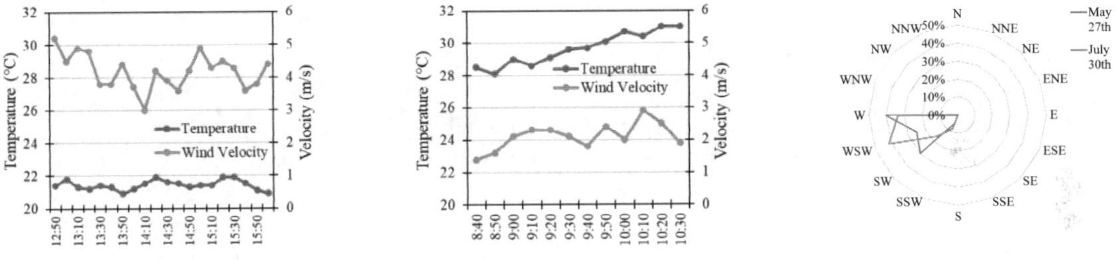

a. May 27th b. July 30th c. Wind direction

Figure 9. Meteorological condition

4. Field Measurement results and analysis

4.1. Carbon dioxide concentration analysis
Based on CO2 concentration measurement result in N113 classroom (Figure 10), the CO^2 concentration level exceed 1000 ppm at measurement point ③, ④, and ⑥ after around 40 minutes measurement started, 30 minutes after all smoke exhausts windows closed, while other CO^2 concentration at other measurement points is under 1000 ppm. According to N113 seating positions distribution map (Figure 5), there were more denser students seated in these 3 points compared to other positions. After 130 minutes from the measurement started, smoke exhaust windows in N113 were opened. It can be seen from the graph that CO^2 concentration were drastically declined to 450-550 ppm.

Based on N107 CO_2 concentration result (Figure 11), the CO_2 concentration level did not exceed 1000 ppm at all measurement points, so that the smoke exhausts windows were closed from class began until finished. It can be presumed that CO_2 concentration in this classroom was not high because the class type when measurement conducted was theory class which is discussion or talk is not allowed.

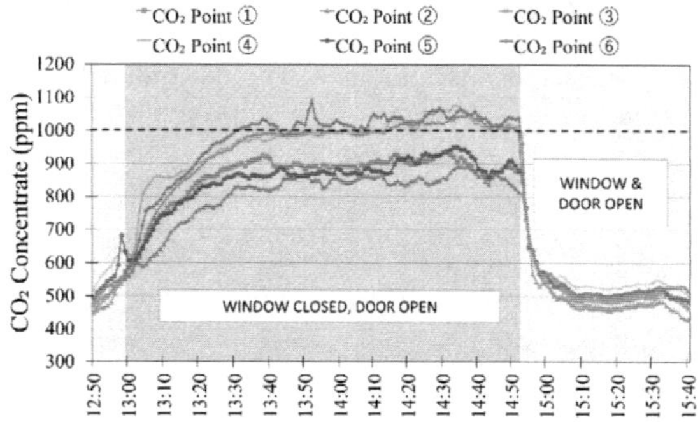

Figure 10. N113 CO_2 concentration.

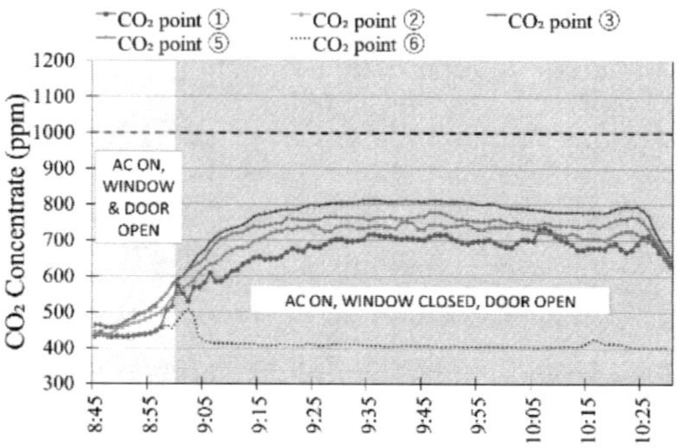

Figure 11. N107 CO_2 concentration.

4.2. Room air temperature and relative humidity analysis

Both N113 and N107 air temperature measurement results show (Figure 12 and Figure 13) that the average value of the room temperature was still in range 17°C-28°C, which is the room temperature standard for classroom based on Japan School Environmental Hygiene management standards [10]. According to the N113 air temperature result (Figure 12), after smoke exhaust windows opened, air temperature down by 3°C.

N113 measurement, which was conducted in spring season, the relative humidity result shows (Figure 12) that the average value of the relative was still in range 30%-80%, which is the room relative humidity standard for classroom based on Japan school environmental hygiene management standards [10]. The measurement result of relative humidity in N113 measurement shows that after smoke exhaust windows were opened, the relative humidity was increased but still in the range of the standard. In the other hand, N107 measurement, which was conducted in summer season, the relative humidity result shows (Figure 13) that relative humidity result in point ③ was above 80%, which is the upper limit of Japan school environmental hygiene management standards [10].

Although the relative humidity results of the measurements are still in range Japan school environmental hygiene management standards, the relative humidity results, especially in N107 (summer season) is looser than the Building Sanitation Law standard which can be seen in psychrometric chart figure 14, 40% to 70% [5] is set as an appropriate range of relative humidity in the building. When the temperature is high and the humidity is high, the risk of heat stroke increases, so it is necessary to pay attention to the humidity. The risk of heat stroke will be discussed in "5.2. WBGT".

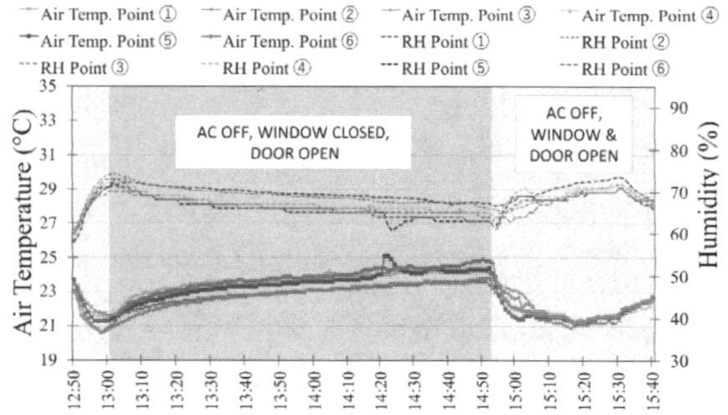

Figure 12. N113 Air temperature and relative humidity.

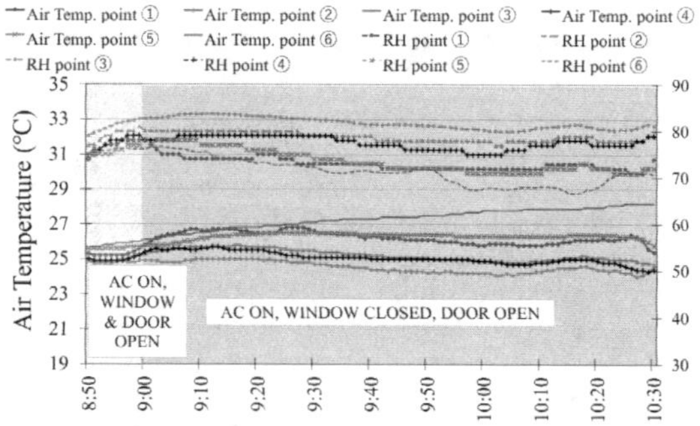

Figure 13. N107 Air temperature and relative humidity.

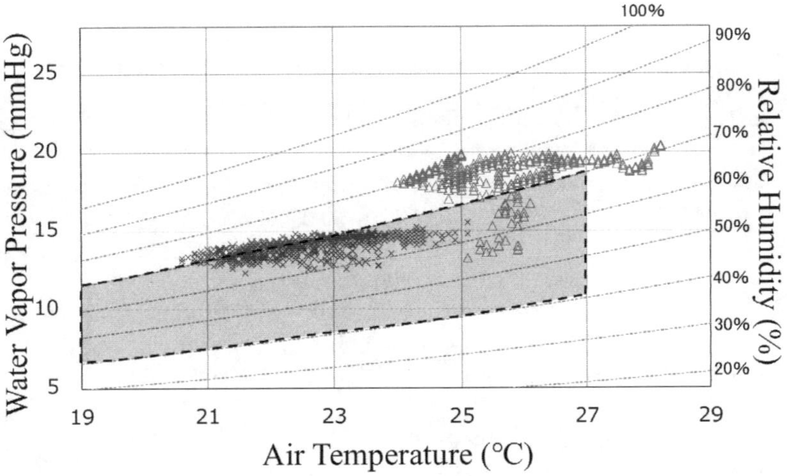

Figure 14. Psychrometric chart of indoor humidity distribution. (×: N113, △: N107)

5. Analysis with secondary data

5.1. Ventilation volume Text

By using the measured CO_2 concentration, the ventilation volume rate is estimated. Ventilation volume rate was calculated using the following formula derived by Seidel [11].

$$Q = \frac{M \cdot n}{C - C_0} \cdot 10^6 \tag{1}$$

Where,

Q: Ventilation volume rate [m³/h]
M: CO_2 exhalation rate [m³/h・person]
n: total people in room [°C]
C: Indoor CO_2 pollutant
C_0: Outdoor CO_2 concentration

CO_2 concentration data used for calculation is the data when window was closed. Based on JIS A1406, 1974 [12], CO_2 exhalation rate (M) based on energy metabolism rate during "seated office work" is 0.0129~0.023 m³/h, and during "slow walking" is 0.023~0.033 m³/h. In this calculation, type of class in N113 was experimental and discussion class, CO_2 exhalation rate used is 0.023 m³/h while type of class in N107 was theory class, CO_2 exhalation rate used is 0.016 m³/h. From the calculation result, both class ventilation volume rate is almost same when the CO_2 exhalation rate is categorized by different energy metabolism rate.

Table 2. Ventilation volume rate.

Class	Date	Time	Q (m³/h)	Q`(m³/h/person)	Frequency (times/h)
N113	May 27th, 2021	13:00-14:51	2509.6	45.6	6
N107	July 30th, 2021	09:00-10:30	251	49.4	6

5.2. WBGT

WBGT (Wet Bulb Globe Temperature) is calculated from the measured values to evaluate the risk of heat stroke. In this research, *WBGT* indoor was calculated using formula (2) proposed by B. Lemke *et al.,* 2012 [13], where *Tpwb* and e_a calculation was calculated using formula (3), (4) proposed by Bernard *et al*., 1999 [14] summarized by Carter et al., 2020 [15].

$$WBGTid = 0.67Tpwb + 0.33Ta \quad (2)$$

$$Tpwb = 0.376 + 5.79e_a + (0.388 - 0.0465e_a)Ta \quad (3)$$

$$e_a = \left(\frac{RH}{100}\right) \times \left(0.6107exp\left[\frac{17.27Ta}{Ta+237.3}\right]\right) \quad (4)$$

Where,

WBGTid: Wet bulb globe temperature Indoor
Tpwb: Psychometric wet bulb temperature [°C]
Ta: Room temperature [°C]
e_a: Ambient vapor pressure

The calculated WBGT of two measurements are shown in figure 15 and figure 16 WBGT in N113 during measurement in spring season was under 21°C, which was in "almost safe range" (Table 3). In the other hand, WBGT in N107 during measurement in summer season was 22.6°C-23.9°C, which was generally in caution range. Although heat store risk level is not high, yet it is stated that WBGT of above 21°C, there is the danger of fatal accidents due to heat illness, so that caution is advised, and water replenishment should be promoted during exercise [16].

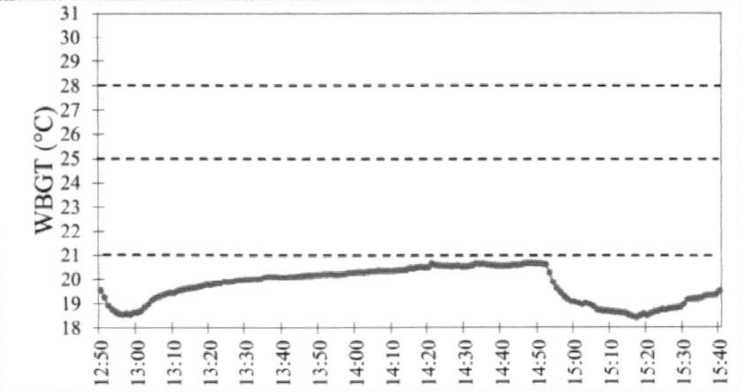

Figure 15. N113 WBGT.

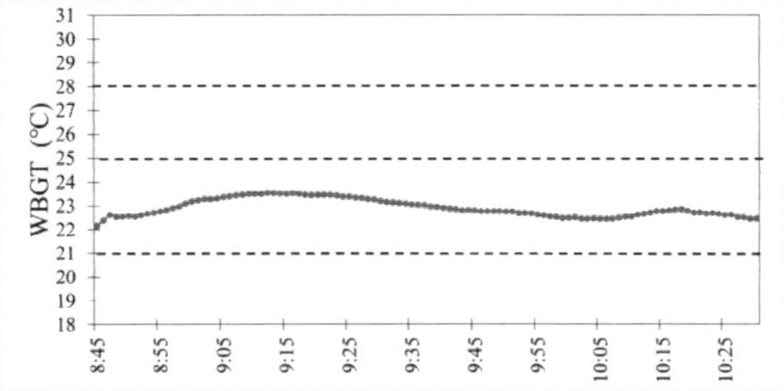

Figure 16. N107 WBGT.

Table 3. Heat store risk guidelines (WBGT) [16].

	Danger	**Severe warning**	**Warning**	**Caution**	**Almost safe**
Temperature standard	Above 31°C	28-31°C	25-28°C	21-25°C	Under 21°C

6. Conclusion

From this experimental measurement results, we conclude as follows:

1. By analyzing wind direction in both days when measurement occurred and the ventilation volume rate calculation in both conditions are same with different CO_2 exhalation rate based on energy metabolism rate, it can be concluded that CO_2 concentration will be high and exceed 1000 ppm after 30 minutes when smoke exhausts windows are closed in the discussion type of class in the classroom with 419.9 m^3 volume area and total number of people 55.

2. It also can be concluded that in classroom of theory class (talking or discussion is not allowed) with 419.9 m^3 volume area and total number of people 51, the CO_2 concentration in the classroom with opened doors and closed windows do not exceed 1000 ppm.
3. When the daily average outside air temperature is close to 30°C, air conditioner turned on with temperature setting 25.5°C and all entrance door opened, indoor air temperature in classroom do not exceed the school hygiene management standard of 28 ° C in summer season, yet mostly humid and relative humidity was higher than the Building Sanitation Law standard for indoor thermal comfort.
4. When the daily average outside air temperature is close to 30°C, air conditioner turned on with temperature setting 25.5°C and all entrance door opened, WBGT in the classroom is above 21°C, which is in "caution" range.

From this study result, we suggest to open smoke exhaust windows to control CO_2 concentration at discussion type of class. For the principle, it can be suggested to strictly prohibit talking in the classroom, but if it is not possible due to the type of the class, it is recommended to open the windows from the beginning until finish or at least every 30 minutes. Although it is not necessarily dangerous because there is no exercise activity in the classroom, hydration during class, especially in the summer season is suggested to prevent heat stroke illness. The result of this study can be expected to become useful for regulations for thermal comfort and infection spread prevention in the classrooms. Indoor thermal comfort in classroom with AC turned on and opened door in summer and winter season will be the focus in the further research.

7. References

[1] MEXT Document 2-1 2020 *Current status, issues, and future direction of corona support* https://www.mext.go.jp/content/20200924-mxt_keikaku-000010097_3.pdf. Accessed on 10 February 2021
[2] Gao W and Novianto D 2018 Eco-House in Kitakyushu, Japan. *In Sustainable Houses and Living in the Hot-Humid Climates of Asia* (Singapore: Springer) pp 483-490
[3] Action guidelines of the University of Kitakyushu to prevent the spread of new coronavirus infection: https://www.kitakyu-u.ac.jp/uploads/a7608254a15f2d8806e9a9c05b39130b.pdf
[4] Infection control measures related to the start of face-to-face classes in the first semester of 2021 (Hibikino version): https://www.kitakyu-u.ac.jp/uploads/2a2d8e529cb4c0712ef69baf18cdd540.pdf. Accessed on 22 March 2021.
[5] Building environmental hygiene management standards, Japan Ministry of Health, Labour, and Welfare: https://www.mhlw.go.jp/bunya/kenkou/seikatsu-eisei10/. Accessed on 20 February 2021
[6] Peng Z and Jimenez J L 2021 Exhaled CO2 as a COVID-19 infection risk proxy for different indoor environments and activities *Environmental Science & Technology Letters* **8**(5) 392-7
[7] Sekartaji D, Novianto D and Ryu Y 2016 Survey on Indoor Thermal Environment of Japanese Minka in Kitakyushu *JAILCD* **147** 50.
[8] Novianto D, Nuffida N E, and Gao W 2021 A Review of Architecture Response to the Pandemic towards New Normal Behavior *Int. Conf. and Works. Re-thinking the HIBIKINO Campus 2001-2021 Kitakyushu (Japan)--(Architectural Design) JAILCD* p 13
[9] Japan Meteorological Agency: http://www.jma.go.jp/jma/index.html. Accessed on 15 February 2021.
[10] Japan School environmental hygiene management manual: https://www.mext.go.jp/component/a_menu/education/detail/__icsFiles/afieldfile/2018/07/31/1292465_01.pdf. Accessed on 2 April 2021.
[11] Tsujihara M, Imamura S and Tanaka M 2009 やさし建築環境 p 89 (Friendly building environment: in Japanese)

[12] JIS a14061974 室内換気量測定法(炭酸ガス法) 2013.11 (Indoor ventilation volume measurement method (carbon dioxide gas method) in Japanese)
[13] Lemke B Kjellstrom T 2012 Calculating workplace WBGT from meteorological data: a tool for climate change assessment *Industrial Health* **50**(4) p 267
[14] Bernard T E 1999 Prediction of workplace wet bulb global temperature. *Applied occupational and environmental hygiene*. Jan 1;**14**(2) p 126
[15] Carter A W, Zaitchik B F, Gohlke J M, Wang S, Richardson M B 2020 Methods for estimating wet bulb globe temperature from remote and low-cost data: A comparative study in Central Alabama *GeoHealth*. **5** e2019GH000231
[16] Heat Illness Prevention Information: https://www.wbgt.env.go.jp/en/wbgt.php
[17] Novianto D, Hidayat AS, Hazrati FY, Rahmavani AM, Fadhila AR, Jaya AM, Koerniawan MD 2021 Rethinking sustainability during WFH: A survey on living environment quality and energy use. *IOP Conf. Series: Earth and Env. Sci.* (Vol. 881, No. 1, p. 012005)

IOP Conf. Series: Earth and Environmental Science 1007 (2022) 012010 doi:10.1088/1755-1315/1007/1/012010

Analysis on the effect of home appliances use to electricity consumption during WFH

S F Hutama[1], M Y A Ghifari[1], I T Ula[1], S D Yong[1,2], B Soemardiono[1], D Septanti[1], D Sekartaji[3], D Novianto[1,4]

[1] Department of Architecture, Institut Teknologi Sepuluh Nopember, Kampus ITS, Sukolilo, Surabaya 60111, Jawa Timur, Indonesia
[2] Interior Design Department, Petra Christian University, Surabaya, Indonesia
[3] Departement of Architecture, Environmental Engineering, The University of Kitakyushu, Japan

Abstract. Various forms of quarantine and social restrictions to prevent the spread of the COVID-19 pandemic has impacted lifestyle. WFH has emerged as a growing trend in this new normal era. The implementation of WFH results in all work-related activities becoming home-centred. This study aims to analyse the pheasant and find out the effect of household appliance use on energy consumption during the implementation of WFH. The questionnaire survey was conducted in April 2020 and it collected 215 responses from residents of Indonesia, during the pandemic which was distributed online. The data were analysed using a multiple regression analysis with SPSS to determine the relationship between variables. The results showed that the length of WFH time affected an increase of energy consumption especially from the use of home computer, internet, rice cooker, AC, and water needs. In addition, the increase that happened also depends on the population living and the size of the building they live in. If this continues, it could lead to the risk of an energy crisis. Strategies related to proper distribution and energy demand are needed to build resilience and form a sustainable society.

1. Introduction

The global crisis of the COVID-19 pandemic that has occurred since the end of 2019 has caused health emergencies around the world. This particular situation has led to many significant changes to city life, about work, recreation, public spaces use, safety and security of transportation, both public or private. As a precautionary measure, governments around the world began enforcing quarantine measures such as social restrictions, face-to-face dismissals in learning, and the closure of public spaces to avoid physical contact.

Positive cases in Indonesia were first identified in March 2020 leading to a fairly rapid spread [1, 2]. Until the paper was started, the increased cases of infection are still significant. In response to this situation, the Indonesian government re-enacted the Enforcement of Restrictions on Community Activities (PPKM) in various regions in Indonesia where have high infection spread points. These restrictions include closing public spaces, restricting transport, traveling ban, online learning, restaurants and café operating hours, and limiting as much as possible by implementing Work From Home (WFH) or Stay At Home (SAH).

[4] Corresponding author: didit.novianto@its.ac.id

The concept of working at home is not a new thing, this concept has existed for the last 10 years [3]. Increasingly advanced technology provides the option that work can be done virtually, leading to efficiency and changes in workplace concepts [4]. Because of the pandemic, this concept is applied in the work, especially in offices because to prevent the spread of viruses. And this is something that inevitable. However, the implementation of WFH also has side effects and risks such as in terms of time management, lack of time communicating with colleagues, and social isolation that leads to reduced motivation and performance in work [5]. Even working from home often leaves employers with no clear boundaries between work and personal life, which is results in excessive workload, increasing stress, anxiety [6], and job satisfaction. Working from home is also reported to disrupt family balance and work-life resulting in stress [7]. Studies conducted in Japan also prove the application of working at home makes the hours and days for work increased, which affects behavior and physical activity [8,20]. In addition, isolation and social restrictions cause drastic changes in life and result in various adverse effects on mental health such as anxiety, depression, insomnia, and emotional fatigue [9,10]. Thus, the coronavirus pandemic resulted in sudden changes in daily routines and working conditions for many employees.

As work and online schooling were done more at home, that makes people are starting to adapt to new lifestyles. Thermal comfort at home is a very important concern [11]. Many experts also recommend that people need to maximize openings for good air circulation. However, it is predicted that Indonesians will still turn on the electric fan and or open the window by keeping the air conditioner turned on because of the Indonesian hot climate [12]. Therefore, during the period of restrictions and WFH is expected to increase the amount of electricity consumption which at the same time the energy demand also increases.

Before the pandemic, the use of energy in households was said to be high so often the electricity supply was inadequate. Associate VP Markplus Inc. reported that there has been an increase in energy consumption by the community since the WFH was implemented. There was an increase in household gas use by 11.19% and electricity by 7.5% [13]. Indonesia is the fourth most populous country in the world, predicted to increase many times during the pandemic. This needs to be underlined. On the other side of the increasing consumption of electricity and gas in the household, the existence of this WFH reduces mobilization and transportation because work does not have to be done in the office [14,15]. After the pandemic ends, it is still uncertain whether energy consumption can return to normal due to the formation of habits during this new normal.

Previous research has discussed how WFH harms well-being, leisure time, lack of motivation, and job satisfaction, as well as the relationship of energy consumption and comfort to socioeconomic factors [5,6,8,11,12]. However, discussions about how household appliances affect increased energy consumption during WFH are still little discussed. Seeing the pandemic conditions in Indonesia that are increasingly spreading, WFH is again applied to a large number of office workers. Therefore, this study aims to analyse energy consumption factors during the implementation of WFH or SAH. It is hoped that this research will be a new source of knowledge for readers and further developments in the future related to lifestyle changes during the COVID-19 pandemic.

2. Research Methods and Scope

The methods used in this research are quantitative-based with descriptive-analytical design. Data collection in this study using online questionnaires through Google Forms that are distributed to various communities with criteria are: (1) the general public living in Indonesia, (2) have access to Google Form, and (3) people who can participate in this study are those who fill out questionnaire sheets that we shared. Questionnaires consist of questions divided into four sections: (1) social-demographic characteristics, (2) about WFH (i.e., whether and how long has been implementing WFH), (3) expenditure levels before and during WFH, and (4) changes in intensity use of electronic devices and households during WFH or SAH. The questionnaires were distributed to the entire community and the total of the respondent that participated in the study is 215 people. The division aims to ascertain whether the instrument variables used in this study are valid and reliable.

This study used dependent variables and independent variables. The variables are processed in IBM SPSS software using multiple regression analysis to test the influence between variables. Then, the result will be interpreted, it can be known the amount of independent variable contribution to dependent variables. So the results of the interpretation can conclude whether the variable has a positive influence on energy consumption during the implementation of WFH or not.

3. Results and Discussion

3.1. Sociodemographic Characteristics

From the distributed questionnaires, the feedback obtained was 215 respondents who participated in this study. Female participants (63.7%) were higher than male participants (36.3 %). This study was distributed to productive ages, most of the participants aged 31 – 35 years (36.7%) participated in this study and others who participated were those aged 21 – 25 years (15.3%), 17 – 20 years (14.9%), and 26 – 30 years (10.7%). Most of the participants worked as staff company (27.4%), civil servants (21.9%), teachers (15.8%), self-employment (14.4%), medical personnel (1.9%), and other professions (6.5%). Not only that, but there were also students (7.4%), retired (2.8%), and unemployed (1.9%) participating in research. A total of 88.8% of respondents implemented the WFH system during the pandemic. About 42.8% of respondents lived in single-storey homes and 39.1% lived in two-story homes. As well as the number of lengths they occupy residence based on classification does not differ much, and the dominating are those who have lived for 4 – 6 years (15.8%) and 7 – 10 years (16.3%). The information can be found in Table 1.

Table 1. Socio-demographic characteristics of respondents

Categories	Variable/N/Percentage									
Gender	Male	Female								
	78	137								
	36,3%	63,7%								
Age	> 17	17-20	21-25	26-30	31-35	36-40	41-45	46-50	50-60	< 60
	2	32	33	23	79	14	9	9	13	1
	0,9%	14,9%	15,3%	10,7%	36,7%	6,5%	4,2%	4,2%	6,0%	0,5%
Building Type	One-storey house	Two-storey house	More than two-storey house	Rental House/ dormitory	Row House	Apartment/Flat				
	93	84	4	15	10	9				
	43,3%	39,1%	1,9%	7,0%	4,7%	4,2%				
Building Direction	West	South	East	North	Others					
	44	52	52	52	15					
	20,5%	24,2%	24,2%	0,0%	7,0%					
Floor Size (m^2)	> 20	20-30	30-40	40-50	51-60	61-70	71-80	81-90	< 90	
	20	19	12	16	22	14	13	19	80	
	9,3%	8,8%	5,6%	7,4%	10,2%	6,5%	6,0%	8,8%	37,2%	
Length of Stay (year)	> 1 year	1-3	4-6	7-10	10-15	16-20	20-30	< 30		
	19	40	34	35	22	26	29	10		
	8,8%	18,6%	15,8%	16,3%	10,2%	12,1%	13,5%	4,7%		
Occupation	Staff	Civil Servant	Teacher	Self-employment	Student	Unemploy ment	Retired	Medical Personnel	Others	
	59	47	34	31	16	4	6	4	14	
	27,4%	21,9%	15,8%	14,4%	7,4%	1,9%	2,8%	1,9%	6,5%	
WFH before Covid-19	Yes	No	Sometimes							
	155	38	19							
	72,1%	17,7%	8,8%							
WFH since Covid-19	Yes	No								
	191	24								
	88,8%	11,2%								

In short, the respondents who participated tend to be female and adult, with a higher proportion of respondents being aged 31 – 35 years, a small percentage under 17 years of age, and over 60 years of age. In addition, most of them are staff, civil servants, teachers, and employers with a total of 80%, as well as 84.2% of participants living in their own homes (not rent).

3.2. Work From Home

Questionnaires also obtained information related to work style before and during the COVID-19 pandemic. 72% of them are families with incomes coming from parents (both husband and wife, father and mother), and the rest are single-income families. Before the pandemic, about 72.1% of respondents had to work in the office/workplace, 17.7% could do work at home, and as many as 8.8% could do work flexibly i.e., sometimes could do it at home. After the pandemic entered Indonesia which resulted in the situation changed. As mentioned above, about 90% of respondents did WFH during the pandemic, and most of them (68%) report that conducting WFH within an uncertain time. This is because of the restrictions on activities imposed by the government to reduce the number of victims of infection. And 11.2% of respondents did not apply it for various reasons such as dominating is because work cannot be done at home which is related to their profession. Another reason is that those who cannot always do the WFH system due to the company's policy of setting working hours/employee shift system and the company system that must continue to do work at the office, they are afraid if they suffer losses in business, and part of them work as medical personnel, as shown in Figure 1.

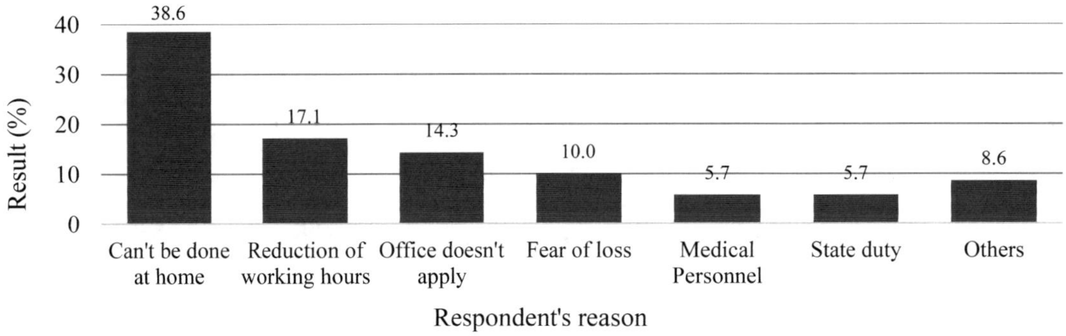

Figure 1. Reasons for not working from home.

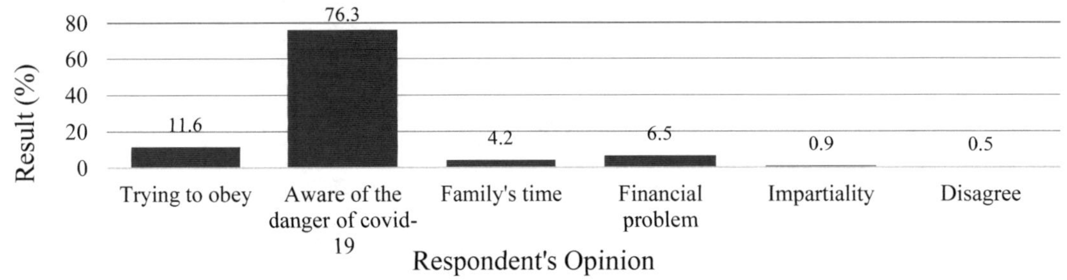

Figure 2. Respondent opinion about working from home.

In the questionnaire, respondents were also asked about their responses to the implementation of WFH during the pandemic. The results showed that about 97% of respondents agreed with the implementation of the WFH. The reason is that most of them replied that they agreed to do WFH because they were aware of the dangers of COVID-19 and obeyed the rules (75%). In addition, 11.6% of them tried to comply with the rules applied, 4.2% agreed because they had more time with family, 6.5% of them admitted that the system applied greatly impacted their economy, and some of them profess to disagree and do not care much about the WFH being implemented (Figure 2).

Overall, according to the results described above WFH activities are inevitable. And that makes the whole activity will always be centred at home. As a result, unorganized time management due to flexible working hours, as well as decreased motivation and performance in work may occur [5].

3.3. Energy Consumption during WFH

Five items (Likert scale 3 levels, 1 = not increased, 3 = more increased) are used to test the impact of increasing energy consumption during WFH. That items are habit change during WFH (working hour, sleeping hour, productivity, work concentration, TV usage, AC usage, and internet), electricity and gas expenditure levels (electricity, gas, and internet), concern level for energy consumption (air conditioning, lighting, air, electricity, gas, water, and internet), thermal comfort in the house (lighting, air temperature, airflow, and humidity), and the intensity of use of electronic devices in households (air conditioning, TV, rice cooker, oven/microwave, stove, home computer/laptop, and electric fan). The connectedness between variable studies is tested using multiple regression tests to see how much influence between the variables. The results can be seen in Table 2 below.

Table 2. Regression test results

Model	Unstandardized Coefficients		Standardized Coefficients	t	Sig.
	B	Std. Error	Beta		
(Constant)	7,900	1,276		6,193	0,000
Intensity of computer/laptop usage	0,993	0,223	0,281	4,460	0,000
Internet usage habits	1,130	0,389	0,181	2,900	0,004
Intensity of rice cooker usage	0,693	0,223	0,191	3,109	0,002
Intensity of AC usage	0,425	0,199	0,135	2,133	0,034
Water spending	0,483	0,235	0,126	2,056	0,041

The statistical significance value must be < 0.05 to see how much influence it is. Of all the variables tested, which had an influence on energy consumption variables were the intensity of home computer/laptop usage (X1) by 0.000 < 0.05, internet use (X2) by 0.004 < 0.05, rice cooker usage intensity (X3) by 0.002 < 0.05, air conditioning usage intensity (X4) by 0.034 < 0.05, and expenditure on water needs (X5) by 0.041 < 0.05 (Table 2). This is quite natural because lifestyle and work style that changes during pandemics cause activities such as school and work were done at home so that it can interfere with the management of a person's time at work [5]. So, the regression equation model based on the results of the analysis can be written:

$$Y = 7{,}900 + 0{,}993X1 + 1{,}130X2 + 0{,}693X3 + 0{,}425X4 + 0{,}483X5 \quad (1)$$

The equation can be interpreted, namely the contribution of the independent variable to the dependent variable on the intensity of computer/laptop use (X1) of 0.993, Internet use (X2) of 1.130, the intensity of rice cooker use (X3) of 0.693, the intensity of AC use (X4) of 0.425, and expenditure on water needs (X5) of 0.483. All the results of the analysis of each variable show a positive number. This means that the longer or more frequent the use of household electronic devices such as home computers/laptops, air conditioners, rice cookers, internet, and water needs, will be followed by an increase in energy consumption.

Simultaneously, Table 3 shows the value of significance for the influence of changes in habits during WFH, the level of electricity and gas expenditure, concern level for energy consumption, the thermal comfort of a house, and the intensity of simultaneous use of electronic devices in households to energy consumption is 0.00 < 0.05, so it can be concluded that there is a simultaneous influence on energy consumption (Table 3).

Table 3. ANOVA analysis's result

	Model	Sum of Squares	df	Mean Square	F	Sig.
1	Regression	500,167	29	17,247	3,011	.000[b]
	Residual	1053,871	184	5,728		
	Total	1554,037	213			

Overall, the current results show that changes in activity due to WFH cause changes in daily habits, especially increase in internet use, rice cooker, air conditioning, home computer/laptop use, and the need for water that causes an increase in spending per month. Continuous and simultaneous use can certainly have a major influence on the energy released. Tools used daily such as television, air conditioning, home computer/laptop, electric fan, rice cooker, lighting, and internet will affect the consumption of electrical energy. Respondents believe that significant improvements are in the use of home computers, rice cookers, air conditioners, and the internet. On the other hand, the level of expenditure on water bills there is increased which means that the water needs will happen during WFH.

This indicates that the changing style of working at home requires them to continue to use personal electronic devices. Activities that are mostly done at home make people turn on the air conditioner and or electric fan to provide thermal comfort. Family members also become more often at home which will automatically increase the cooking needs. So the longer the period of implementing WFH can also increase the energy consumption. If this continues, it can lead to energy poverty, which will directly affect the health and welfare of the population [11].

3.4. Energy consumption on family members, household, and electricity

In this section, we analyze the number of family members, the age of the building, and the building size to explore the potential influence on the increase in energy needs that occur.

Table 4. Correlation analysis's result

	Model	β	Sig.	R	R Square
	(Constant)	7,072	0,000		
1	FAMILY MEMBERS	0,745	0,001	.496[a]	0,246
	(Constant)	1,215	0,000		
	Intensity of rice cooker usage	0,297	0,000		
	Internet spending	0,341	0,001		
	Airflow comfort	0,315	0,005		
	Intensity of electric fan usage	0,174	0,020		
2	BUILDING SIZE	1,073	0,000	.383[a]	0,147
	(Constant)	1,690	0,000		
	Intensity of AC usage	0,252	0,027		
	Airflow comfort	0,348	0,047		
3	BUILDING AGE	0,062	0,571	.352[a]	0,124
	(Constant)	4,091	0,000		
	Soundproof comfort	-0,247	0,030		

Based on a statistical analysis of family members ($\beta 1 = 0.745$), it shows that the more family members who live in one house, the energy demand will increase by 24.6%. The main impact comes from the use of rice cookers, internet/Wi-Fi, and the need to achieve comfortable airflow, of course, an electric fan is needed considering the hot climate in Indonesia. Building size ($\beta 2 = 1.073$) and building age ($\beta 3 = 0.062$) also show that the wider and older their building will increase the need for thermal comfort energy consumption by 14.7 % and 12.4 %. For example, the intensity of air conditioning use will be increased and soundproofing will decrease as the performance of the building decreases. Meanwhile, simultaneously, family members have an impact on increasing energy consumption by 49.6%, house size by 38.2%, and building age by 35.2%. From the three items that were tested, the one that has the biggest impact on daily energy consumption is the family members (Table 4).

Figure 3 explains the average amount of household electricity consumption per population living and the increase in electricity when doing WFH in a percentage. For a house occupied by one person/single, the average electricity consumption will be 195 kWh/month. In terms of lifestyle attributes, a house with a single member will use artificial ventilation such as air conditioning and electric fans to cool the room, cook, and use the home internet all the time due to work or other needs. Compared to others, single members had the highest electricity consumption usage from the other clusters, and it was revealed that there was an increase reaching 75% after they did WFH. Meanwhile, households with 9 members have the lowest monthly electricity consumption, 23 kWh/month. It can be understood because the air conditioning and electric fans used at home are more often used together, not specific to certain people or members. The increase occurred when WFH and SAH were 10% from before because all family members tended to do activities at home. In terms of increase, households with 9 members are no different from households with 7 members. However, the average electricity usage per person is 73 kWh/month. Which is the family members will share the air conditioning and electric fans use with other members, rice cookers can be used more than two times a day, and the internet/Wi-Fi is needed more than usual because some family members mostly do their work at home or studying online. This also applies to households with other multi-members or more. Looking at the graph, it can be seen that the more members, has the smaller the electricity consumption requirement per kWh/person/month. The rate of increase will also be felt more in households with single members because they have to bear the electricity needs alone, compared to households with multiple members or more, who shared the ventilation or cooling needs with other members.

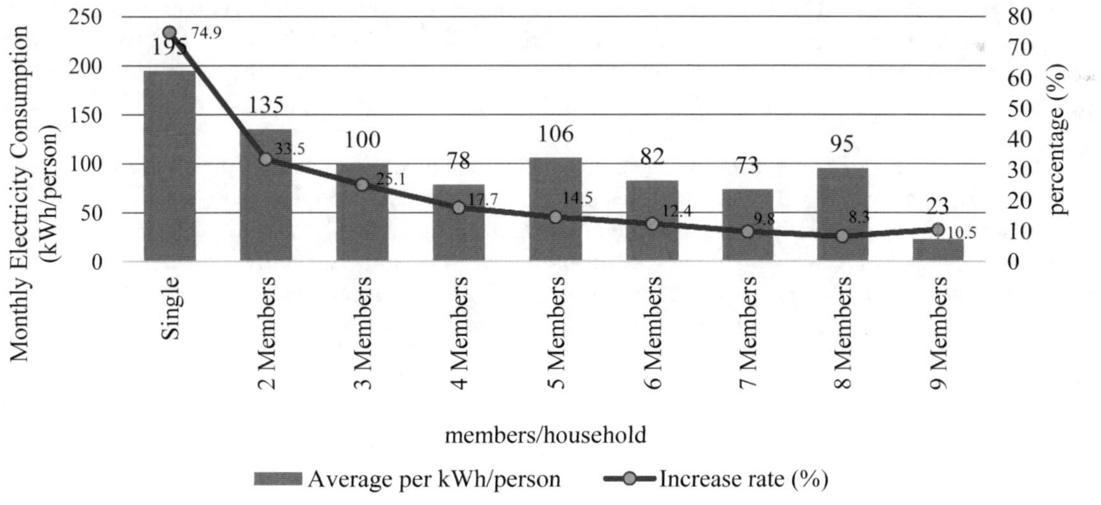

Figure 3. Electricity use consumption change

However, electricity demand and increase are highly dependent on the house size and population living. It means that if a small or narrow house occupied by many family members automatically will use a lot of electricity, while a large house occupied by few people the possibility of an increase in electrical consumption is not too significant or may not even change. In this case, respondents agreed that as long as the WFH was enacted it could increase energy needs, judging by the increase in electricity and gas usage per month.

4. Conclusion

Health problems occurring around the world related to pandemics are inevitable [16]. Various efforts have been made to suppress the spread of the virus. Such as the closing of public spaces, and even restrictions on office activities by implementing a WFH. In this study, 215 people living in the various city in Indonesia were investigated. The results showed that many citizens during this pandemic implemented WFH and most of them used air conditioning to cool their homes. It is also confirmed the studies about the implementation of WFH has negative effects such as time management, decreased motivation and performance in work, as well as impacting the health and well-being of the community [5, 6, 9, 10, 12].

It is known in this study that about 90% of respondents believe that the implementation of WFH is necessary during a pandemic. This shows that public awareness of the dangers of pandemics is increasing. Therefore, people will often spend more time indoors than outside. And comfort will always be an aspect of being noticed. To achieve comfort at home, people tend to use an electric fan or air conditioner, because of the hot climate in Indonesia. The study highlights the long-time influence of WFH, the number of family members, and the extent of the home on energy consumption. During WFH that has a significant influence is the intensity of the use of home computers/laptops, internet, rice cooker, air conditioning, and water needs. This is natural considering lifestyle changes that require activities such as learning and work to be done at home. Then it can also be interpreted that the longer the time WFH can increase energy needs in households [12].

In addition, other influences of increasing energy consumption also come from the number of family members who live in one house and the area of the house. The more people who live in a certain area of the house automatically the need for electricity follows each individual. Evidenced by the increase in electricity and gas consumption per month in households.

This problem needs to be considered because if it happens in the long run, it can lead to bad situations such as energy poverty due to lack of supply [17]. It will directly affect the health and well-being of the community [18]. Even after the pandemic ends, the majority of people will tend to return to their original needs [13]. However, it does not close the possibility that there will be people who still follow the habit during the pandemic period.

Therefore, measures such as considering the real needs of the community are needed to reduce and prevent the occurrence of adverse situations and the risk of energy poverty due to lack of supply [19]. Proper strategies on the distribution and use of energy are also needed to build a sustainable society.

Although this study has limitations in terms of information obtained from respondents, and the scarcity of studies on energy consumption during WFH. This research is expected to be a new science for its readers. As well as more research is needed for the depth of knowledge.

5. References

[1] Ratcliffe R 2020 *First coronavirus cases confirmed in Indonesia amid fears nation is ill-prepared for an outbreak.* The Guardian. Available online: https://www.theguardian.com/world/2020/mar/02/first-coronavirus-cases-confirmed-in-indonesia-amid-fears-nation-is-ill-prepared-for-outbreak (accessed on 11 October 2021)

[2] Susilo A, et al. 2020 *Coronavirus disease 2019: Tinjauan literatur terkini.* Jurnal Penyakit Dalam Indonesia **7**(1): pp 45-67

[3] Parthasarathy A 2020 *Coronavirus Challenge-Propelling a New Paradigm of Work from Home.* NISCAIR-CSIR. SR **5**(04): p. 14-19.

[4] Holland S J, et al 2016 *I can't steal from a coworker if I work from home: Conceptual and measurement-related issues associated with studying counterproductive work behavior in a telework setting.* Human Performance. **29**(3): pp 172-190

[5] Martin, B.H. and R. MacDonnell 2012 *Is telework effective for organizations? A meta-analysis of empirical research on perceptions of telework and organizational outcomes.* Management Research Review

[6] Liu H -L and V -h Lo 2018 *An integrated model of workload, autonomy, burnout, job satisfaction, and turnover intention among Taiwanese reporters.* Asian Journal of Communication. **28**(2): pp 153-169

[7] Irawanto D W, K R Novianti, K Roz 2021 *Work from home: Measuring satisfaction between work–life balance and work stress during the COVID-19 pandemic in Indonesia.* Economies. **9**(3): p 96

[8] Javad Koohsari M, et al. 2021 *Working from Home After the COVID-19 Pandemic: Do Company Employees Sit More and Move Less?* Sustainability. **13**(2): p 939

[9] Fofana N K, et al. 2020 *Fear and agony of the pandemic leading to stress and mental illness: an emerging crisis in the novel coronavirus (COVID-19) outbreak.* Psychiatry Research. **291**: p 113230

[10] Yang S, et al. 2020 *The mental health burden of the covid-19 pandemic on physical therapists.* International Journal of Environmental Research and Public Health. **17**(10): p 3723

[11] Cuerdo-Vilches T, M Á Navas-Martín, I Oteiza 2021 *Behavior Patterns, Energy Consumption and Comfort during COVID-19 Lockdown Related to Home Features, Socioeconomic Factors and Energy Poverty in Madrid.* Sustainability. **13**(11): p 5949

[12] Novianto D, et al. 2021 *Rethinking sustainability during WFH: A survey on living environment quality and energy use.* IOP Conference Series: Earth and Environmental Science. **881**(1): p 012005

[13] Novika S 2020 *Pemakaian Gas hingga Listrik Meningkat Tajam Selama WFH, BBM Anjlok.* Detik Finance. Available online: https://finance.detik.com/energi/d-4976787/pemakaian-gas-hingga-listrik-meningkat-tajam-selama-wfh-bbm-anjlok (accessed on 11 October 2021)

[14] Novianto D, W Gao, S Kuroki 2015 *Review on people's lifestyle and energy consumption of Asian communities: Case study of Indonesia, Thailand, and China.* Energy and Power Engineering. **7**(10): p 465

[15] Novianto D, Nuffida N E, Gao W 2021 *A Review of Architecture Response to the Pandemic towards New Normal Behavior* (International Conference and Workshop: Re-thinking the HIBIKINO Campus 2001-2021: Kitakyushu (Japan))--(Architectural Design). JAILCD.13-8

[16] Blanco H, et al. 2009 *Hot, congested, crowded and diverse: Emerging research agendas in planning.* Progress in Planning. **71**(4): p 153-205

[17] Karpinska L and S Śmiech 2020 *Conceptualising housing costs: The hidden face of energy poverty in Poland.* Energy Policy. **147**: p 111819

[18] Castaño-Rosa R, J Solís-Guzmán, M Marrero 2020 *Energy poverty goes south? Understanding the costs of energy poverty with the index of vulnerable homes in Spain.* Energy Research & Social Science. **60**: p 101325.

[19] Hesselman M, et al. 2020 *Mapping COVID-19 Emergency Measures on Household Energy Services.* Engager Cost. Available online: http://www.engager-energy.net/covid19/ (accessed on 2 April 2021)

[20] Gao W, Novianto D 2018 *Eco-House in Kitakyushu, Japan.* In Sustainable Houses and Living in the Hot-Humid Climates of Asia (pp 483-490). Springer, Singapore.

Acknowledgments

This research is part of Penelitian Kerjasama Antar Perguruan Tinggi Institut Teknologi Sepuluh Nopember 2021, No. 1315/PKS/ITS/2021. The authors gratefully acknowledge this financial and technical support.

IOP Conf. Series: Earth and Environmental Science 1007 (2022) 012011 doi:10.1088/1755-1315/1007/1/012011

Analysis study of light shelves, light pipes, mirror ducts utilization and implementation in Indonesia

E S Rumondang[1,2], Y V Regina[1], Y R Angela[1] and D Larasati[1]

[1] Department of Architecture, Bandung Institute of Technology, Bandung 40116, Jawa Barat, Indonesia

Abstract. This research was conducted because of the issue of exhaustible energy in Indonesia, particularly in the architecture and construction division. One of the methods to reduce energy-use in buildings is by using passive lighting techniques. This research examined several types of passive lighting techniques to determine the advantages and disadvantages of this type of lighting, as well as their implementation in Indonesia. Three types of passive lighting that will be analyzed in this study are light pipes, mirror ducts, and light shelves. Literature studies and precedents are used as a comparison for analysing the types of passive lighting and as precedents to determine the variables and parameters in the simulation. Simulations are carried out to calculate effectiveness based on the parameters that have been chosen. In addition, a survey in the form of a questionnaire was conducted to see the preferences and implementation of passive lighting among practitioners in the architectural field in Indonesia. This research is expected to be a source of information and can be used by architects to understand more deeply about passive lighting and its effect on the energy efficiency of buildings.

1. Introduction

In this era of globalization, energy consumption in the world has increased drastically. This occurs in all sectors, including industrial, household, commercial, transportation, and others [1]. Therefore, the issue of non-renewable energy is increasingly being discussed in the world today. Whereas non-renewable energy is only available in limited quantities and cannot be replaced in a short time.

In addition, the impact of globalization has an especially strong influence on the process of designing and constructing buildings, which makes buildings one of the largest energy users for the environment. In Indonesia, building sector energy consumes about 18-20% of the total energy consumption of all sectors in general [2].

Energy consumption in the building category in Indonesia is still wasteful. Buildings in Indonesia will be the third largest energy user, accounting for around 30% of total national energy consumption [3]. If not managed properly, energy consumption from buildings has the potential to increase up to 40% by 2030. For this reason, it is necessary to reduce energy consumption in buildings in Indonesia. The Indonesian government itself has already issued a regulation to maintain energy consumption about the use of electricity saving [4].

Lighting is a very important aspect of human life. Humans can only use their visual ability with the presence of light. In result, artificial lighting has always been one of the largest energy consumption that humans use. According to Alrubaih, in recent years, it has been proven that 40% of energy use in

[2] Corresponding author: evelynsarahrs@gmail.com

buildings is used in the lighting sector [5]. Various lighting alternatives such as passive lighting were developed to reduce the energy consumption.

Traditional methods of lighting generally only enter a maximum of 4.6 meters of light from the opening. However, expanding the area of the opening has been shown to be ineffective because only a small amount of light can reach the inside of the room and also increases the radiation in the part of the room close to the opening [6].

The concept of energy-efficient lighting by optimizing lighting systems (known as passive lighting techniques) could be a way to reduce energy use in buildings. By using passive lighting, absorbed sunlight can be reflected and be able to reach the deepest parts of a room. In addition, passive lighting techniques do not require special mechanical equipment and energy sources.

As a solution to reduce energy in buildings, the researcher wants to examine the application of a few passive lighting types in Indonesia. This can be started by examining the effectiveness of different types of passive lighting and comparing them with the preferences among academics and architectural professionals. The researcher will also examine the application of the use of these types of passive lighting in Indonesia.

This research is limited by examining three types of passive lighting, namely light shelves and light tubes (light pipes and mirror ducts). These three passive lighting were chosen because of a precedence of a building: Zero Energy Building at the BCA Academy Singapore. This building is called a live demonstration of an energy efficient building and received many international awards including ASEAN Energy Awards, IES Award, etc. [7] and they use all three (light shelves, light pipes, mirror ducts) of the tools.

The formulations of the problems in this study are (1) How is the effectiveness of types of passive lighting (light pipes, mirror ducts, light shelves? (2) What is the preference of practitioners in architecture for these types of passive lighting? And (3) How is it applied in Indonesia?

Therefore, this study aims to (1) Find out the comparison of energy use in passive lighting design of light pipes, mirror ducts and light shelves, (2) Knowing the preferences of practitioners in the field of architecture regarding passive lighting, and (3) Ascertain the real implementation of passive lighting by people in the architectural field in Indonesia.

2. Research methods

There are three research methods in this research (1) literature study, (2) passive lighting simulation, and (3) passive lighting application questionnaire survey.

The literature study method is a qualitative method to determine and further review the types of passive lighting in case studies. Case studies are used to analyze and study the features of the three selected passive lighting. The results of these studies will be used as design considerations in conducting simulations.

The second research method in this study is a quantitative method to do simulations on buildings using the three alternatives of passive lighting. The building is designed with an area of 324 m^2 consisting of 5 classrooms. Any passive lighting will be compared to a room that only uses windows as lighting.

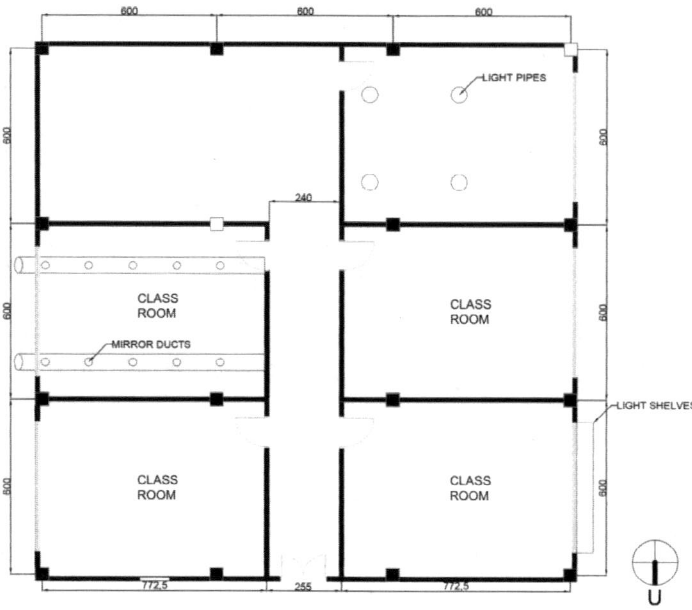

Figure 1. Classroom plan for simulation.

Fixed variables used in this building are size of the space, opening area, material, direction of light, and time. The size of the classrooms made is 7.8 x 6 m. The openings in each class are 11.57 m^2. The material used for the walls is brick. The time used to perform this simulation is on January 21 at 14.00. The classroom which will be used as in the simulation is the room on the west side.

For the simulation of light shelves, it used 2 different locations, namely outside the building and inside the building. Reflective boards with a size of 60 x 445 cm are used in the simulation. The two positions will be simulated further to determine which passive lighting is more optimal.

In the simulation with mirror ducts, the simulation used 2 pipes that have a diameter of 45 cm (surface area 0.15 m^2). Pipe diameter considerations are determined based on a case study at the BCA Academy. The area of space used in the classroom to be simulated is half of the area of the BCA Academy, then the pipe surface area used is half of the duct surface area at BCA Academy. Each pipe has 5 apartments with an apartment diameter of 20 cm. The distance between the apartments is 150 cm obtained from the angle of reflection of sunlight entering the pipe at an angle of 35°.

For the simulation using light pipes, 4 light pipes with a pipe diameter of 55 cm and a pipe length of 120 cm will be used. Light pipes are placed at a distance of 3 m from the outer wall and the distance between the light pipes is 3 m.

The third method used is a qualitative research method, which is a survey in the form of an online questionnaire distributed to 209 respondents from various types of work. Respondents' occupations were mostly students or architecture graduates at 72.2%. In addition, there are also architects, educators, employees / employees, interior design students, urban designers, interior designers, and others. The distribution of respondents' locations varies widely, some from West Java, DKI Jakarta, East Java, Banten, Central Java, Riau Islands, North Sumatra, Bengkulu, Riau, East Kutai, and many other cities. This questionnaire is a qualitative questionnaire to determine the extent to which Indonesians are familiar with passive lighting and to find out designer preferences in the use of passive lighting and their reasons.

3. Literature studies

According to Kadir, passive lighting or daylighting is a technique to lighten up building interiors with natural light without artificial lighting in daytime [8]. Window is the most commonly used daylighting

IOP Conf. Series: Earth and Environmental Science 1007 (2022) 012011 doi:10.1088/1755-1315/1007/1/012011

tool as it lets side lightning from the sun come into a building but there are many other daylighting tools other than window. There are many benefits for using natural lights, such as (1) influencing a human's well-being physiologically [9], (2) increasing productivity of a human, and energy consumption [8].

Energy consumption of minimum 10% can be reduced by using passive lighting strategies. Daylighting offers higher luminous efficacy compared to the majority of artificial lighting. According to Mardaljevic, in daylight, sunlight offers 70-10 lm/w, clear blue skylight, overcast skylight, and global skylight offers 130, 110, 105 lm/w respectively [10]. This is pale in comparison with artificial light which only offers 15 lm/w for an incandescent bulb. Both compact fluorescent and T5 linear fluorescent offer 57-72 and 70-100 lm/w respectively [10].

These are some passive lighting strategies according to Brandon Gullotti about Passive Daylighting Systems [11]:

1. Building orientation;
2. Window;
3. Sky lights;
4. Clerestories;
5. External shadow system;
6. Light shelves;
7. Solar tubes;
8. Light wall colors;
9. Parametric modeling, daylight simulation and artificial intelligence (AI).

Passive lighting that will be discussed furthermore are:

3.1. Light pipes

According to Jadhav, Light pipes or light tubes work as a light distributor to illuminate the room. The light pipes consists of two essential components [12]:

1. A light collection dome that work similar to skylight, but the shape is hemispherical to bring light inside the tube from various angles;
2. A long hollow tube is attached on one end of the light dome and the other end may look like a light fixture that emits light into the desired space in the room or ceiling. Inside the tube is made up of reflecting material such as metal that helps to reflect and transport light along the tube.

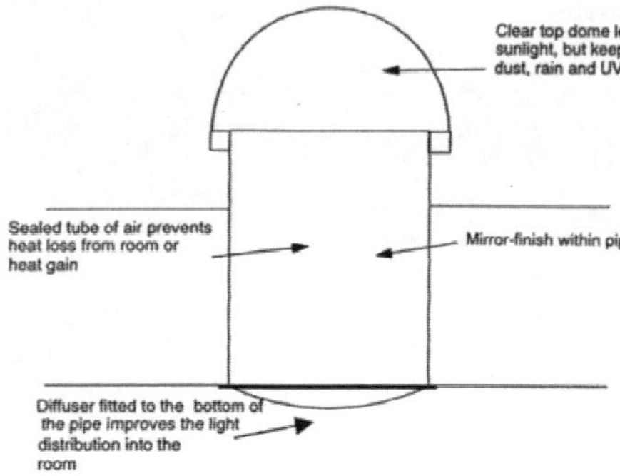

Figure 2. Schematic of typical light pipe [13].

3.2. Mirror ducts

Mirror ducts are another variation of the light pipes, but on horizontal axis and mounted on the ceiling in the room [12]. Mirror ducts are designed to catch daylight through external collectors and transferred into horizontal reflective ducts that are incorporated within the ceiling. The light then exits through the ceiling apertures or illuminating units into the workspace below. The main advantage of the mirror ducts is usually glare free.

Two main components on mirror ducts are [12]:

1. Collecting Unit, located on the exterior of the building which serves to capture light;
2. Radiation Unit, located in the interior of the building which functions to emit light into the room.

Various designs of mirror ducts according to Beltran:

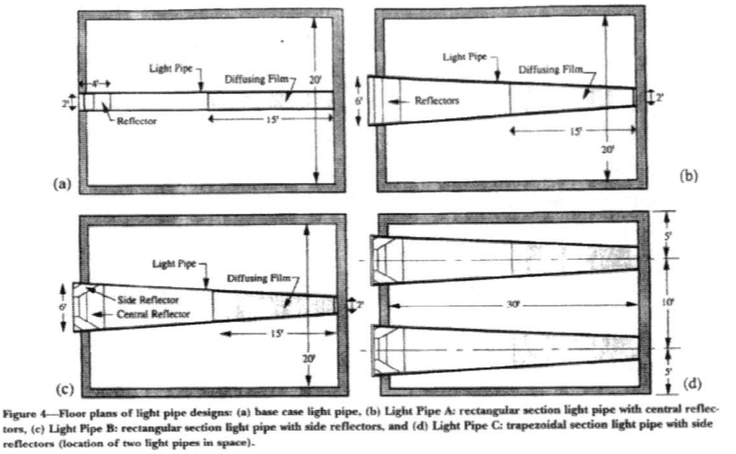

Figure 3. Various design of *light pipes* (horizontal) [6].

1. Base case light pipe;
2. Rectangular section light pipe with central reflectors;
3. Rectangular section light pipe with side reflectors;
4. Trapezoidal section light pipe with side reflectors.

3.3. *Light shelves*

Light shelves are reflective surfaces that can reflect light into the room. The reflective surface is placed on top of the window. There are 4 types of light shelves designs according to Beltran [6]:

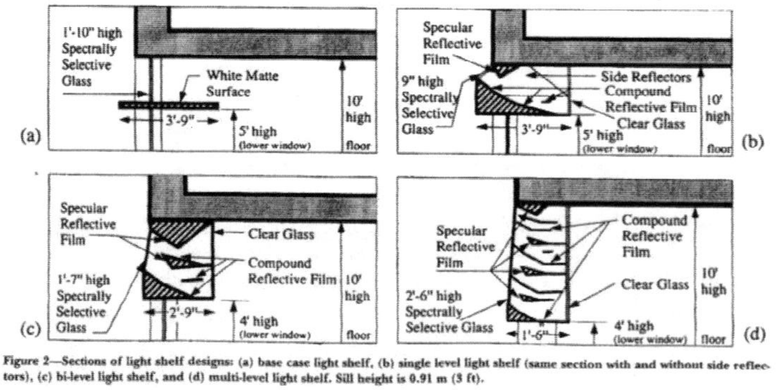

Figure 4. Section of the various design of the *light shelves* [6].

1. Base case light shelf;
2. Single level light shelf;
3. Bi-level light shelf;
4. Multi-level light shelf.

IOP Conf. Series: Earth and Environmental Science 1007 (2022) 012011 doi:10.1088/1755-1315/1007/1/012011

4. Case study

Case study conducted by reviewing the BCA Academy that is located at the National University of Singapore, or well known as NUS in Singapore. This building becomes one of the buildings that raises the concept of "Zero Energy Building" where the energy produced is surplus, or in other words, more than what is needed. This building dimension is 20 x 76 metres. This building applied three passive lighting that will be discussed further, light pipes, mirror ducts, and light shelves [14]. Several aspects in this case study that are taken into consideration in designing passive lighting are the placement and size of the passive lighting, as well as the ratio to the area of the room.

This building's width is 20 metres and the single loaded plan becomes the challenge in designing the room's lighting. To fulfill the "Zero Energy Building" concept, this building uses these passive lightings. Three of the passive lighting are placed on different sides of the building. Light pipes placed at the upper floors of the building, mirror ducts (horizontal light pipes) placed on the floors below, and light shelves placed on the east side of the building [15].

The use of the light pipes in this building because it is more effective than skylights. Skylights have smaller surfaces, so that the light is more focused and doesn't spread anywhere. Light pipes are useful for keeping the inside of the building stay bright. The diameter of the light pipes are 1 m with 4 pieces placed in the middle of the multipurpose room with an area of 400 m^2. Light pipes are channeled from the roof to the top of the building (third floor). There are two types of light pipes installed, which are mirrored and not mirrored [15].

Mirror ducts (horizontal light pipes) are used to illuminate the inner building on the 1st floor to 3rd floor. Mirror ducts are placed on the east facade of the building. It has a cross sectional area of 0,3 m^2 with a total area of 2 pieces on the east side of the classroom with an area of 100 m^2. Sunlight is channeled through a reflective horizontal pipe of the ceiling of the room [15].

Light shelves are applied all over the west facade of the building. It uses materials that have high reflectivity that can reflect lights. The width is in the range 50-70 cm along the opening. In addition, light shelves also function as shading to prevent direct light entering the inner building. The effectiveness of light shelves is also increased by using a ceiling with a reflective coating material [15].

5. Result and discussion

5.1. Simulation results

Simulation results for ordinary window and three types of passive lighting (light pipes, mirror ducts, and light shelves) are presented in the following:

5.1.1. Light pipes. The points taken for analysis are the points close to the window, in the center of the room, and in the deepest part of the room.

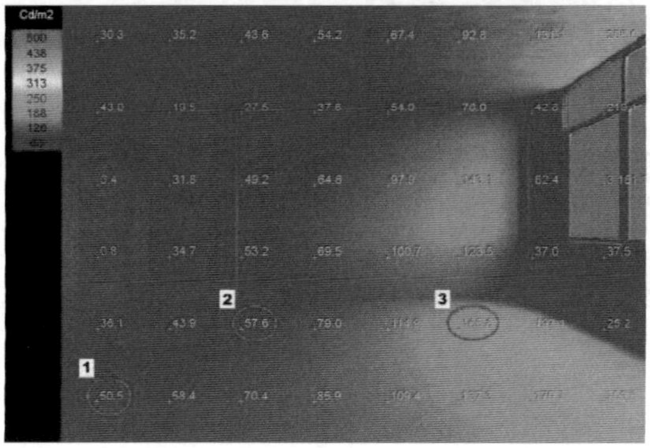

Figure 5. Ordinary window simulation results.

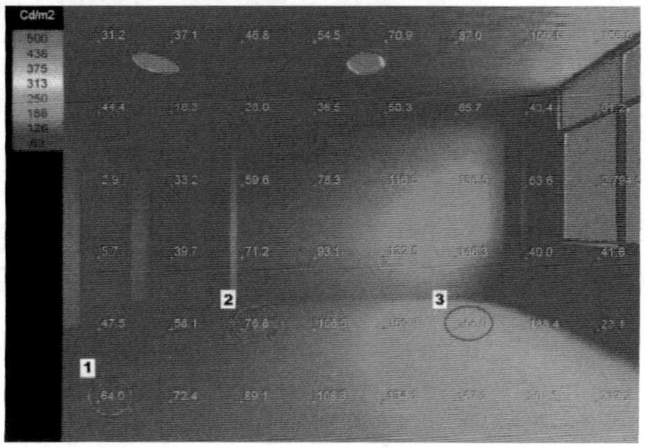

Figure 6. Light pipes simulation results.

Table 1. Light pipes simulation results.

Types	Point 1 left (lux)	Point 2 middle (lux)	Point 3 right (Lux)
Ordinary Window	50.5	57.6	165.8
Light Pipes	64.0	76.8	206.9

5.1.2. Mirror ducts

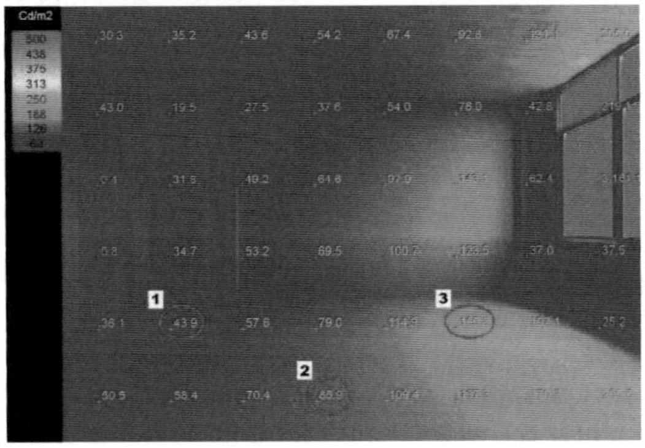

Figure 7. Ordinary window simulation results.

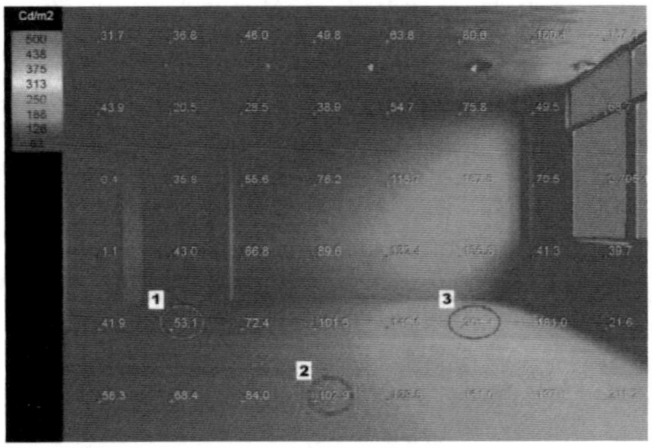

Figure 8. Mirror ducts simulation results.

Table 2. Mirror ducts simulation results.

Types	Point 1 left (lux)	Point 2 middle (lux)	Point 3 right (Lux)
Ordinary Window	43.9	85.9	165.8
Mirror Ducts	53.1	102.9	201.4

5.1.3 Light shelves

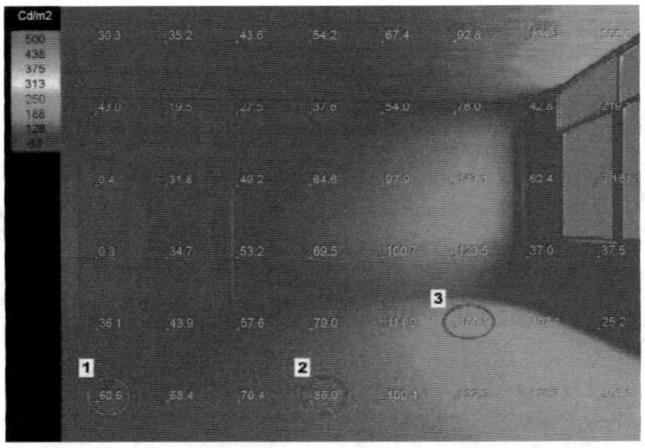

Figure 9. Ordinary window simulation results.

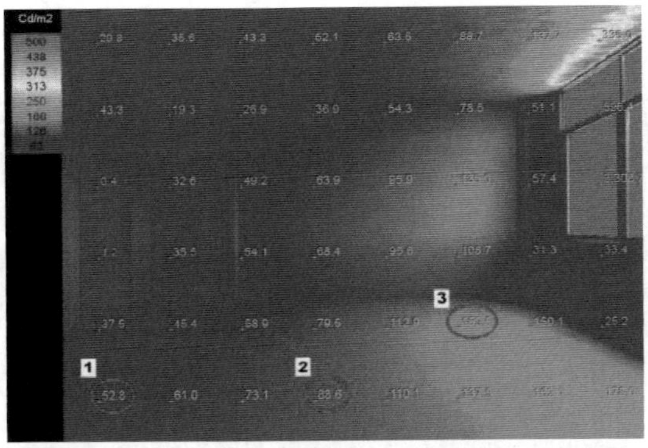

Figure 10. Light shelves (outer) simulation results.

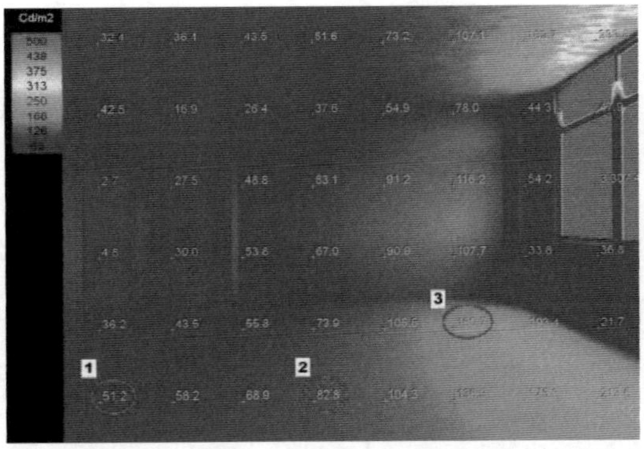

Figure 11. Light shelves (inner) simulation results.

Table 3. Lights shelves simulation results.

Types	Point 1 left (lux)	Point 2 middle (lux)	Point 3 right (Lux)
Ordinary Window	50.5	85.9	165.8
Light Shelves (outer)	52.8	88.6	152.5
Light Shelves (inner)	51.2	82.8	160.7

5.2. Simulation analysis

As stated in Table 1, the differences of the room which used ordinary windows and used light pipes are significant. This could happen because sunlight entered the hole built in the roof and reflected using reflective pipes. Using light pipes, sunlight can enter through the deepest part of the room.

The weakness of using light pipes is it can only be used on the highest floor of a building. Light pipes also need a big space between the ceiling and roof as a place to put reflective pipes. Also, by using light pipes, the structure of the building, especially the roof structure, will be affected.

In Table 2, mirror ducts made greater light intensity than ordinary windows. But the differences of the intensity are not as significant as light pipes (table 1). This is caused by the direction of sunlight, as in light pipes, the sunlight enters directly from the roof. In mirror ducts, the sunlight entered from the side of the building as the angle is lower and made the sunlight intensity which entered the room lower than the case while using light pipes. Furthermore, while using mirror ducts, sunlight is hindered by obstacles such as other building components.

Light pipes and mirror ducts have the same system with different axes (mirror ducts have horizontal axis, light pipes have vertical axis). These two types both can access deep parts of the room which can't be reached by ordinary windows. The advantage of using mirror ducts over light pipes is the system can be used in every level, as light pipes can only be used in the highest level of the building.

Table 3 showed the results of light shelves (inner and outer types) and ordinary windows simulation. The simulation showed that the differences by using light pipes and ordinary windows are not significant. As it can be seen in point 3 (figure 6 and 7) that in the part of the room that is close from the window, the intensity of light is lower while using light shelves rather than using ordinary windows (point 3 in figure 4 and 5). Although in the middle part of the room (point 2 in figure 6 and 7), the room with the outer light shelves is brighter than the room which used ordinary windows. Last, at the inner part of the room (point 1 in figure 6 and 7), the room which used light shelves is brighter than the one which used ordinary windows.

The principle of light shelves is depending on reflection of sunlight which makes the intensity even in every part of the room. This also can be used to diminish sun radiation exposure. From the simulation result, it can be observed that the part of the ceiling that is close to the windows is brighter while using light shelves. This fact can be used to determine materials (which have more reflective abilities) used for ceilings in design.

In general, it can be concluded that the light intensity which is produced by light pipes have the most significant differences and also can reach the deepest part of the room compared to other types of passive lighting. Although there are limitations by using light pipes, such as it can only be used at the highest level floor of a building and can affect building systems especially roof parts.

Mirror ducts produce less significant light intensity compared to light pipes. Although there isn't any limitation of the use and also able to reach the deepest part of the room.

Light shelves produce the most insignificant number of light intensity compared to others, also the lights are not able to reach the deepest part of the room. But the usage of light shelves is easy and flexible. Furthermore, the application of light shelves can diminish sun radiation because the sunlight is reflected by the shelves in the building facade.

5.3. *Questionnaire results*

From 209 respondents, 59.8% consider energy consumption when designing buildings. Respondents answered that the consideration is used to save energy, save operating costs, and to achieve the 'green building' concept. While the respondents who don't consider energy building consumption answered that they didn't know, didn't understand, never thought about it, and felt that energy consumption wasn't their field of job. On the other hand, 97.7% of respondents consider natural lighting in their building design.

Table 4,5, and 6 showed an average score (1-5 scale) that respondents gave about their preferences in respectives aspects: aesthetic aspect, natural lighting utilization aspect, energy building reduction aspect.

Table 4. Respondent preferences based on aesthetic aspect.

Types of Passive Lighting	Score (1-5 Scale)
Light shelves	4.03
Mirror Ducts	3.8
Light Pipes	3.04

IOP Conf. Series: Earth and Environmental Science 1007 (2022) 012011 doi:10.1088/1755-1315/1007/1/012011

Table 5. Respondent preferences based on natural lighting utilization aspect.

Types of Passive Lighting	Score (1-5 Scale)
Light shelves	4
Mirror Ducts	3.91
Light Pipes	3.37

Table 6. Respondent preferences based on energy building reduction aspect.

Types of Passive Lighting	Score (1-5 Scale)
Light shelves	4.03
Mirror Ducts	3.8
Light Pipes	3.04

5.3.1. Respondents which has experience designing building with passive lighting. 81 respondents (38.75%) have experienced using passive lighting in their building design. Among these respondents, there were 22 respondents respectively who answered 50% and 75% of the designs that they had made considering passive lighting. The answers given by all respondents show that the considerations are thought based on space needed, typology, design concept, efficiency, budget, level of conveniences, and client desires.

5.3.2. Respondents which never experience designing building with passive lighting. There are 128 respondents (61.25%) who have never applied passive lighting in their designs. The following are respondents' answers regarding the reasons for never applying passive lighting: don't know, don't understand, and haven't thought about it before.

A total of 70 respondents (54.68% of those who answered never) answered “Yes” to apply passive lighting to their next design. While the rest (58 respondents, 45.31%) answered "maybe".

Respondents who answered “Yes” reasoned that passive lighting saves energy, is environmentally friendly, enhances the concept of sustainability, and is interesting to try. Meanwhile, respondents who answered “Maybe” felt they still didn't know, wanted to learn more, and designed the building according to requirement needs.

5.4. Questionnaire analysis

The questionnaire respondents fairly consider energy consumption enough while designing buildings. Especially in the lighting aspect, most of the respondents pay attention to natural lighting.

In table 4,5, and 6, the preferences of respondents could be seen in three aspects (aesthetic, natural lighting utilization, energy building reduction). Respondents have high preference for light shelves as the best passive lighting in all three aspects, followed by mirror ducts and light pipes as the worst.

The results of questionnaire respondents contrast with simulation results. Simulation results concluded that light pipes are the most effective, followed by mirror ducts and light shelves. This can be explained for aesthetic aspects, as it is subjective. But for the other aspect (natural lighting utilization and energy building reduction), the contradiction could be shown by the differences of the answers and it proves that the respondents still haven’t understood the characteristics of each type of passive lighting.

Most of the respondents (61.25%) never applied passive lighting. This happens because the respondent does not know about light pipes, mirror ducts, and light shelves. After completing this questionnaire, none of the respondents who had never previously applied passive lighting directly refused to use passive lighting. Respondents who answered 'maybe' mostly answered with the reason they wanted to get information and know and understand more about the features of light pipes, mirror ducts, and light shelves.

For respondents who have applied passive lighting: light pipes, mirror ducts, and light shelves are still unknown compared to other passive lighting. Most of the respondents (60.49%) used passive lighting apart from those three types.

5.5. *Further discussion*

In general, it can be concluded that respondents haven't understood light pipes, mirror ducts, and light shelves. This shows how the answers of the respondents' preferences contrast with simulation results. Moreover, among respondents which have experience using passive lighting, the amount of respondents who have experience using light pipes, mirror ducts, and light shelves are very few. Furthermore, most respondents who never use passive lighting answer that they never knew about passive lighting technology before.

6. Conclusion

Based on literature studies and simulation that have been done, it can be concluded that light pipes produce the largest intensity, followed by mirror ducts and light shelves. Light pipes and mirror ducts can both reach deeper parts of the room than light shelves. Application of light shelves become the least that disturbs the structural and facade systems of the building. Meanwhile, according to the results of the questioner, light shelves have the highest preference, followed by mirror ducts and light pipes.

In addition, according to the results of the questionnaire, application of the light pipes, mirror ducts, and light shelves are still under-applied by Indonesians. This is due to lack of knowledge about the types of passive lighting.

Suggestions from this study are distribution of information about passive lighting, especially light pipes, mirror ducts, and light shelves. Information dissemination can be in infographics, seminars, or lectures at the universities.

7. References

[1] Omer A M 2008 Energy, environment and sustainable development *Renewable and Sustainable Energy Reviews* **12** pp 265-300

[2] World Business Council for Sustainable Development 2016 *EEB laboratory Jakarta energy efficiency in building* http://docs.wbcsd.org/2016/11/EEB_Lab_Jakarta.pdf

[3] Green Building Council Indonesia 2019 *Studi IFC dan GBC Indonesia: Bangunan gedung hijau 30-80% lebih hemat air & listrik* https://pressroom.ifc.org/all/pages/PressDetail.aspx?ID=24900

[4] Pemerintah Indonesia 2012 *Peraturan Menteri ESDM no. 13 tahun 2012 tentang penghematan pemakaian listrik* (Jakarta: Sekretariat Negara)

[5] Alrubaih M S, Zain M F M, Alghoulb M A, Ibrahima N L N, Shameria M A and Elayebb O 2013 Research and development on aspects of daylighting fundamentals *Renewable and Sustainable Energy Reviews* **21** pp 494–505

[6] Beltrán L O, Lee E S and Selkowitz S E 1997 Advanced optical daylighting systems light shelves and light pipes *Journal of The Illuminating Engineering Society*, *Annual Conf. on IESNA* vol 26 pp 91-106

[7] Building and Construction Authority 2020 *Zero energy building* https://www1.bca.gov.sg/buildsg/sustainability/zero-energy-building

[8] Kadir A A, Ismal L and Kasim N 2015 *Optimization of daylighting system by using light pipe system in a building* https://www.semanticscholar.org/paper/Optimization-of-daylighting-system-by-using-light-a-Kadir-Ismail/fc3333ce82e2fc5ccacf73923d1f419955c5d7ce

[9] Mayhoub M S and Carter D J 2009 The cost and benefits of using daylight guidance to light office buildings *Building and Environment* **46** pp 698-710

[10] Mardaljevic J, Heschong L and Lee E 2009 Daylight metrics and energy savings *Lighting Research and Technology* **41** pp 261-283

[11] Gullotti B 2019 *Passive daylighting systems could transform the architecture of natural light* https://hmcarchitects.com/news/passive-daylighting-systems-could-transform-the-architecture-of-natural-light-2019-05-24/
[12] Jadhav N Y 2016 *Green and smart building: Advanced technology option* (Singapore: Springer) pp 54-57
[13] Oakley G, Riffat S and Shao L 2000 Daylight performance of light pipes *Solar Energy* **69** pp 89-98
[14] Building and Construction Authority 2013 *Singapore: Leading the way for green buildings in the tropics* https://bca.gov.sg/greenmark/others/sg_green_buildings_tropics.pdf
[15] Yudelson Associates 2011 *Building and construction authority zero energy building Bradell road campus, Singapore* http://www.solaripedia.com/files/1009.pdf

Arranging *here* and there: A microlens observation on domestic waste separation

A Karimah[1], R Suryantini[1,2] and M M Y Harahap[1]

[1] Department of Architecture, Faculty of Engineering, Universitas Indonesia, Depok 16424, Jawa Barat, Indonesia

Abstract. This paper explores ease of access in the spatial practice of domestic waste separation. Such an ease was observed by investigating how foodstuff is arranged, *here* and *there*, as an attempt to *bring order*. In the process of separating the foodstuff waste, the inhabitant's body–her position, movement, and technical gestures–interrelates with the foodstuff, utensils, and the surrounding spaces become crucial aspects. Arguably, performing such a spatial practice boosts the inhabitant's convenience, thus encouraging sustainable waste separation in domestic settings. In particular, this paper employs a microlens observation with a qualitative approach to dismantle the spatiality of the everyday domestic waste separation practice. The findings suggest dimension and adjustability are the essential measurable aspects in obtaining the desired convenience. Such a role raises the awareness for considering ergonomics for obtaining ease of access in separating waste. It drives the utilisations of particular portable objects and working surfaces, to assist in separating waste conveniently throughout the phases of bringing order. Notably, the temporary placement of wanted and unwanted foodstuff parts signifies the arrangement of here and there in such ways. Therefore, these findings expand the discussion of ergonomics, towards a sustained practice of waste treatment.

1. Introduction

This paper aims to reveal the spatiality that supports the inhabitant's ease of access, which may contribute to a sustained treatment of waste in domestic settings. This aim arises from the phenomenon of inhabitant's daily activities toward the foodstuff parts to provide the daily meal for households, that generates the production of organic waste [1,2]. It contributes a heap of organic waste to the waste processing plant, especially from the process of sorting, removing, and classifying the wanted or unwanted foodstuff parts [3]. Separating organic waste from other kinds of waste is a recommended practice that could reduce the workload on further waste treatment [4]. Arguably, once made to be performed on a broader scale, such practice would contribute to a more sustained treatment of waste.

The idea becomes an entry point for us on the importance of understanding and exploring the possible ways, in demonstrating the everyday spatial practice of waste separation in domestic settings. Unfortunately, practising waste separation requires inhabitants to devote their time, when their hands are already full in processing foodstuffs and cooking them [1,5]. Which is probably why, while encouraged, such treatment of waste is still rarely being practised in many places [4]. Such a practice can be seen only in countries where waste management has been systematically integrated into

[2] Corresponding author: r.suryantini@ui.ac.id

households and rooted in inhabitants' daily life; such as Japan, Germany, South Korea, and Vermont [6–8].

Studies on the practice of waste separation and alike in domestic settings, mostly revolved around observing the inhabitant's knowledge, intention, habit, and facilitating conditions that drive a sustained practice of waste separation [9,10], along with the employment of such understanding to suggest a strategic policy recommendation on waste management [11,12]. In addition, other studies focus to discuss the role of external actors in promoting waste separation by local authorities or volunteers; through doing the educational campaign, the programmed intervention with monitoring or win-win outcomes, declaring a statement of policies, also providing the physical facilities [12–15].

However, a study that looks at waste treatment from the perspective of spatial practice, which situates them toward a sustained treatment of waste, is still rarely discussed [2,16,17]. This perspective allows a shift from the study of waste management that uses high-end technology or tends to explore the middle to end phase of the waste management chain [18], to an exploration that is rather micro. This opportunity is more relevant to explore the sustained treatment of waste, practised by the inhabitants in domestic settings. Nonetheless, this paper argues, looking from this perspective will gain an in-depth understanding of the spatiality that contributes to the ease of doing such waste separation.

2. The ease of access: encouraging a sustained treatment of waste at home

This paper situates the daily practice of waste separation in the domestic setting, as part of an attempt toward a sustained waste treatment. During the daily process of *bringing order* on foodstuff, the organic waste is produced in the domestic settings, from parts of foodstuff that are categorised as *out of place* [2]. It occurs on sorting and removing the foodstuff parts that are inedible or unappetising, to maintain *good order* in food consumption [1–3] since the very beginning. The practice of waste separation concerns on, even though the organic waste is considered as *unwanted* material on food consumption *here*, it might be a *wanted* material *there* within a process of waste treatment, i.e., recycling or composting [3].

The process of bringing order on foodstuff also consists of classifying foodstuff parts, as the wanted or the unwanted, then belong here or there [3]. Such processes are practised by the inhabitants in the domestic settings, including keeping the wanted foodstuff parts for meal creation and at the same time removing aside the unwanted parts [1,19]. The inhabitants create *orders* so that only the wanted material to be here, then the unwanted material goes there [1,3].

There are several spatial strategies developed in the domestic space that can be identified as part of ordering. Implying that the use of domestic spatial elements are able to be adjusted, spatial fluidity becomes significant as one of the strategies that support the inhabitant's health and well-being throughout occupations [20]. In addition, a related study on food flow states that particular ways of regulating the dirt from foodstuff would generate particular patterns of circulation and localisation of the dirt in certain areas of the domestic space [21]. Such domestic spatial practices construct interrelations between the inhabitant's body and foodstuff parts, to the available spaces and objects [1,16], which further become parts of a machinic architectural system that is conducive for bodily comforts [22]. Importantly, such orders are supported by various utilisations of the available domestic spaces and objects, through changes of activities [23] while employing the order of the foodstuff. Further, it brings the need to consider the inhabitant's ease [16] in practising waste separation regularly at home, prior to, brings a more "personalised and healthy approach" [24]. Therefore, designing house spaces that carefully aids the inhabitant's activities is crucial [25], suggesting further the spatiality of waste treatment [26] at the household level.

This paper observes the inhabitant's ease of access that conducts convenience throughout the spatial practice of waste separation; particularly in utilising the available spaces and objects in domestic settings. Arguably, the spatiality that enables such ease of access has a contribution to the sustained practice of waste treatment at the household level [11,27]; especially caused by the shorter distance between the facilities, as well as the ease in operating them [28]. This raises the importance to reveal how the inhabitant's body position, movement, and technical gestures are interrelated with wanted-unwanted foodstuff parts, then to the available working surfaces and portable objects [1,16,19,29]. Related studies regarding the ease of access—from the discourse of ergonomics—are emphasising dimension and adjustability as the measurable aspects of space or furniture to accommodate the user's need [30]. Or, as an input to modify the design of a mass product to attain optimum uses, along with creating its aesthetics [31,32]. Implying that studies on ergonomics have the potential to encourage a sustained practice that is operated manually by a human; even though the discussion is still limited [4,5,33]. This paper presents an observation of the spatial practice of waste separation in domestic settings, particularly throughout the process of bringing order on foodstuff parts. In referring to the above study, this paper emphasises the observation of the inhabitant's ease of access, related to particular spaces and objects during such practice; therefore, expanding the discussion of a sustained treatment of waste within the discourse of architecture.

3. Methodology

This paper observes the everyday practice of separating organic waste, while preparing food in domestic settings, particularly on the inhabitant's ease of access to the available spaces and objects. It adopts an approach of micro-spatial analysis from [34] to gain an in-depth understanding of the interrelation between people, spaces, and objects. It then applied to observe how the inhabitants have been treating the unwanted foodstuff parts conveniently, especially on separating them with other kinds of waste, by utilising the available working surfaces and portable objects.

The spatial practice was observed within a household that has already started separating the organic waste from other kinds of waste for about three years and continuously improving the ease of doing it. Three female inhabitants were being observed, as the main actors that have been conducting such waste treatment in the observed household. The house is located within a neighbourhood in Depok City, Indonesia, that requires each household to separate their organic and non-organic waste disposal but allows each to develop its way in separating both materials. This household utilises the available basic domestic utensils and working surfaces, to separate the organic waste during the particular phase of bringing order on foodstuff. Data gained from the direct observation of such practice in this household are being mapped into a set of diagrams, to gain an understanding of how inhabitant's bodies are interrelated to spaces and objects, which provides support in performing ease of access throughout the practice. Arguably, by observing the spatial practice of waste separation through a microlens, the understanding of a sustained treatment of waste from the very beginning could be revealed.

4. Performing the ease of access

In this paper, separating the organic waste at the household level is seen as a spatial practice that is conducted by the inhabitant. This spatial practice constructs such an interrelationship of the body with wanted-unwanted foodstuff parts, to the available spaces and objects. This study observes its spatiality by particularly looking at the inhabitants' ergonomics that is related to the ease of access. Briefly, at the observed household, such spatial practice occurs within three phases of bringing order on foodstuff. Each phase differs in a manner on how order on foodstuff will be brought; either will be stored (figure 1), will be used for particular meal creation (figure 2), or will be removed entirely from the cutleries or utensils (figure 3). In general, separating the disposal of organic waste in different phases, takes a particular role of the inhabitant's body parts, along with their movements and technical gestures. The organic waste separation on each phase occurs in a particular area, involving particular portable objects and working surfaces, to assist the inhabitant's ease of access in doing it.

4.1. A spatial practice: utilising the available objects and spaces

The observation begins by mapping the practice of organic waste separation in a household, into a series of diagrams (figure 1-3); therefore, enables a more micro look on the spatiality that assists the inhabitant's ease of access during such practice. The diagram highlights the interrelation between body, spaces, and objects, within each phase of bringing order on foodstuff. It points out the utilisation of portable objects throughout the occurrences of organic waste separation, marked by different kinds of the coloured-lines. The diagram also points out the utilisation of working surfaces with different heights (Y-axis, in metre), marked by the thin grey lines. In addition, it states the inhabitant's sequence of bodily position or movement (from left to right in X-axis, they pointed out with the arrows and texts), while transferring the organic waste between the portable objects in sequence. The end-temporary container inside the kitchen is holding the collected unwanted foodstuff parts—organic waste only—for a half-day. It is placed at the corner, some steps away from the working areas where the process of bringing order on foodstuff occurs. By looking at the treatment of waste from the perspective of spatial practice, this paper argues that practising such separation with ease depends on the inhabitant's current working range. Arguably, the temporal placement of wanted and unwanted foodstuff parts also signifies here and there [2,3] inside the kitchen. Therefore, a more microlens observation becomes essential to gain an understanding of an easy manner of waste treatment at the household level.

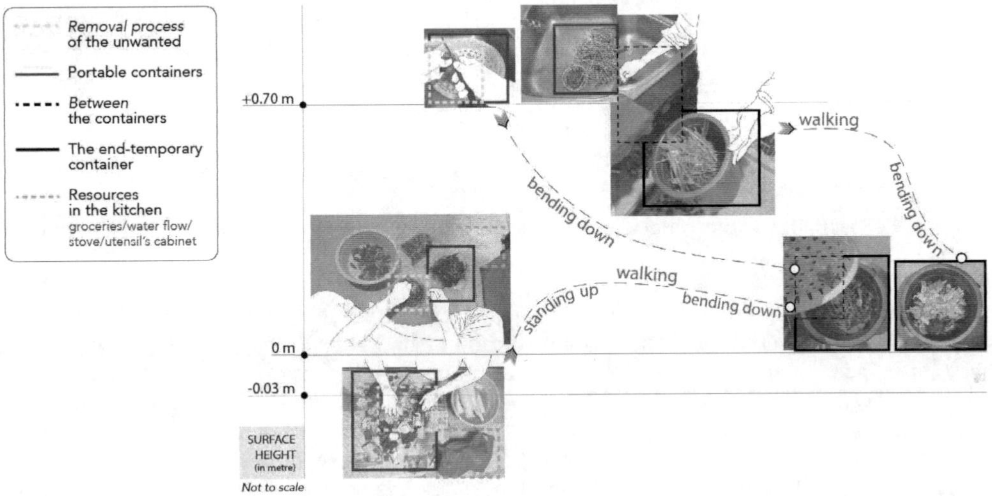

Figure 1. Throughout arranging the foodstuff piles that just be bought, waste separation is practised by utilising the basic kitchen utensils, and occurs on grounded surfaces or within the sink area.

Practising the organic waste separation on each phase, are assisted by the different roles of portable objects and working surfaces. The first phase of bringing order on foodstuff is conducted in a manner how piles of foodstuff from the groceries will be stored (see figure 1). In this phase, the organic waste separation is incorporated throughout the process of sorting the rotten pieces of vegetables, peeling their outer skin or cutting the root and stem, getting them rid-off from the soil or dust, or trimming the excess meat fat. Such processes only utilise the basic kitchen utensils, e.g., using a regular knife or scissors. Afterwards, the portable containers, e.g., small basket, colander, used-newspaper, and sink drainer, are utilised to assist the classification of foodstuff parts; either collecting the wanted parts for further meal creation, or holding the unwanted foodstuff parts temporarily. In this first phase, the practice of organic waste separation occurs immediately after each unwanted foodstuff part has been removed, then being held temporarily by the portable containers that are located just beside the removal area. Such processes may occur on the grounded surfaces, or occur on the surfaces of the sink area when needed to employ the water flow.

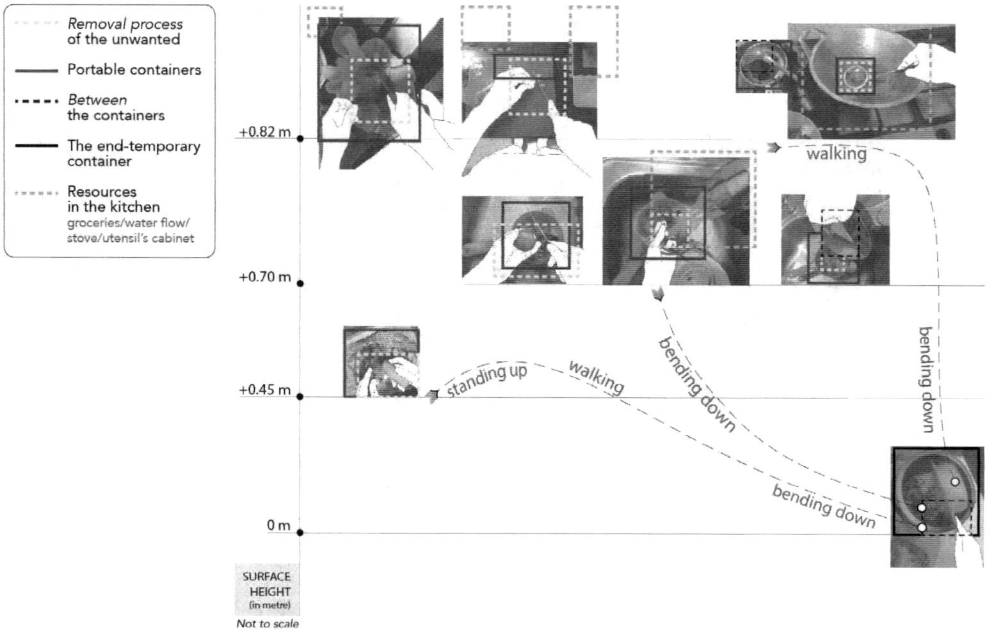

Figure 2. Practising waste separation in a more careful manner, by utilising wider range of utensils and working surfaces.

The second phase of bringing order is conducted more carefully, in a manner how the selected foodstuff parts will be used for particular meal creation (see figure 2). The practice of organic waste separation is incorporated throughout the removal process of the unwanted foodstuff parts; such as peeling to a certain depth of vegetable flesh, cutting its unappetising or crusty parts, taking out the seeds inside, or getting rid-off the tiny crumbs from hot oil. Such more careful processes utilise a broader range of kitchen utensils, e.g., kinds of knife, scissors, sieve, mini strainer, even cutleries like forks or chopsticks. The wanted and unwanted foodstuff parts are classified using a broader range of portable containers, e.g., small bowls or chopping boards, used wrappers, plates, or strainers. Nonetheless, in this second phase, the practice of organic waste separation occurs on the various spots of working surfaces; following the working flow on meal creation, as long as the surface's height is taller than the floor but lower than the inhabitant's elbow.

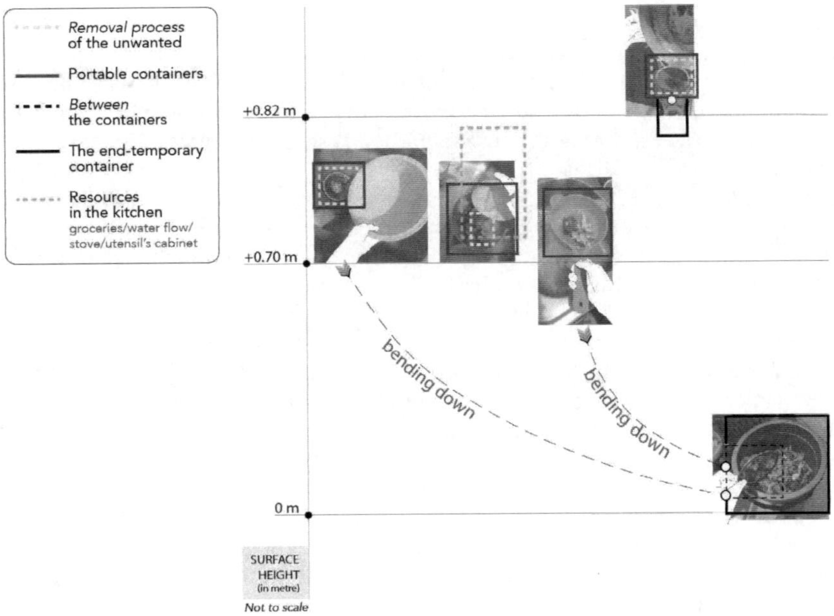

Figure 3. Practising waste separation while removing unwanted foodstuff from cutleries or utensils, occurs on and around the source of water flow.

The last phase of bringing order is conducted after the eating or cooking process has been done, in a manner how the unwanted foodstuff will be removed entirely from the cutleries or utensils (figure 3). Throughout rinsing or washing activities, the separation of waste is incorporated. The perforated portable containers, e.g., strainer or sink drainer, are utilised to segregate the unwanted foodstuff parts from liquid materials; while at the same time, to prevent the foodstuff from going to the sink's gut together with the water flow. In this last phase, separating the organic waste heavily interrelates with the position of inlet and outlet of the water flow, therefore it occurs within and around the sink's working surface.

4.2. Here and there: the inhabitants bodily aspects

The previous subsection explains that the inhabitant's ease of access in practising waste separation is enabled by the presence of portable objects and working surfaces within his current reach. It is found that the current reach differs for each phase of bringing order, and was generated by the body parts' movement and technical gestures, needed in a particular phase. Throughout the practice of organic waste separation from other kinds of waste (figure 1-3), here and there [2,3] are arranged by the inhabitants, by utilising the available portable objects on particular working surfaces. These findings raise the importance to have an in-depth look at the inhabitant's ergonomics that drive such utilisation of objects and spaces to assist the ease of access throughout practising the organic waste separation in domestic settings.

Notably, the organic waste separation is practised immediately after the removal of unwanted foodstuff parts, which happened several times in a phase of bringing order on foodstuff. This means the waste separation is practised within the inhabitant's current working range. It is found that the longer devoted time and a heavier workload needed in a phase, trigger the inhabitant to practice such separation within the minimum working range. It is only as far as the movement of his hip and round-arm in the third phase. Or even smaller, only as far as the movement of his round-arm in the first phase. Especially, in the first phase, it drives the inhabitant to settle on the working surfaces that are able to support his leg and back body parts—for example, settling on the floor adjacent to the wall, thus enabling him to lay on eventually. Only when needed to employ the water flow, the inhabitant then gets up and moves closer to the source of water flow.

Meanwhile, another finding has been learned from the second phase of bringing order, which emphasises the utilisation of working surfaces. Separation practice in this phase occurs on various spots of the working surfaces, near to the current needs of kitchen resources, i.e., stove, grocery bags, utensil cabinet, and water flow; that are continuously changed in need, following the working flow and pace in bringing order on foodstuff. In this phase, separating the organic waste is practised in a more expanded working range, demonstrated by the movement of the inhabitant's footsteps between resources in the kitchen. Up till here, it can be understood that the working range of waste separation may vary; depending on the different working load, the duration or pace, the working flow, also the resources needed. These conditions are interrelated to the inhabitant's body position, movement, and more specifically the technical gestures, to the particular portable objects and working surfaces. In addition, the current working range is also based on the need for a shorter distance between the specific body parts to the working surfaces or portable objects. Either to provide a clearer vision during a careful way of removal, or to accommodate the looping movement between kitchen resources at a fast-working pace, or to assist in transferring the unwanted foodstuff parts between the portable containers. These findings imply that considering the dimension and adjustability are needed to perform ease of access, which will conduct convenience in separating waste. Some of the examples consider the working surfaces' height and position to the kitchen resources, or, the container's placement and portability; therefore, performing the ease of access to separate waste here immediately after the process of removal.

Furthermore, this paper demonstrates that the placement of wanted and unwanted foodstuff parts, signifying the here and there [3] that are temporary within the domestic kitchen. At the observed household, the unwanted foodstuff parts are being removed mostly using the inhabitant's right hand, while the wanted foodstuff parts are detained or gripped by the left hand. Therefore, the unwanted is mostly placed at the right side of the working area, which signifies there; while the wanted goes to the left side, which signifies here. Also, here is mostly placed on a higher surface, than the placement of there. Such a placement is assisted by the portable objects only, or combined with working surfaces in such ways. For example, utilising the portable container with a holder, assists in transferring the unwanted foodstuff from here (e.g., colander) to there (e.g., the end-temporary container). Another example, directly separating on the sink's working surface is allowed, if combined with the utilisation of a sink drainer to hold the tiny particles—avoiding the unwanted parts to stay here. Such utilisations are assisting the arrangement of here and there to become clear yet flexible enough, adapted to the current working range. Therefore, implies the importance of performing inhabitant's ease of access throughout practising the organic waste separation at the household level.

5. Conclusion

This paper demonstrates a microlens observation on separating waste in domestic settings—as a spatial practice of bringing order here and there. The study took a closer look at the practice throughout phases of bringing order on foodstuff, particularly on the spatiality that performs in assisting inhabitants doing their task more conveniently. It highlights how the inhabitant's bodily aspects are interrelated to the available objects and spaces. Also, the observation of the separation practice through micro perspectives [34] along with mapping analysis were useful in dismantling how the interrelationships enabled ease of access that could promote sustainable waste separation practice in domestic settings.

The findings highlight that the inhabitant's ease of access throughout the practice is performed by the utilisation of particular portable objects and particular working surfaces, which differs on each phase of bringing order. Such utilisations are driven by the measurable aspects of ergonomics—i.e., dimension and adjustability—that are related to the inhabitant's body parts while separating waste in each phase. Moreover, such utilisations create ways of arranging both the wanted and unwanted foodstuff parts. They signify the temporary arrangement of here and there [17], as either placed on the left and right side, or, on the higher and lower surface of the current working area.

Overall, the findings suggest the role of ergonomics in promoting a sustained domestic practice of waste separation that could be performed regularly. The study further positioned the measurable aspects of ergonomics as the essential convenient factor that would support the sustained practice of separating waste manually in domestic space, that is rather comfortable and healthy. It also implies the importance to consider the performance of objects and surfaces for addressing the ergonomics of waste separation practice: (1) they exist separately yet inseparable and interrelated one another; (2) they work together through a certain treatment variedly depending on the inhabitant's phase of activities. However, this study only looks at a domestic setting and is limited to the acts of waste separation observed in a household. A more comprehensive study that also considers the type of household, such as shared houses, apartment complexes, or households with adjacent household business programmes, might deepen the understanding of sustainable waste separation. There is also an opportunity to look at the importance suggested in this paper from its quantitative side, draw the study at the different aspects of ergonomics, use other perspectives than spatial practice, or look at the other occurrences of waste separation in different contexts.

6. References

[1] de Certeau M, de Giard L and Mayol P 1998 *The practice of everyday life volume 2: Living and cooking* (Minnesota: University of Minnesota Press)

[2] Douglas M 1996 *Purity and danger: An analysis of concept of pollution and taboo* (New York: Routledge)

[3] Strasser S 2000 *Waste and want: A social history of trash* (New York: Holt Paperbacks)

[4] Barbosa L and Veloso L 2014 Consumption, domestic life and sustainability in Brazil. *J. Clean. Prod.* **63** 166–72

[5] Bernstad A 2014 Household food waste separation behavior and the importance of convenience *Waste Manag.* **34** 1317–23

[6] Lee S and Paik H S 2011 Korean household waste management and recycling behavior *Build. Environ.* **46** 1159–66

[7] Liu C, Hotta Y, Santo A, Hengesbaugh M, Watabe A, Totoki Y, Allen D and Bengtsson M 2016 Food waste in Japan: Trends, current practices and key challenges *J. Clean. Prod.* **133** 557–64

[8] State of Vermont 2020 *Food scraps* https://dec.vermont.gov/waste-management/solid/materials-mgmt/organic-materials

[9] Issock P B I, Robert-Lombart M and Mpinganjira M 2020 Understanding household waste separation in South Africa: An empirical study based on an extended theory of interpersonal behaviour *Manag. Environ. Qual. Int. J.* **31** 530–47

[10] Loan L T T, Nomura H, Takahashi Y, Yabe M 2017 Psychological driving forces behind households' behaviors toward municipal organic waste separation at source in Vietnam: A structural equation modeling approach *J. Mater. Cycles Waste Manag.* **19** 1052–60

[11] Knickmeyer D 2020 Social factors influencing household waste separation: A literature review on good practices to improve the recycling performance of urban areas *J. Clean. Prod.* **245** 118605

[12] Zhang Z and Wang X 2020 Nudging to promote household waste source separation: Mechanisms and spillover effects *Resour. Conserv. Recycl.* **162** 105054

[13] Bonrood K, Towprayoon S, Bonnet S and Tripetchkul S 2015 Enhancing organic waste separation at the source behavior: A case study of the application of motivation mechanisms in communities in Thailand *Resour. Conserv. Recycl.* **95** 77–90

[14] Jamal M, Szefler A, Kelly C and Bond N 2019 Commercial and household food waste separation behaviour and the role of Local Authority: A case study *Int. J. Recycl. Org. Waste Agric.* **8** 281–90

IOP Conf. Series: Earth and Environmental Science 1007 (2022) 012013 doi:10.1088/1755-1315/1007/1/012013

Influence of a facade design on thermal and visual comfort in an elementary school classroom

P P Suradhuhita[1,2], E Setyowati[1], E Prianto[1]

[1] Department of Architecture Faculty of Engineering, Diponegoro University, Semarang 50275, Jawa Tengah, Indonesia

Abstract. Thermal and visual comfort are important elements in the learning process. The object chosen was the classroom of the Sawah Besar 01 public elementary school, Semarang where the building tends to have a homogeneous facade design but with openings that have varying orientation directions. This affects the level of comfort both thermal and visual in the classroom. Data analysis was carried out quantitatively, namely to obtain a thermal comfort index and natural lighting comfort index according to the Indonesia National Standard. The thermal comfort conditions of all selected rooms on the 1st and 2nd floors did not meet the comfortable thermal comfort standard for classrooms. Rooms 1 and 2 on the 1st floor tend to be above the minimum level of classroom lighting. Room 3 on the 2nd floor tends to be stable and all of them are above the minimum limit. In room 4 on the 2nd floor, the general lighting level has exceeded the minimum limit. New façade opening designs and mechanical ventilation aids need to be applied to improve thermal and visual comfort according to the Indonesia National Standard.

1. Introduction

There are comfort factors in buildings that must be considered while designing a building. In the case of this research, there are visual comfort and thermal comfort. Visual and thermal comfort in buildings is needed so that people can do their activities properly for a prolonged time. The level of productivity and human health is influenced by local climatic conditions. That is why it is important to accommodate human comfort [1]. One of the factors that influence visual and thermal comfort is the building façade [2]. The building facade is a part of the building that is often exposed to solar radiation. The direction of building orientation has a major influence on the effectiveness of the function of the building facade. One of the tasks of the building facade is to regulate the conditions around the outside of the space, which aims to ensure comfortable conditions inside the room.

1.1. Recommended level of thermal comfort

The thermal comfort limit for equatorial conditions ranges from a lower limit of 19°C TE to 26°C TE. The most suitable comfort limit for Indonesians used in this study is the Mom-Wiesebron comfort limit as follows [3].

[2] Corresponding author: interior.purnama93@gmail.com

IOP Conf. Series: Earth and Environmental Science 1007 (2022) 012013 doi:10.1088/1755-1315/1007/1/012013

Table 1. Mom-wiesebron comfort zone criteria.

Criteria	Effective Temperature (TE)
Cool Comfortable	20,5°C TE - 22,8°C TE
Optimal Comfortable	22,8°C TE - 25,8°C TE
Warm Comfortable	25,8°C TE -. 27,1°C TE

1.2. Recommended level of lighting

The main function of Sawah Besar 01 Elementary School Semarang classroom is for reading and writing activities. However, classrooms are also used for scientific experimentation and artistic activities such as playing musical instruments, singing, and drawing. This affects the recommended minimum lighting level so that the user's visual comfort is achieved as described between 250-750 Lux [4].

2. Methods

The object chosen by the researcher in the classroom of Sawah Besar 01 Semarang state elementary school. Public elementary school buildings tend to have a homogeneous building facade design with openings on both sides of the wall as shown in figure 1. Facade designs that tend to be the same but have varied orientation directions affect the level of comfort both thermal and visual in the classroom.

Field measurements for the room temperature were carried out in the time range from 07.00 – 16.00 WIB on Wednesday, June 2, 2021. Data collection points are air movement, dry temperature, and air humidity by opening window openings in the room and without artificial air conditioning. By using the data that has been obtained then an analysis is carried out, obtaining effective temperature data. The measurement point for the level of lighting in the selected classroom is carried out as shown in figure 2 [5]. The measurement results obtained are then compared with the recommended lighting levels [4].

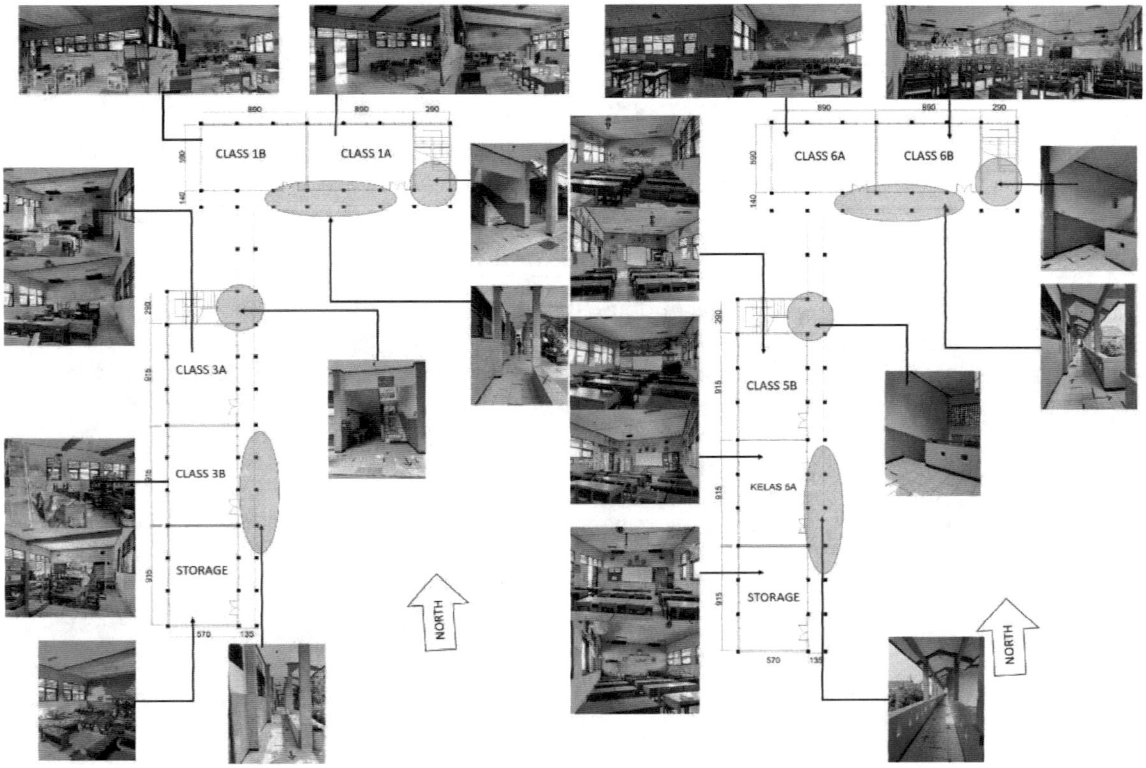

Figure 1. Sawah Besar 01 elementary school layout 1st floor (left) and 2nd floor (right).

IOP Conf. Series: Earth and Environmental Science 1007 (2022) 012013 doi:10.1088/1755-1315/1007/1/012013

Data collection was taken in 2 classrooms on the 1st floor, Class 3B or Room 1 and Class 1A or Room 2, with openings facing the same direction east and west. And also 2 rooms on the 2nd floor, Class 5A or Room 3 and Class 6B or Room 4 with openings facing north and south.

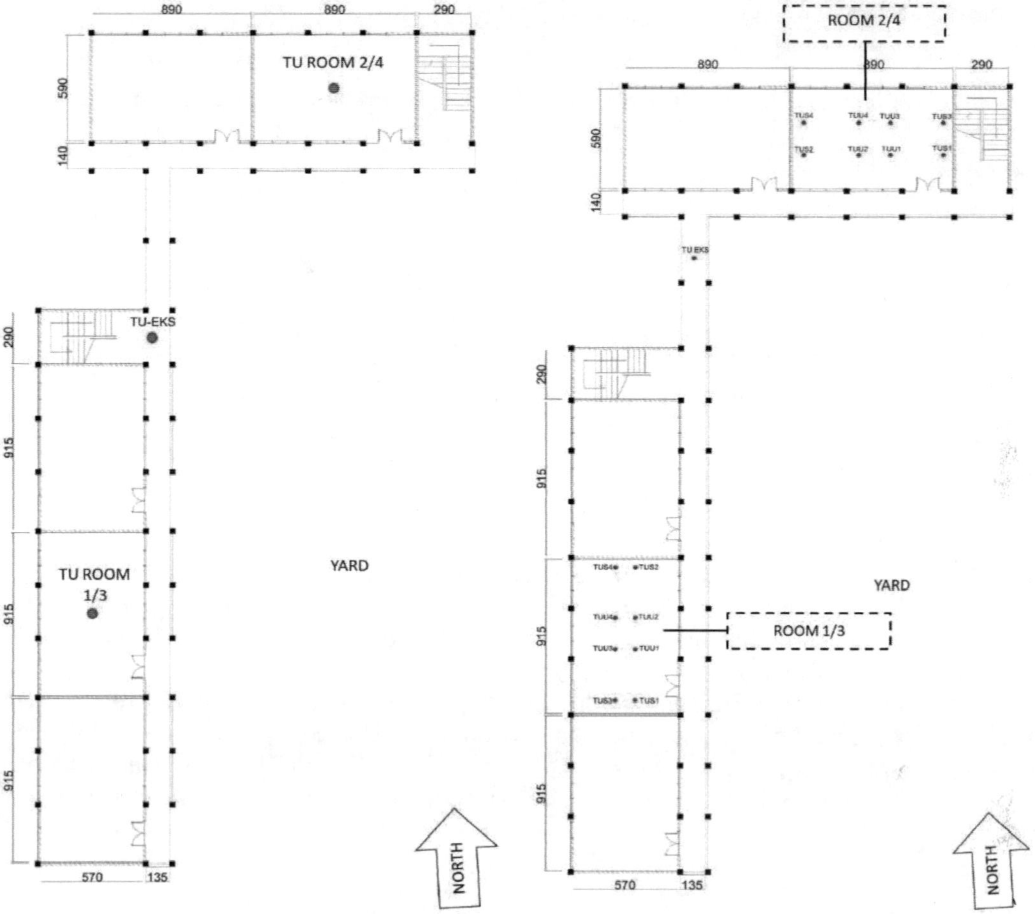

Figure 2. Measuring point for thermal comfort (left) and visual comfort (right).

Measuring instruments as shown in table 2 for the implementation of direct measurements on objects in accordance with the required research variables. Quantitative data analysis to obtain a thermal comfort index by Indonesian standards and a natural lighting index using measurement theory according to Indonesian standards [4]–[6].

Table 2. Measuring instruments used in the current research.

Tools	Type	Function
Hot Wire Anemometer	Lutron AM 4234SD	Measure wind speed, temperature, and humidity
Lux Meter	AEMC Instruments CA813	Measuring lighting levels
Laser Distance Meter	KXL-E100	Measuring the dimensions of space

doi:10.1088/1755-1315/1007/1/012013

3. Result

3.1. Air movement

Air movement is influenced by the shape and orientation of the opening in each room, the elevation of the room, and the wind speed outside the room. The wind speed itself is influenced by differences in air pressure, location, altitude, and time [7].

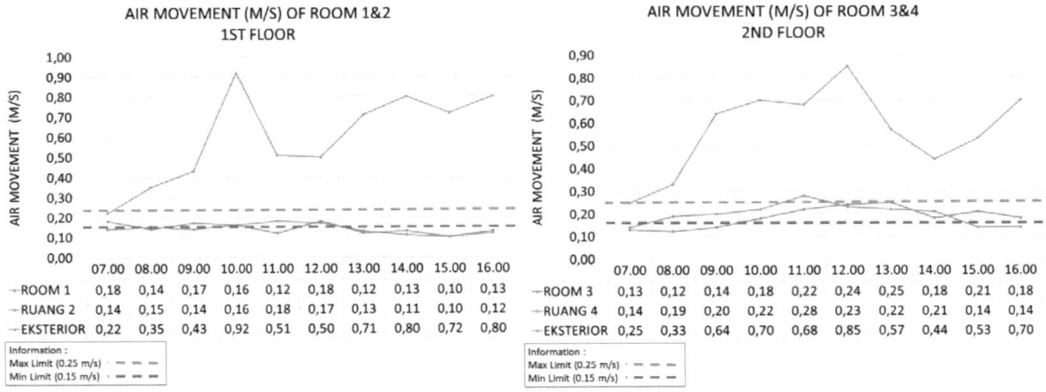

	07.00	08.00	09.00	10.00	11.00	12.00	13.00	14.00	15.00	16.00
ROOM 1	0,18	0,14	0,17	0,16	0,12	0,18	0,12	0,13	0,10	0,13
RUANG 2	0,14	0,15	0,14	0,16	0,18	0,17	0,13	0,11	0,10	0,12
EKSTERIOR	0,22	0,35	0,43	0,92	0,51	0,50	0,71	0,80	0,72	0,80

	07.00	08.00	09.00	10.00	11.00	12.00	13.00	14.00	15.00	16.00
ROOM 3	0,13	0,12	0,14	0,18	0,22	0,24	0,25	0,18	0,21	0,18
RUANG 4	0,14	0,19	0,20	0,22	0,28	0,23	0,22	0,21	0,14	0,14
EKSTERIOR	0,25	0,33	0,64	0,70	0,68	0,85	0,57	0,44	0,53	0,70

Figure 3. 1st floor and 2nd floor air movement chart.

To maintain comfortable conditions, the air velocity is between 0.25 m/s to 0.15 m/s [6]. At the measurement point with a height of 1m from the floor, the air velocity in rooms 1 and 2 on the 1st floor tends to be under the comfortable category. On the other hand, in rooms 3 and 4 on the 2nd floor, air movement is categorized as comfortable. Air movement is affected by the shape and orientation of the opening, the elevation of the room, and the speed of the wind outside. Wind speed itself is influenced by differences in air pressure, location, altitude, and time [7]. This condition creates air movement in the 2nd floor classroom which tends to be higher than the 1st floor classroom. This is because the openings on the 2nd floor tend not to be blocked by plants or other buildings so that outside air can enter the classroom efficiently.

3.2. Air humidity

For tropical areas, it is suitable for rooms with a dense number of people such as classrooms, the relative humidity is still allowed to range between 55-60% [6]. From the results of field measurements, the humidity is very high in the morning and tends to decrease until 14.00 WIB but rises again until 16.00 WIB. Figure 4 shows that all selected rooms on the 1st and 2nd floors do not meet the comfortable humidity standards for classrooms.

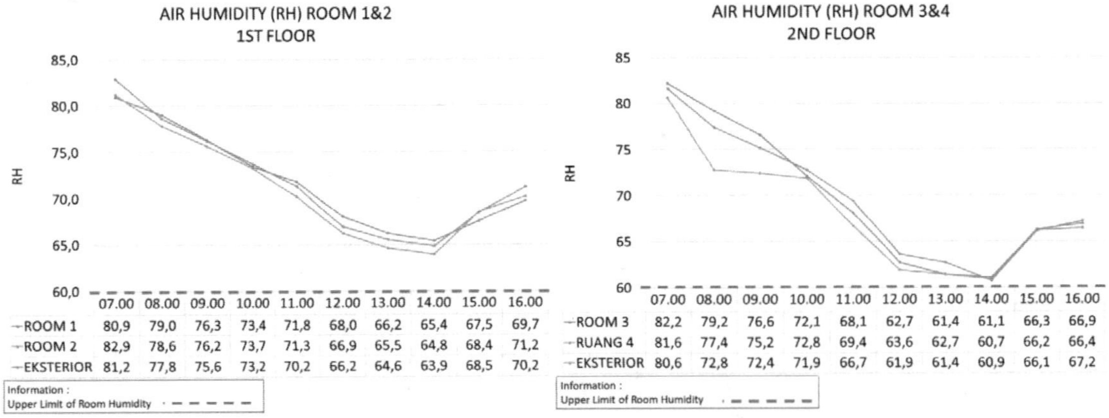

	07.00	08.00	09.00	10.00	11.00	12.00	13.00	14.00	15.00	16.00
ROOM 1	80,9	79,0	76,3	73,4	71,8	68,0	66,2	65,4	67,5	69,7
ROOM 2	82,9	78,6	76,2	73,7	71,3	66,9	65,5	64,8	68,4	71,2
EKSTERIOR	81,2	77,8	75,6	73,2	70,2	66,2	64,6	63,9	68,5	70,2

	07.00	08.00	09.00	10.00	11.00	12.00	13.00	14.00	15.00	16.00
ROOM 3	82,2	79,2	76,6	72,1	68,1	62,7	61,4	61,1	66,3	66,9
RUANG 4	81,6	77,4	75,2	72,8	69,4	63,6	62,7	60,7	66,2	66,4
EKSTERIOR	80,6	72,8	72,4	71,9	66,7	61,9	61,4	60,9	66,1	67,2

Figure 4. 1st floor and 2nd floor air humidity chart.

IOP Conf. Series: Earth and Environmental Science 1007 (2022) 012013 doi:10.1088/1755-1315/1007/1/012013

Dry temperature is inversely proportional to air humidity. Figure 4 explains that the air humidity in the morning is higher than the humidity in the afternoon because the water vapour content in the air in the morning is higher than in the afternoon. This condition occurs because of the low heat of solar radiation and dry temperatures in the morning [7]. The dry temperature on the 2nd floor tends to be higher than on the 1st floor. This is caused by several factors, namely the angle of incidence of sunlight, altitude, wind direction, ocean currents, clouds, and duration of irradiation [7].

3.3. Dry temperature

In determining thermal comfort, one of the influencing factors is dry temperature. In measuring the dry temperature in the selected classroom, there is a slight difference in dry temperature on the 1st floor and 2nd floor room as shown in Figure 5. The temperature in the room increases due to the heat of solar radiation entering the room, thus affecting the temperature in the room.

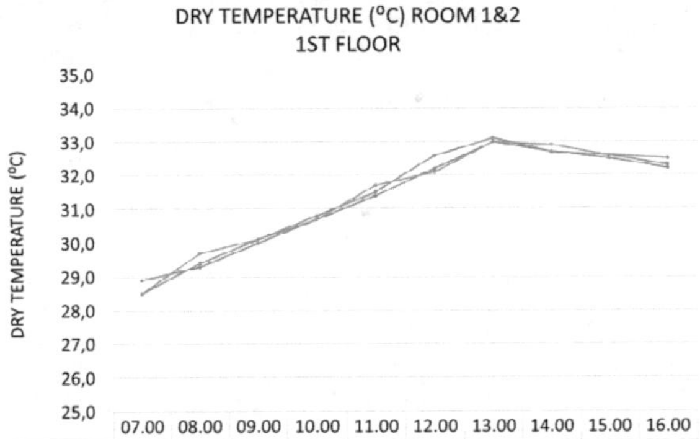

	07.00	08.00	09.00	10.00	11.00	12.00	13.00	14.00	15.00	16.00
ROOM 1	28,9	29,3	30,0	30,7	31,4	32,2	33,0	32,7	32,5	32,2
RUANG 2	28,5	29,4	30,1	30,8	31,5	32,6	33,1	32,7	32,6	32,3
EKSTERIOR	28,5	29,7	30,1	30,7	31,7	32,1	33	32,9	32,6	32,5

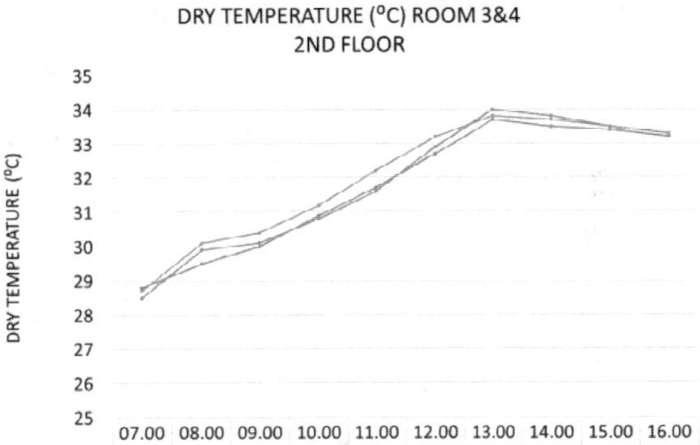

	07.00	08.00	09.00	10.00	11.00	12.00	13.00	14.00	15.00	16.00
ROOM 3	28,8	29,5	30	30,9	31,7	32,7	33,7	33,5	33,4	33,2
RUANG 4	28,5	29,9	30,1	30,8	31,6	32,9	34	33,8	33,5	33,3
EKSTERIOR	28,7	30,1	30,4	31,2	32,2	33,2	33,8	33,7	33,5	33,3

Figure 5. 1st and 2nd floor dry temperature charts.

The direction of solar radiation and protective measures against it greatly affect the entry of heat into the room [7] This condition can be seen from figure 5 where the dry temperature of Rooms 3 and 4 which are on the 2nd floor is higher than that of Rooms 1 and 2 which are on the 1st floor.

3.4. Natural lighting level

The level of natural lighting in Room 1 on the 1st floor tends to be higher than in Room 2. This occurs because the opening on Room 1 faces west and east without being blocked by buildings or trees and it's facing the sun path so it receives direct sunlight. Whereas opening in Room 2 faces north to another building and south to a small park, this causes the room to receive very little direct sunlight. Field measurement of Room 3, 2nd floor tends to be stable until 12.00 WIB but then increases drastically until late afternoon. For field measurements in Room 4 on the 2nd floor, the lighting level of the classroom tends to be high in the morning and lower in the afternoon.

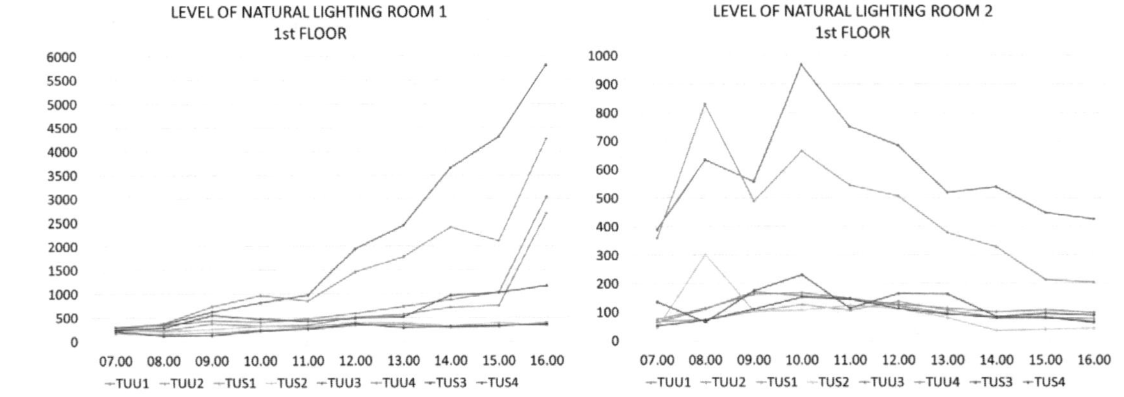

Figure 6. Room 1 and room 2 of natural lighting charts.

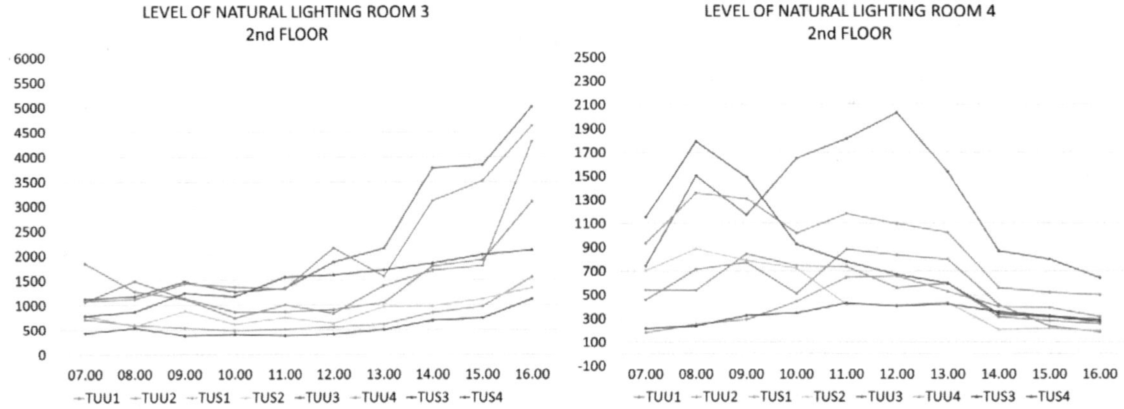

Figure 7. Room 3 and room 4 of natural lighting charts.

Room 1 and Room 3 has the same opening but different altitude, this situation causes some difference in natural lighting level because the opening on Room 3 did not get obstacle from buildings and plants around it, this allows sunlight to enter the room to the maximum. This situation also occurs in Room 2 and Room 4, where Room 4 has a higher level of natural lighting because it lacks obstacle.

4. Discussion

4.1. The effective temperature analysis

Thermal comfort for the tropics can be divided into cool comfortable between 20.50C ~ 22.80C TE, optimal comfort between effective temperature 22.80C ~ 25.80C TE, and warm comfortable between 25.80C ~ 27.10C TE [3]. For a room to be considered comfortable, the room must be included in Mom-Weiseborn category, if it is not included in the category then there needs to be a change in the facade design or additional mechanical assistance.

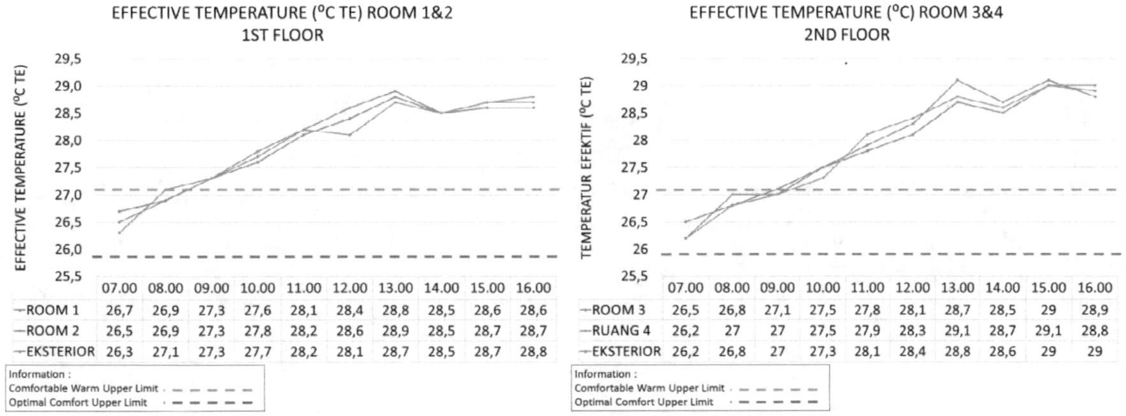

Figure 8. 1st and 2nd floor thermal comfort charts.

From the results of field measurements, it can be seen in figure 8 that only at 07.00 WIB, both at the measuring point on the 1st and 2nd floors, is in the warm comfortable category. Meanwhile, from 08.00 to 16.00 WIB, it is not included in the comfortable category. Due to the condition of the room that only relies on openings in the form of windows as effective temperature controllers in the classroom, the temperature outside the room which tends to be above comfort level, greatly affects the effective temperature in the room. The direction and strength of the wind, also the elevation of the façade opening greatly affect the amount of wind that enters the room. This event can be seen from Room 1 on the 1st floor that has an average effective temperature slightly higher than Room 3 on the 2nd floor. In room 1, the east side is blocked by plants, and the west side is blocked by a building while both openings in Room 3 are not blocked by buildings or plants. This causes that the wind to enter Room 3 effectively and lower the effective temperature in the room [7].This temperature difference also occurs in Room 2 on the 1st floor which is hotter than Room 4 which is on the 2nd Floor. From the results of this field analysis, it can be concluded that the humidity and dry temperature outside are very influential on the effective temperature in the room especially when using natural ventilation. To increase the thermal comfort of the room, it is necessary to increase air velocity with the help of mechanics such as fans or by applying cross ventilation with changing the design opening of the façade to increase wind for outside to enter effectively, another method including blocking direct sunlight coming into the room to minimize the heat that occurs from solar radiation using solar shading device both fixed such as vertical louver or movable such as a curtain [8,9,10].

4.2. Natural lighting analysis

The measuring point in Room 1 on the 1st floor tends to be above the minimum level of classroom lighting, as described in figure 9. This is because the window openings in Room 1 face east and west so that they get good natural lighting from sunlight. However, it can be seen that in the afternoon the lighting level tends to exceed the maximum limit of 750 lux that has been set and it is not ideal for students' activities. This condition happened because the opening on the west side receives direct sunlight without being blocked by buildings and plants. The geometric shape of the window that spans the length of the room with an overhang type canopy also has a significant impact on performance concerning daylight penetration, because it can not prevent the entry of direct sunlight into the room [11]. This can result in glare thereby reducing visual comfort in the classroom. In Room 2 on the 1st floor, the lighting level of the classroom tends to be below standard. This is because the window openings in this room face north and south. On the south side, the opening faces a canopied hallway and a garden full of trees so that it blocks the entry of sunlight into the room. On the north side, the opening in Room 2 on the 1st floor is opposite to a 1-story building opposite. This condition blocked the sunlight from entering the room.

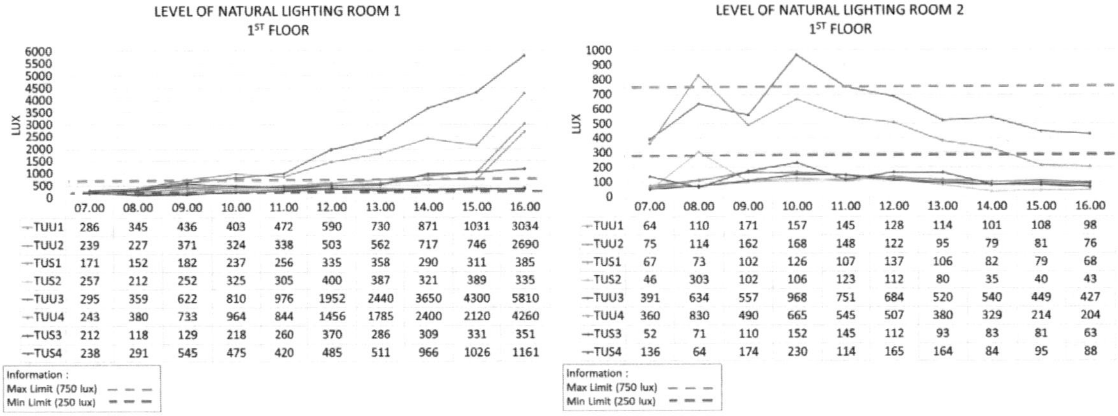

Figure 9. Graph of natural lighting levels for room 1 and room 2 on 1st floor.

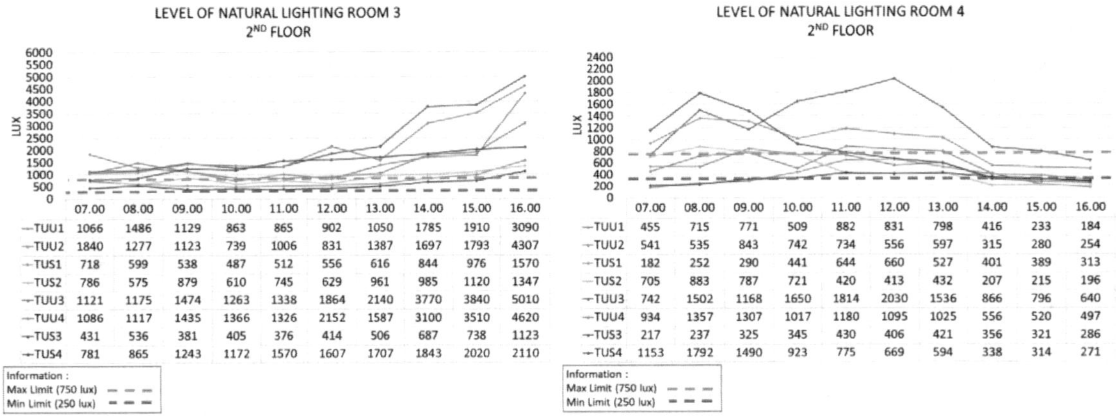

Figure 10. Graph of natural lighting levels for room 3 and room 4 on 2nd floor.

Figure 10 shows the measuring point in Room 3 tends to be stable and all of them are above the minimum limit of 250 lux until 12.00 WIB then increase drastically until late afternoon. This is due to the fact that the window openings in Room 3 face east and west without being blocked by buildings or surrounding trees so that they get natural lighting from sunlight all day long. However, it can be seen that the measuring points TUU1, TUU2, TUU3, and TUU4, figure2, which are located in the middle of the room tend to be far above the maximum limit of 750lux that has been set. At this point, it even tends to be very high and can cause glare thereby reducing visual comfort in the classroom. Windows that are not optimally protected by shading devices barely meet the acceptance criteria and thus require additional control devices such as vertical louver canopy or curtains [12]. Room 1 on the 1st floor and Room 3 on the 2nd floor have the same direction of openings but the level of natural lighting in Room 3 on the 2nd floor tends to be higher than Room 1 on the 1st floor. This is because the eastern opening of Room 1 faces the hallway and garden so that the morning sun tends to be blocked by trees.

Infield measurements Room 4, the lighting level of the classroom tends to be high in the morning and down in the afternoon. The lighting level, in general, has exceeded the minimum limit of 250 lux, this is because the window openings facing north and south are not blocked by buildings or trees so that the room can receive sunlight well. With the condition of openings facing north and south that I not obstructed, the lighting level tends to be within the 250-750 lux range so that the level of visual comfort is better than Room 2 on the 1st floor which has the same direction of opening. Meanwhile, for field measurements in Room 2 on the 1st floor, the natural lighting level of the classroom tends to be below standard [4]. On the south side, the opening faces a canopied hallway and a garden full of trees so that it blocks the entry of sunlight into the room. To assist in increasing visual comfort, Room 2 needs to be assisted with artificial lights or redesigned by using larger windows. Even though it helps increase the level of natural luminance, direct sunlight should always be avoided because it brings heat into the building that can reduce the thermal comfort in the classroom [10]. The amount of illumination in each zone based on the required light level needs to be controlled to improve occupant visual comfort and also reduce lighting energy consumption. Natural lighting strategies must be able to reduce and control the level of solar radiation, especially from side and top lighting, to overcome the problem of heat distribution by providing shading and adjusting the size and location of windows to get indirect sunlight and also prevent exposure from east or west. In the building area facing East-West, new shading devices should be applied to reduce solar radiation such as egg crate and vertical louver [13].

5. Conclusions

Based on the discussion above, it can be concluded that with the current opening and facade design, all selected classrooms do not fit the comfortable criteria according to the Mom-Wiesebron standard. Different designs are needed in the form of larger openings with cross ventilation models, different canopy designs from existing conditions, and mechanical assistance to help maximize air movement. In addition, with the current situation, the selected classrooms did not meet the criteria of visual comfort according to Indonesia National Standard with east-west openings tend to experience glare while openings with a south-north direction are less than optimal in receiving natural lighting. In redesigning the facade, it is necessary to design according to the need to let natural light enter the room but also protect it from solar radiation heat so visual and thermal comfort can be achieved.

6. References

[1] Olgyay V 1963 *Design With Climate* (Princeton: Princeton University Press)
[2] Mangunwijaya Y 1997 *Pengantar Fisika Bangunan* (Jakarta: Djambatan)
[3] Soegijanto 1998 *Bangunan di Indonesia dengan Iklim Tropis Lembab Ditinjau dari Aspek Fisika Bangunan* (Jakarta: Departemen Pendidikan dan Kebudayaan)
[4] Standar Nasional Indonesia 2000 Konservasi energi pada sistem pencahayaan *SNI 03-6197-2000* p 17
[5] Standar Nasional Indonesia 2001 Tata cara perancangan sistem pencahayaan alami pada bangunan gedung *SNI 03-2396-2001*
[6] Standar Nasional Indonesia 2001 Tata cara perancangan sistem ventilasi dan pengkondisian udara pada bangunan gedung *SNI 03-6572-2001* pp 1–55
[7] Lippsmeier G 1994 *Bangunan Tropis* (Jakarta: Erlangga)
[8] Ching F D K 1997 *Arsitektur Bentuk, Ruang, dan Susunannya* (Jakarta: Erlangga)
[9] Reza F and Kojima S 2020 Thermal comfort investigation based design considerations for the tropical studio type classroom *Int. J. Sustain. Dev. Plan.* vol 15 no 8 pp 1179–1185
[10] Szoklay S V 1980 *Environmental Science Hand Book for Architect and Builders* (London: The Construction Press Lancaster)
[11] Ochoa C E, Aries M B C, Loenen E J van and Hensen J L M 2012 Considerations on design optimization criteria for windows providing low energy consumption and high visual comfort *Appl. Energy* vol 95 pp 238–245
[12] Teitelbaum E *et al* 2020 Membrane-assisted radiant cooling for expanding thermal comfort zones globally without air conditioning *Proc. Natl. Acad. Sci. U. S. A.* vol 117 no 35 pp 21162–21169
[13] Egan M D 1975 *Concept in Thermal Comfort* (Engelwood Cliffs: New Jersey: Prentice Hall Inc)

Acknowledgment
The author would like to thank Sawah Besar 01 Elementary School, Semarang City, Central Java, for allowing me to carry out this research at his institution. We also thank the Department of Architecture, Faculty of Engineering, Diponegoro University for providing facilities in data processing.

IOP Conf. Series: Earth and Environmental Science 1007 (2022) 012014 doi:10.1088/1755-1315/1007/1/012014

The sustainability of the Pegat Lake as the centre for the orientation of the Giri kingdom, socio-cultural studies

M Koderi[1,2], **B Soemardiono**[1], **M Faqih**[1]

[1]Architecture Department, Faculty of Civil, Planning, and Geo-Engineering, Institut Teknologi Sepuluh Nopember. Kampus ITS, Sukolilo, Surabaya 60111, Indonesia

Abstract. The synergy between environmental sustainability and technological progress is the key to sustainable development. The dependence of the people of the Giri Kingdom on the Pegat Lake as a source of clean water as well as the orientation centre of the Giri Kingdom's downtown area, has shifted due to the development of the city and advances in clean water technology. The potential for socio-cultural sustainability which is based on the existence of Pegat Lake is the strength of the meaning of local wisdom which is a process of harmonious relationship between nature, humans and architecture is starting to be forgotten. This paper explores the role of Pegat Lake from time to time based on artefacts, data and history as well as notes on the results of previous research. The important finding of interdependence between the Pegat Lake area, the Giri Palace area and the Tomb and Ainul Yaqin Mosque area is a strength of sustainability to continue to be developed according to the socio-cultural demands of the community as well as the development of religious and historical-based ecotourism as an answer to sustainable regional development.

1. Introduction

The World Commission on Environment and Development contains the idea of sustainable development [1]. This idea is realised in the form of development solutions that can meet the needs of the present by taking into account the needs of future generations. Meanwhile, Holmberg suggests three aspects of sustainable development: economic, ecological and social [2]. The ecological aspect is intended as a system to protect the environment and natural resources. The social aspect is the provision of social facilities and adequate access to health, education, gender equality, as well as a clean and participatory political system, while maintaining the sustainability of the economic aspect in the utilisation of renewable natural resources. This sustainable development includes food production, alternative economic and environmental policies [2]. Meanwhile, Agenda 21 defines culture as a process of dynamic development of individual and group identities [3]. Active participation involving the community in cultural activities. Improved quality of life for the better as well as their life choice opportunities. Socio-cultural factors in the community environment play an important role in the development of development [3]. In Undang-Undang no. 23 tahun 2014 that formulate a comprehensive and integrated approach to development that is able to generate creativity, heritage, knowledge, and differences in it [4]. The potential as a centre for regional arrangement orientation and tourist destinations that are excavated from historical heritage is the basic reason in research exploring Pegat Lake as a cultural heritage whose current condition tends to be poorly maintained and its role is starting to be forgotten by the community. The existence of Pegat Lake, which was once the foundation for the lives of the people of the Giri Kingdom city centre in the past, will be extinct and damaged even though the great potential for regional development makes Pegat Lake as a centre of orientation in sustainable landscape arrangement can be an important answer in the future.

[2]Corresponding author: m.koderi.mt@gmail.com

2. Literature review

2.1. Meaning in architecture

2.1.1. Definition of Meaning. Philosopher Runes in Arias understands meaning as a very ambiguous term with four meanings, namely: 1) purpose, 2) designation or reference, 3) definition or translation, 4) consequences. Meanwhile, Arias defines meaning and use as two different phenomena. Meaning is related to communication such as the idea to be conveyed while usage is related to how architecture is used [5].

Use and meaning are interrelated, use is a manifestation of effective behaviour and meaning is affective behaviour. As Quin in Noth meaning is explained in substitution and cognitive equivalence [6]. Meaning basically can be derived from observable contexts such as the opinion of Lech in Noth which states that contextual meaning theory follows a syntactic approach when considering the meaning of a function sign from its relationship with other signs in its context [6]. It has an understanding that meaning is pragmatic when defining meaning as a function of its situational context. Meanwhile, Rapoport distinguishes the levels of meaning in the context of the built environment, namely: 1) High-level meanings are related to cosmology, cultural schemas, worldviews, philosophical and secret systems; 2) Middle-level meaning, communicating about identity, status, wealth, power; 3) Low-level meaning, every day and instrumental meaning, for example accessibility, use of settings as intended and influenced by social situations, expected behaviour and so on [7]. According to Hershberger in Broadbent, meaning can be divided into two, namely: 1) representational meaning (objective meaning) is meaning that arises from outside and is related to objects and events; 2) responsive meaning (subjective meaning), is everything related to internal factors and is only owned by the observer and captured by his own feelings [8].

The understanding of meaning above shows that meaning is seen from affective assessment and responsive (subjective) meaning. The nature of this assessment depends on the judgement of the observer himself and cannot be the same as that of other observers. In contrast to the representational (objective) meaning, the assessment is based on the condition of the object. Gibbs states that the notion of meaning in the context of communication has six (6) definitions, namely: 1) something like a symptom or sign that can help; 2) something that states its importance, significance and consequences in a certain way; 3) the meaning of an object can be in the form of the object itself, its purpose and justification; 4) Something that someone wants to say or communicate and in this case what is meant by that person is the expression of the speaker's intent; 5) meaning means if someone wants to do something; 6) Expressions or symbols are expressions or symbols that contain what is expressed, represented and presented [9]. Meaning in language requires a way of conveying understanding directly, contextually and in conclusion. Gibbs' statement shows that understanding meaning can be reviewed based on direct understanding without any other reference, based on contextual and real information based on direct and contextual understanding but requires conclusions from the results of understanding meaning.

Meanwhile, according to Rapoport, the meaning of the environment can be understood in three main ways, namely: 1) A semiotic approach (the science of meaning) [10]. Semiotics is the process by which things serve as signs. Semiotics contains three main components, namely: the sign vehicle (what acts as a sign), the designation (for what refers to a sign), interpretation (the influence on the interpreter based on a thing, namely a sign). Semiotics as a study of the importance of structured system elements, can also be understood as three main components, which help in understanding some problems with semiotics, including: a. Syntax (the relationship between the sign and what must be signed as a sign system, namely the understanding of a system structure); b. Semantics (the relationship of the sign with something that must be signed, namely how the sign that leads to the meaning is owned by the element); c. Pragmatics (the relationship of signs with people's behaviour, where the results are interpreted as part of the unity of behaviour); The second approach is the symbolic approach, namely the traditional elements used to solve the mysteries of high-style architecture and vernacular environments. This approach is very useful in situations especially in traditional cultures. This approach is usually used in traditional contexts. The third approach is a nonverbal communication approach, namely communication that involves the observer (human) and the object of observation (object). Understanding the meaning of the environment according to Rapoport can be done through a language approach, a symbolic approach and a non-verbal communication approach [10].

Moore in Arias, states that the symbolic meaning in architecture is the human experience through the mediation of arches [5]. According to Capon, architecture is arranged systematically in 3 (three) main categories, namely form, function and meaning (meaning) [11]. Form refers to the visible presence of architectural objects, function refers to how well they can operate (how well they perform) and meaning refers to how well we know architectural objects. A different view from specifically Rapoport which explains that architectural products have 3 (three) namely buildings, urban space and landscape culture [7]. City buildings and spaces are the result of community designs that involve elements of community culture. The most important thing about landscape culture is the meaning for users (i.e. society) of architectural products. From the statements above, it shows that in architecture, meaning is one of the main things in architectural assessment. The difference between the two views between Capon and Rapoport is the meaning as part of the architecture and the meaning as the meaning of the cultural landscape.

From some of the meanings mentioned above, it can be concluded that 1) meaning in architecture is related between humans and their architecture or the built environment, 2) meaning is related to the interpretation of signs or signals produced through one's experience with architectural representations, 3) architectural representation is related to the general meaning that related to the purpose and justification of something as well as the meaning of non-verbal communication between the observer and the architecture.

2.1.2. Denotative and connotative. According to Eco in Broadbent, denotative meaning shows the function of architectural objects while architectural objects can have different meanings depending on their connotations [8]. The meaning can be denotative or connotative and the ordering of these two categories is highly dependent on the observer. Connotative meaning can be denotative for different observers. However, if the two observers have made quite deep observations, then the same meaning is obtained, because understanding meaning can be learned through experience and deepening. Jencks in Noth explains that connotation is a key concept of semiotics, style and literary theory [6]. The concepts of connotation and denotation were first used for analysis in distinguishing between abstract and concrete. Leech in Noth defines denotative meaning as conceptual meaning and distinguishes six kinds of meaning, namely: 1) connotative, 2) stylistic, 3) affective, 4) reflection, 5) collocative/conceptual, 6) thematic [6]. Bloomfield in Noth, connotation as a supplementary value that cannot be calculated and determined where its scope is, as a subject of more systematic research in linguistics than language varieties, registers and styles [6]. Eco asserts a typology of connotations which basically consists of a primary meaning and a secondary meaning [6].

Architectural properties of buildings that reflect the mental image of the building and the association's representation of the building, as Lang in Arias defines the term symbolic meaning as a result of cognitive processes in which objects acquire connotative evaluations outside of their use as objects. People can infer emotional and other qualities from style or schemes. Pleasant connotative meaning, one can talk about symbolic aesthetic experience [5]. Meanwhile, Rapoport calls it a pragmatic meaning, the meaning of style for the wearer [10].

The conclusion from the understanding above is that individual experience of an object depends on the interaction between object features and individual representations to remember past experiences related to the type of building.

2.2. Culture and sustainability

Meanwhile, according to Dessein, there are three roles of culture in sustainable development, among others; the first representation considers culture as if it has an independent role in sustainability: it

[15] Nguyen T T and Watanabe T 2019 Win-win outcomes in waste separation behavior in the rural area: A case study in Vietnam *J. Clean. Prod.* **230** 488–98
[16] Franck K A and Lepori R B 2000 *Architecture inside out* (Chichester: Wiley)
[17] Pink S 2012 *Situating Everyday Life: Practices and Places* (Los Angeles: SAGE)
[18] Engkvist I-L 2010 Waste and recycling – A challenge for society *Appl. Ergon.* **41** 335
[19] Pollan M 2014 *Cooked: A natural history of transformation* (New York: Penguin Books)
[20] Suryantini R, Paramita K D and Yatmo Y A 2021 A healthy machine for living: Investigating the fluidity of open spaces in the domestic environment during the pandemic *AIP Conf. Proc.* **2376** 040018
[21] Suryantini R, Atmodiwirjo P and Yatmo Y A 2021 Toward a healthy home: Investigating food flow and the shift in domestic spatial practice during the COVID-19 pandemic *Hous. Soc.*
[22] Atmodiwirjo P and Yatmo Y A 2015 Architecture as machine; Towards an architectural system for human well-being *Int. Cong. Le Corbusier, 50 Years Later*. 18th-20th November 2015 (Valencia)
[23] Paramita K D and Schneider T 2018 Passage Territories: Reframing Living Spaces in Contested Contexts *Interiority* **1** 113–29
[24] Marco E, Williams K and Oliveira S 2021 Prioritising Storage Practices: A New Approach to Housing Design Thinking *Interiority* **4** 223–48
[25] Marco E, Williams K and Oliveira S 2020 Too Much 'Stuff' and the Wrong Space: A Conceptual Framework of Material Possessions *Interiority* **3** 219–42
[26] Yatmo Y A, Atmodiwirjo P and Paramita K D 2013 Whose Waste Is It Anyway? *J. Urban Des.* **18** 534–52
[27] Aini M S, Fakhru'l-Razi A, Lad S M and Hashim A H 2002 Practices, attitudes and motives for domestic waste recycling *Int. J. Sustain. Dev. World. Ecol.* **9** 232–8
[28] Zhang S, Zhang M, Yu X and Ren H 2016 What keeps Chinese from recycling: Accessibility of recycling facilities and the behavior *Resour. Conserv. Recycl.* **109** 176–86
[29] Harman G 2018 *Object-oriented ontology: A new theory of everything* (London: Pelican Books)
[30] Nadadur G and Parkinson M B 2013 The role of anthropometry in designing for sustainability *Ergonomics* **56** 422–39
[31] Akita M 1991 Design and ergonomics *Ergonomics* **34** 815–24
[32] Dianat I, Molenbroek J, Castellucci H 2018 A review of the methodology and applications of anthropometry in ergonomics and product design *Ergonomics* **61** 1696–720
[33] Martin K, Legg S and Brown C 2013 Designing for sustainability: Ergonomics – carpe diem *Ergonomics* **56** 365–88
[34] Tine P, van Acker M and de Vos E 2015 The Public Interior: The meeting place for the urban and the interior *IDEA J.* **15** pp 44–55

Acknowledgments
The authors would like to thank Prof. Yandi Andri Yatmo for his supervision throughout this research. The research is part of the research funded through the Penelitian Dasar 2021 grant from Kementerian Riset dan Teknologi/Badan Riset dan Inovasi Nasional (Kemenristek/BRIN) Republik Indonesia (Grant Number NKB-039/UN2.RST/HKP.05.00/2021).

becomes the fourth pillar of sustainability [12]. This representation represents culture in sustainability, and sees cultural sustainability as parallel to ecological, social, and economic sustainability; The second representation refers to culture that has a mediating role to achieve economic, social and ecological sustainability. This representation, shows that material and immaterial culture is seen as an important resource for local and regional economic development. It also implies that cultural values and perceptions need to be considered when aiming for ecological or social sustainability; the third representation considers culture as a necessary foundation for meeting overall sustainability goals. This representation, which culture refers to as sustainability, encompasses the other pillars of sustainability and becomes an overarching dimension of sustainability. In other words, sustainability becomes embedded in culture and leads to an eco-cultural civilization.
The three pillars of sustainability holistically according to Korten [13] consist of:

2.2.1. Economically. Viable is an economic pillar that becomes an important aspect of the development of certain cities or regions, including; Sustainable economic development and increasing people's income; Increasing employment and equal distribution of business opportunities and income; Development based on local resource economy, but export-oriented to regional, national and global levels.

2.2.2. Environmentally. Friendly is an environmentally friendly pillar which is an important aspect of the development of certain cities and areas, including; Environmental conservation and maintenance of soil nutrients; efficient use of resources, with efforts to use renewable resources.

2.2.3. Socially. Politically Acceptable and Culturally Sensitive are Pillars that are socially and politically accepted by the community and sensitive to culture which is an important aspect of the development of certain cities and areas including: Democratisation of spatial planning and management; Community empowerment can be interpreted as an effort to increase the dignity of the people from the inability to escape from alienation from the system of life, underdevelopment and poverty; Decentralisation of spatial planning and management; Utilisation of indigenous knowledge of the area, usually in the form of traditional technology, customs, beliefs and taboos of society which is often underestimated because it is not scientific. Social equity, integration between physical issues with social issues and cultural integrity. The conclusions drawn from the understanding of culture and sustainability are related to the understanding that culture is the basis for understanding sustainability which is inspired by the process of ideas and thoughts that underlie it.

2.3. The Royal City Centre as a cultural existence

2.3.1. Downtown landscape. Armstrong et al. in Benson Landscape planning and urban form are determined by Land Use Plans, Water Plans, Plant Plans, and Plans for environmentally friendly buildings and land use [14]. Meanwhile, Carlo Vezzoli and Ezio Manzini [15]. stated that design for environmental sustainability is reflected in the life cycle which is influenced by the use of environmentally friendly materials, product life cycle, design cycle, functional approach, objectivity design life cycle, impact of life cycle design, design approach, strategy life cycle design, Linkages between strategies, Priority strategies, Disposal design, Priority environmental costs. In more detail Edwart T White formulates the process of designing a good site must show the scope of influence factors in detail and specifically include: Location, relevance to the Environment, Area and Area, Natural Physical Disability, Artificial Benefits, Circulation, Utilities, Humans and Culture, Climate [16].

2.3.2. Culture and environment. The process of cultural formation according to Rapoport is seen from several things: 1) location, 2) landscape, 3) special elements, 4) special locations, 5) special rooms and types, 6) having names and special ways, 7) special orientation, 8) have colour, texture, 9) have people who will be interesting in special activities [17]. While Altman said that the close relationship between

The high roof of the Tongkonan House also allows for a transition space from the heat of the incoming sunlight. The ability of roofing material is able to reduces the heat of the rays it receives.

In addition, from the data, the rainfall in South Sulawesi can also be categorized as quite high. The rainy season occurs for 5.6 months, from November to April with a greater proportion of the possibility of rain reaching 38%, with a peak reaching 73%. Coupled with the character of the soil in Tongkonan which is difficult to absorb water, resulting in very high humidity levels. Maximizing natural ventilation is also a strategy to create a comfortable shelter for residents.

Table 1. Formatting sections, subsections and subsections [8,9].

Element	Geometric Form	Material	Structure	Climate Response
Foundation		Uru wood, chrysolite wood, jackfruit wood.	Column and beam frame systems, pin and hole joints.	The wooden pillar does not hit the ground directly due to the wet soil and avoids weathering.
Wall	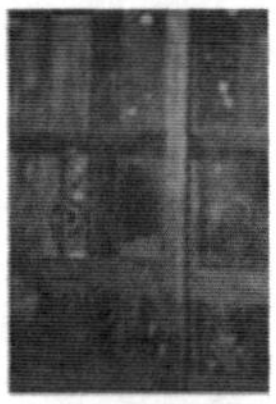	Wooden board	Bearer wall	Optimizing natural light and air circulation
Floor	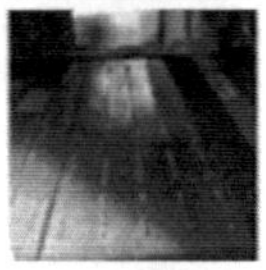	Wooden planks, bamboo	Field frames	Provides gaps between the boards to allow air to enter and exit at a low speed.
Roof		Petung bamboo, thatch roof covering	Roof cover plane frame and beam column frame structure	Drain rainwater and reduce heat. Space on the roof creates volume for the insulation and transitional spaces.

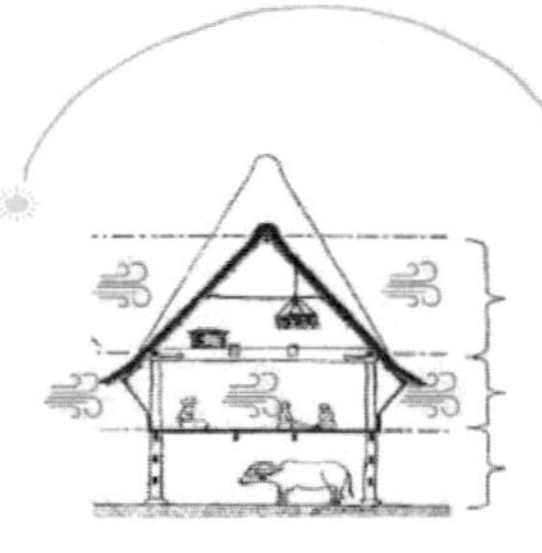

Figure 1. Tongkonan House respons to sun and wind direction.

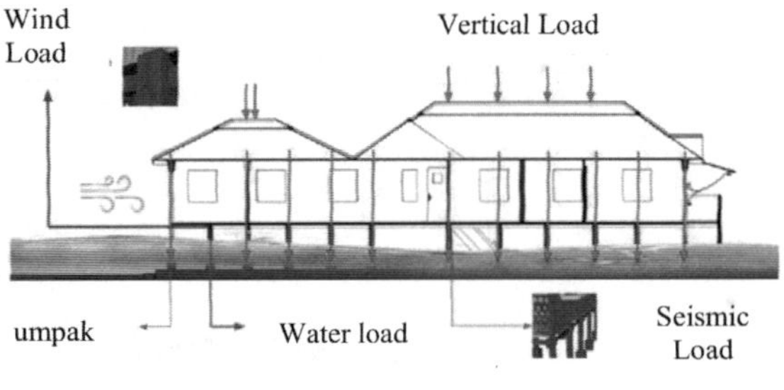

Figure 2. The Betawi Stage House load distribution analysis.

culture and the environment depends on five important factors: 1) natural environment, 2) global orientation and view of the environment related to religion where values and ways of thinking are dominant, 3) cognitive environment, 4) environmental behaviour and processes such as personal space, territorial behaviour, and privacy, 5) environment or the end product of behaviour includes the results of community actions that can be built in environments such as houses, communities and cities, as well as modifications of the natural environment such as agriculture, dams and change climate [18].
From the understanding of the above formulation that the downtown landscape can be realised through a natural and artificial design process as the process of realising the past city centre city centre as an embodiment of the culture that developed in the community.

3. Research method

This study uses a qualitative method using case studies. According to Groat and Wang, qualitative research is the interpretation and meaning of ongoing or ongoing situations, emphasising the role of the researcher as the main part of the objective eye as a research tool and the data can be obtained through field observations, interviews with actual backgrounds [19]. Focused on the respondent's way of thinking. Meanwhile, the case study method, according to Yin in Groat and Wang, implies more than just a simple study of a phenomenon in the field. Moreover, this research includes a case study on the complex dynamics of the past related to the interrelationships that exist in it. In obtaining primary data, this study took it from the field through observation and interviews with respondents [19]. This research was conducted based on a qualitative methodology including observations and surveys at the research location of the Pegat Lake area in Giri, which began with extracting library data related to the past history of Pegat Lake.

4. Analysis and discussion

Qualitative data analysis, according to Moleong in general is a data analysis process which includes, data reduction, categorization, synthesis and summary [20]. Related and more critical matters include, environmental sustainability is determined in the following aspects: Maintaining the vitality and diversity of the earth; Preserving life support systems; Using renewable resources in a sustainable manner; Minimise the use of non-renewable resources; Minimise pollution and damage to the environment and the health of living things; As well as preserving the cultural and historical environment. Analysis related to the cultural and social meaning of Lake Pegat will be able to find the important role of Lake Pegat in supporting the socio-cultural life of the community where Giri worked in the past. The vitality of the role of the Pegat Lake is as a main source of clean water, a centre for socio-cultural orientation of the community as well as environmental conservation in arid and hilly areas. The dependence of the Giri kingdom community on the function and meaning of Pegat Lake in daily life can be found as a potential for sustainability, but the declining role of Pegat Lake from time to time in the cultural and social context for the surrounding community and the Gresik community in general can also be found as a serious problem must be handled carefully and simultaneously in maintaining the meaning and role of Lake Pegat as a sustainable landscape development.

4.1. Telaga Pegat in denotative meaning

Pegat Lake is topographically positioned as a meeting point between the hills of Patireman and Mount Mbah Agung (Bagong). Telaga Pegat was once a basin between the two hills by Sunan Giri with students and the Giri community, it was dug to a certain depth as an effort to accommodate rainwater for daily needs considering this area consists of limestone and is hilly so water is very difficult to get. The existence of Lake Pegat is so important and is supported by its strategic position, so that many travellers and merchants stop by, at the same time making the Lake Pegat area a meeting place and resting place for traders and travellers either by river, sea or land, especially those who will go to the Kingdom Centre. Giri at that time. The location and layout of the Pegat Lake to the Giri Kedaton area is as shown in figure 1.

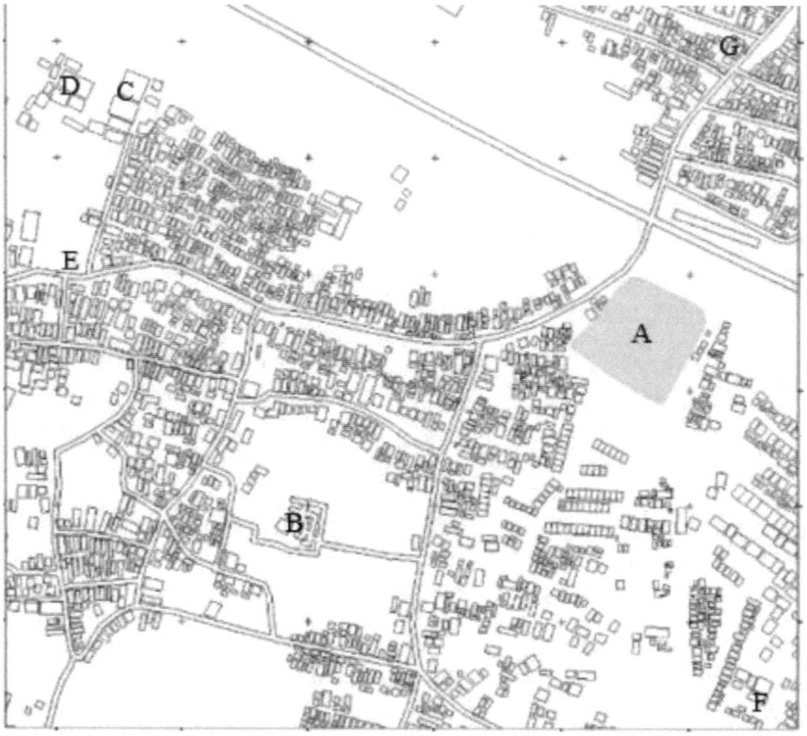

Figure 1. The position of Telaga Pegat and historic buildings in the downtown area of the Giri, among others :A.Pegat Lake, B.Giri Palace, Pesantren Luhur, Makam R. Supeno; C.Ainul Yaqin Mosque, D.Tomb of Sunan Giri E: Market and Field, F: Tomb of Princess Champa and Dewi Sekardadu, G.Market and City in Gresik [21].

4.2. Telaga Pegat in the connotative meaning

This lake was built by Sunan Giri in 1473 AD by mobilising thousands of santri, taken from the word pegat or divorce, indicating that the lake is a separator from the hills between Patireman and Bukit Mbah Agung (Bagong). The area of the Pegat Lake is approximately the size of a football field with a depth of 2 metres. In building the spirit of the Giri people at that time, it became an extraordinary force for the motivation of a great cleric, the Waliyulloh Raden Paku alias Sunan Giri who was considered to have supernatural powers because of religious scholars and the power of prayer to Allah SWT who became the main leader of the Luhur Islamic boarding school at the top of the hill. Giri, which was founded, was filled with divinely inspired meanings that descended on the relatively arid and rocky hilltop of Giri. Telaga Pegat as a source of life in meeting the clean water needs of the Giri Islamic boarding school at that time which accommodated thousands of students from all corners of the archipelago. As shown in Figure 2.

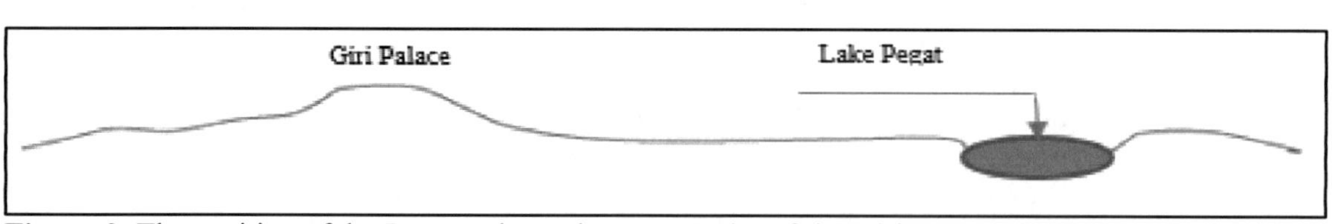

Figure 2. The position of the Pegat Lake to the topography of the Giri Palace area and its surroundings [21].

IOP Conf. Series: Earth and Environmental Science 1007 (2022) 012014 doi:10.1088/1755-1315/1007/1/012014

Efforts to repair Pegat Lake were recorded twice, namely in 1955 and 1977, but the change in the function of Pegat Lake after not being a source of clean water for the city centre has not been maximally carried out, which include being used as a playground for children around, a fishing place with conditions without adequate facilities. With plants and trees that are poorly maintained and some of them even become garbage dumps and are closed by fences and buildings. The land use of the past centre of the Giri kingdom is still recorded today, further land use as in Figure 3.

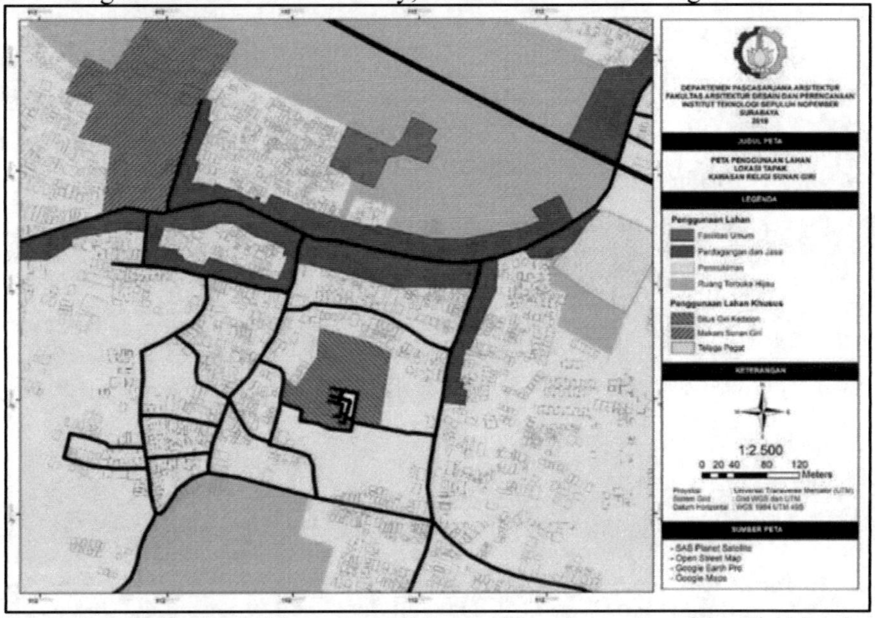

Figure 3. Distribution of the main buildings of the city centre of the Giri kingdom in the past and the development of current land use in the Pegat lake area of the Giri kingdom [21].

4.3. Pegat Lake as a cultural landscape sustainability

Pegat Lake, which has great potential, will be able to become an economic pillar, if the optimization of its former role as a source of clean water for the city centre of Giri's abandoned work is changed as a regional orientation centre considering the strategic position of Pegat Lake in a religious tourism area and cultural heritage can be used as a trigger for economic revival. By changing its role into recreational parks or park tours, pools and other supporting activities such as restaurants, gas houses, etc. The change in the role of the Pegat Lake will become a new regional landmark in the city of Gresik which is able to increase people's income, job opportunities are wider and more equitable. The arrangement of Pegat Lake as the initial movement point for tourism in the Giri royal heritage area which combines other tourist potentials in this area, both religious tourism, history and educational tourism as well as recreation and sports parks as a consequence of changing the role of Pegat Lake must continue to support the concept of environmentally friendly landscapes and be able to become the advantages of the heritage area of the Giri royal city centre in the past, which at the same time took care of the environment properly in environmental conservation and maintenance of soil nutrients, efficient use of natural resources in a sustainable manner, both flora and fauna, in maintaining the beauty of Pegat Lake in the future. Other impacts are expected from the change in the role of the Pegat Lake in the future, which will be felt in democratic community involvement starting from the planning and management of the Pegat Lake spatial planning later. Increasing community involvement is defined as an effort to increase the dignity of the community to live a more decent, healthy and prosperous life. Preservation of traditions and socio-cultural activities will support each other by changing the role of Lake Pegat which is in line with the times and global demands as a form of historical-based ecotourism in the city of Gresik. The examples of suitable planning models are as shown in figure 4.

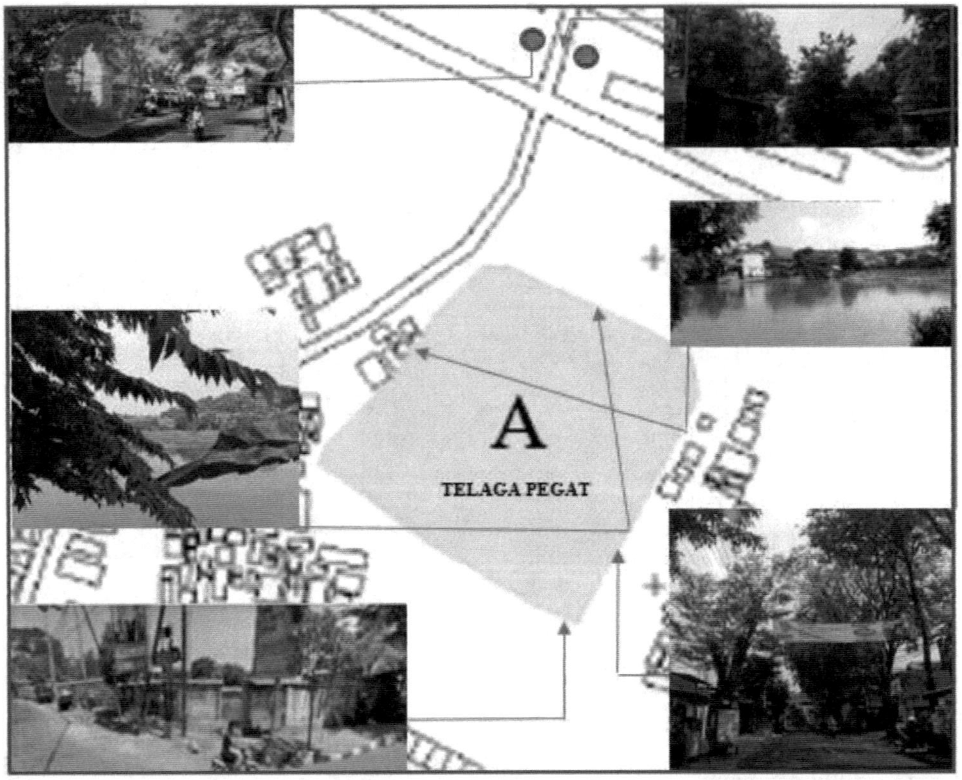

Figure 4. The atmosphere of the Pegat lake and the condition of the buildings and road facilities around it D. Pond Pegat as the main water source of the royal city centre and the Giri Royal Palace [21].

Telaga Pegat was built as an inseparable part of the life of the Giri royal palace which became the centre of control of the Giri kingdom government which was located on the Giri hill which was formerly a Luhur Islamic boarding school building. Giri Palace or known as Kedaton was built since Sunan Giri started building the Luhur Islamic Boarding School in 1478 AD and has been equipped as a Palace building since Sunan Giri was crowned as Prabu Satmata in 1487 AD as a consequence of the Luhur Islamic Boarding School building changing its function as the Giri Royal Palace, whose building is equipped with Taman Sari, Mosque and Islamic Boarding School. Giri Palace was destroyed and burned by the VOC who conspired with Prabu Amangkurat 2 of the Mataram kingdom in the 1612th century AD. The current condition is that the ruins of the foundation artefact which is terraced 7 and is at the top of Giri hill. The sustainability of the life of the royal city centre and the family of the King's Palace is very dependent on the source of clean water from this Pegat Lake. The dynamics of the life of the Giri Palace since the earth scorched Mataram and VOC troops until it was abandoned and vulnerable to being evicted and did not get the attention of the surrounding community, the potential of the Giri Palace area as part of an educational tourism object and sports area with hilly road conditions will be able to revive its role when this, it's just that the difficulty of accessing either four-wheeled vehicles or other facilities makes it possible.

It can be solved if the connectivity with the Pegat lake area becomes the beginning of the ecotourism movement in this area. As for the situation and condition of the artefacts of the ruins of the Giri Palace as shown in figure 5.

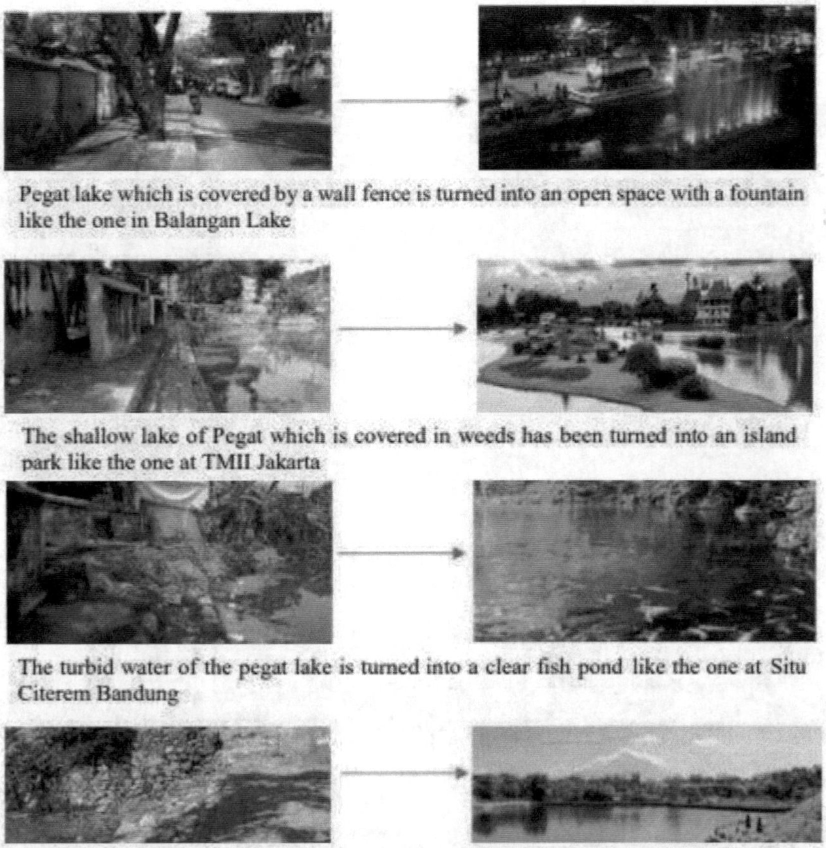

Pegat lake which is covered by a wall fence is turned into an open space with a fountain like the one in Balangan Lake

The shallow lake of Pegat which is covered in weeds has been turned into an island park like the one at TMII Jakarta

The turbid water of the pegat lake is turned into a clear fish pond like the one at Situ Citerem Bandung

The edge of the Pegat Lake which is covered with walls and piles of garbage has been turned into a pedestrian and connecting road to the Kedaton Site area and the Sunan Giri Tomb and Mosque Complex, such as in the Embun Tambak Boyo area, Yogyakarta.

Figure 5. The condition of the Pegat lake and the surrounding buildings, as well as an overview of the changes [22,23,24,25].

4.4. Pegat lake as the main water source of the royal city centre and the Giri Royal Palace

Telaga Pegat was built as an inseparable part of the life of the Giri royal palace which became the centre of control of the Giri kingdom government which was located on the Giri hill which was formerly a Luhur Islamic boarding school building. Giri Palace or known as Kedaton was built since Sunan Giri started building the Luhur Islamic Boarding School in 1478 AD and has been equipped as a Palace building since Sunan Giri was crowned as Prabu Satmata in 1487 AD as a consequence of the Luhur Islamic Boarding School building changing its function as the Giri Royal Palace, whose building is equipped with Taman Sari, Mosque and Islamic Boarding School. Giri Palace was destroyed and burned by the VOC who conspired with Prabu Amangkurat 2 of the Mataram kingdom in the 1612th century AD. The current condition is that the ruins of the foundation artefact which is terraced 7 and is at the top of Giri hill. The sustainability of the life of the royal city centre and the family of the King's Palace is very dependent on the source of clean water from this Pegat Lake. The dynamics of the life of the Giri Palace since the earth scorched Mataram and VoC troops until it was abandoned and vulnerable to being evicted and did not get the attention of the surrounding community, the potential of the Giri Palace area as part of an educational tourism object and sports area with hilly road conditions will be able to revive its role when It's just that the difficulty of accessing either four-wheeled vehicles or other facilities may be overcome if connectivity with the Pegat Lake area becomes the beginning of the ecotourism movement in this area. The situation and condition of the ruins of the Giri Castle artefact is shown in figure 6.

Figure 6. Topographical contours, atmosphere of the Kedaton Palace complex with Stairs and Ruins ex. Giri, Palace [21].

4.5. The Sunan Giri Mosque and Tomb as a regional generator node

The Ainul Yaqin Mosque, which is located east of Sunan Giri's tomb, was built during the time of Sunan Giri 2, this mosque is a form of respect for Kajeng Sunan Giri or Raden Ainul Yaqin as the opening of the Luhur Islamic boarding school and the founder of the previous Giri kingdom. Respect for the ancestors is believed to be able to raise the spirit in continuing the struggle of the ancestors as well as a form of worship. The enthusiasm of the people from all corners of the island of Java and Madura, especially to visit the Kajeng Sunan Giri tomb as part of the Wali Songo is very high in their daily lives, will result in an increase in the density of both vehicles and pilgrims, so that the convenience of access to facility services is often a complaint, especially high traffic jams. This problem will be mutually resolved if the potential for Religious or Pilgrimage tourism is connected to the Pegat Lake area as the initial node of the movement which is connected to the Giri Palace area and the Sunan Giri Tomb and Mosque complex properly supported by the arrangement of traders and guest houses along the area's generating corridor. Simulant and complement each other and overcome problems that arise together. The tomb and mosque complex of Sunan Giri which was built by adopting an architectural form that depicts the spirit of Sunan Giri in filling the spirit of Islamic teachings in the cultural traditions that developed in the Majapahit Hindu community is well illustrated, as shown in Figure 7 and the position of each building as in figure 7 and 8.

Figure 7. The atmosphere of the tomb and mosque complex of Sunan Giri [21].

Figure 8. The atmosphere of the Gate and the tomb complex, the Sunan Giri mosque, the pool and surrounding supporting facilities [21].

4.6. Pegat Lake as Centre for city centre economic orientation

The role of the Giri kingdom and the Luhur Islamic boarding school which is famous in remote parts of the archipelago, its influence is able to attract the arrival of merchants, travellers and students from various directions. The forerunner of this meeting was initiated because of the strategic position of the city of Giri as a stopover port which was spearheaded by a female merchant, namely Nyi Ageng Pinatih who was none other than the adoptive mother of Sunan Giri who was also the pioneer of the trading market in the kingdom of Giri. Meanwhile, the Pegat Lake area is strategically located between the Market, Bandar and the Kingdom Centre so that it is able to attract all newcomers to stop and spend the night. The arrangement of the shopping tourism supporting area between the trade centre and the market

IOP Conf. Series: Earth and Environmental Science 1007 (2022) 012014 doi:10.1088/1755-1315/1007/1/012014

is very important. The market heritage is still functioning and has been developed as shown in Figure 9,10 and 11.

Figure 9. The current atmosphere of the Nyi Ageng Pinatih market complex in Gresik.

Figure 10. The atmosphere of the tourist market that is in the mosque and tomb complex of Sunan Giri Gresik at this time.

Figure 11. The current atmosphere of the Gresik Harbour Gate.

4.7. Telaga Pegat as a meeting and cultural unity

The arrival of tourists, merchants and royal envoys to Giri and mingling in the Pegat lake area makes this area a meeting point between cultures that come, respect for the diversity of cultures is reflected in several artefacts, including tombs of different ethnicities. The picture of multi -ethnic life and culture is depicted by the presence of the tomb of the daughter of Compo and the goddess Sekardadu in the Giri hill area as seen in Figure 12.

Figure 12. The atmosphere of the tomb complex of Putri Compo and Dewi Sekardadu on Giri Hill.

4.8. Potential of Lake Pegat's socio-cultural landscape as sustainability.

From the analysis in the previous section, it shows that the potential of the Pegat Lake area is in accordance with the rapidly growing ecotourism concept at this time. Hijriyati E Ecotourism in question is a combination of various interests that grow from concerns for the environment, socio-culture, and economy [26]. Telaga Pegat has the potential to become an ecotourism-based tourism. Its activities are carried out in a simple way, including maintaining the authenticity of nature, the authenticity of art and culture, customs, habits of life, creating tranquillity, maintaining flora and fauna, and maintaining the environment so as to create a balance between human life and nature surroundings. Development is basically an effort to maintain a balance between the natural environment (biological and non-biological natural resources) and the built environment (human and artificial resources), so that the nature of the interaction and interdependence between the two remains in a balanced harmony.

Socio-cultural sustainability according to Sugandhy and Hakim is an activity that can be carried out through an effort to respect people's knowledge and local wisdom which has been the guide and adaptation [27]. Community participation according to Hadi is a process in which the community participates in taking part in decision making. In terms of quality, participation is an input for policy, strategy, communication, public solving media and social therapy [28]. Empowerment that prioritises customs and culture as well as community participation tends to be more effective. Meanwhile, according to Sedarmayanti, ecotourism activities that attract a lot of tourists have contributed foreign exchange for the country and have also opened up job opportunities for the surrounding community [29]. People not only get jobs and increase their income, but can also create new jobs that support living standards and improve people's welfare. The concept of arrangement in the simultaneous movement of the Giri area by uniting the potential for sustainability of Pegat Lake, Giri Palace and the Mosque Complex as well as the tomb of Sunan Giri as shown in Figure 13.

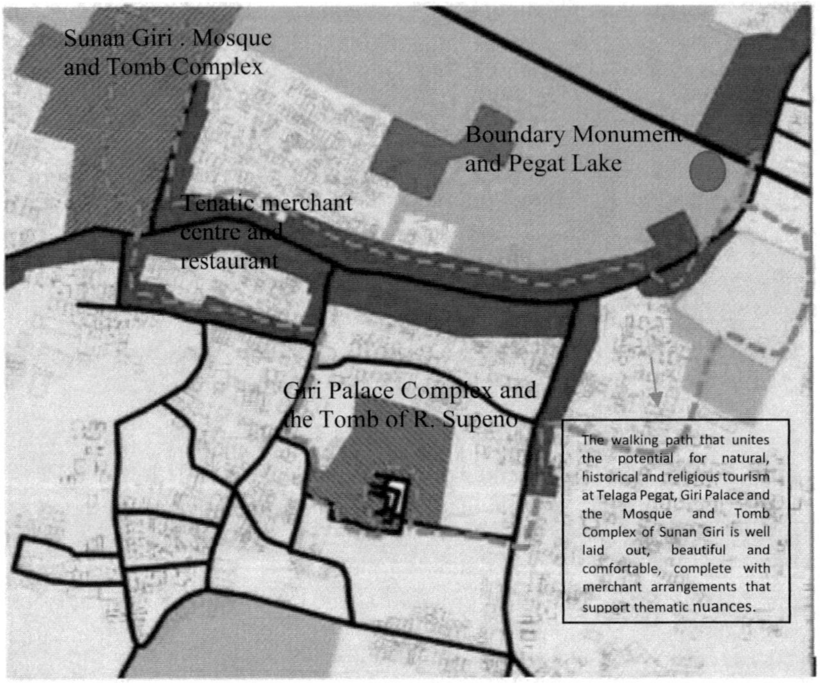

Figure 13. Conception of simultaneous movement of potential sustainability of Pegat Lake, Palace, Tomb and Sunan Giri Mosque as well as arrangement of pedestrian elements and other thematic traders [21].

5. Conclusions and recommendations

The conclusion from the findings of the potential sustainability of the social and cultural landscape of Pegat Lake as from the analysis of historical data, literature and site analysis as well as socio-cultural records of the life of the Giri kingdom community from four different backgrounds, it can be concluded that the meaning of Pegat Lake is currently experiencing a very large decline both denotatively nor connotative. The strength of the meaning and role of the past Pegat lake, especially for the sustainability of the socio-cultural life of the Giri kingdom community, can affect the position of the Giri kingdom against other kingdoms in the archipelago, which include: as the main source of clean water needs; meeting point for traders and preachers; orientation for the surrounding community and relatives who are anchored in the territory of the Giri kingdom either by sea or by land; The orientation centre of the Giri kingdom area. This finding can be concluded that the attachment and dependence between the Pegat Lake area, the Giri Palace Area and the Sunan Giri Mausoleum and Mosque Area which are connected to the movement pathways that are interconnected simultaneously and supported by supporting facilities such as residential areas, markets and open spaces are the key to their respective important roles. Each area including Pegat Lake.

While the recommendations from this paper are the development and arrangement of the Pegat Lake area in particular and the ex. the current Giri palace, must be bound in one simultaneous movement that supports each other by making Telaga Pegat a landmark and the centre of the main landscape orientation which was developed as the entrance to the heritage area of the Giri royal city centre which is adapted to the demands of a tourist area that unites the potential for religious, socio-cultural tourism and nature with a design that supports and complements each other.

6. Reference

[1] World Commission on Environment and Development 1987 *Brundtland Report 1987.* United Kingdom: Our Common Future. https://sustainabledevelopment.un.org/content/documents/5987our-common-future.pdf
[2] Holmberg J 1992 *Policies for a small planet: from the international Institute for environment and development* (London: Earthscan Publications)
[3] United Nations Conference on Environment & Development 1992. *Agenda 21.* Brazil: UNCED. https://sustainabledevelopment.un.org/content/documents/Agenda21.pdf
[4] Undang-Undang no 23 tahun 2014 tentang Pemerintahan Daerah
[5] Arias E G 1993 *The Meaning and Use of Housing International Perspective Approaches and Their Application* (USA: Avebury Ashgate publishing Limited)
[6] Noth W 1995 *Handbook of Semiotics* (USA: Indiana University Press Bloomington and Indianapolis)
[7] Rapaport A 1990 *The meaning of the built environment* (Tucson: University of Arizona Press) pp 256 ISBN: 0 8165 1176 4
[8] Broadbent G, Bunt R and Jencks C 1980 *Sign Symbols and Architecture* (New York: John Wiley & Sons Chichester)
[9] Gibbs R W 2004 *Intentions in The Experience of Meaning* (United Kingdom: Cambridge University Press)
[10] Rapoport A 1982 *The Meaning of the Built Environment* (Tucson: The University of Arizona Press) p 36-43
[11] Capon D S 1999 *The Vitruvian Fallacy* (England: John Willey and Sons Ltd)
[12] Soini K and Dessein J 2016 Culture Sustainability Relation : Toward a Conceptual Framework. *Sustainability Journal,* **8** (2), 167. doi:10.3390/su8020167
[13] Korten 1996 Sustainable Livelihood Social. *Development Review,* Vol. **1** No 2
[14] Benson J F and Roe M H 2000 *Landscape and Sustainability* (New York: Spon Press)
[15] Vezzoli C and Manzini E 2008 *Design for Environmental Sustainability* (London: Springer Verlag)
[16] Edwart T White 1985 *Site Analysis* (Tucson: Architectural Media)
[17] Rapoport A 1977 *Human Aspects of Urban Form: Towards a Man-Environment Approach to Urban Form and Design.* Urban and Regional Planning Series 15 (Oxford: Pergamon Publishing)
[18] Altman I and Chemers M 1984 *Culture and Environment* (Monterey California: Brooks/Cole Pub Company)
[19] Groat L and Wang D 2013 *Research Methods* (New York: John Willey & Sons Inc)
[20] Moleong L J 2010 *Metodologi Penelitian Kualitatif* (Bandung: Remaja Rosdakarya)
[21] https://www.google.com.br/maps/place/Telaga+Pegat
[22] Finopi Indonesia 2020 *Danau Balangan.* Finopi Indonesia. Retrived 01/01/2021. https://www.jasaairmancur.com/jual-air-mancur-di-danau-kab-balangan/
[23] Wonderful Indonesia 2016 *Taman Mini Indonesia Indah.* Indonesia Travel. Retrived 01/01/2021. https://www.indonesia.travel/tw/zh-tw/destinations/java/jakarta/taman-mini-indonesia-indah
[24] Muthalib A 2021 *Situ Cicerem.* Java Ttavel. viewed 2021.https://www.javatravel.net/situ-cicerem
[25] Sobat Jogja 2021 *Embung Tambak Boyo.* Wisata Alam Jogja. Retrived 01/01/2021. https://www.sobatjogja.com/embung-tambakboyo
[26] Fennell D A 1999 *Ecotourism an Introduction* (London: Routledge)
[27] Sugandhy A dan Hakim R 2009 *Prinsip Dasar Kebijakan Pembangunan Berkelanjutan Berwawasan Lingkungan* (Jakarta: Bumi Aksara)
[28] P Hadi 2013 *Community Empowerment Through Community Social Responsibility Activities* (Surakarta: Sebelas Maret University)
[29] Sedarmayanti 2005 *Manajemen Sumber Daya Manusia* (Bandung: Refika Aditama)

IOP Conf. Series: Earth and Environmental Science 1007 (2022) 012015 doi:10.1088/1755-1315/1007/1/012015

Sustainability principle in Nusantara architecture: case study of the Tongkonan House, the Betawi Stage House, the Gadang House, and Lamin House

M B Nabilunnuha[1], S Hervanda[1], G E Xian[1], A Tjong[1], F R Indarti[1], N E Nuffida[1], A Ardhyanto[2], D Novianto[1,3]

[1]Architecture Department, Faculty of Civil, Planning, and Geo-Engineering, Institut Teknologi Sepuluh Nopember. Kampus ITS, Sukolilo, Surabaya 60111, Indonesia
[2]Architecture Department, Environmental Engineering Faculty, The University of Kitakyushu, Hibikino 808-0135, Japan

Abstract. This study aims to analyze and identify the principles of sustainability in case studies of several traditional Indonesian architectures, or more specifically known as Nusantara Architecture. This paper discussed the identification, application of principal thinking, analysis, and observation through literature review and hypothesis. In these three case studies, which are the Tongkonan House, Panggung Betawi House, and Gadang House, we found that each building was represented by the choice of forms, materials, functions, and structures, which used the local resources, and as well as the meaning of architectural elements found in the details. During the modernization, the characteristics of the Nusantara Classical Architecture began to be applied to the current architecture which emerged as the Nusantara Contemporary Architecture. Through this paper, we encourage the complexity of architectural design knowledge that reflect the genius-logic, technological development, identity, lifestyle, ethical and aesthetic values of the Indonesian, which have been developed by the ancestors and it is important to always become a consideration and precedent for the architects, planners, and decision-makers in the future.

1. Introduction

1.1. Nusantara Architecture

Nusantara architecture is an understanding based on architectural philosophy and theory, which explores several aspects that exist in formal, tectonic, structural, representational, and aesthetic principles that influence the form of building composition well [1,2]. Nusantara Architecture is the nickname given to a series of vernacular architecture that can be found in Indonesia or the archipelago, built before 1800 BC and has the same characteristics, due to the similarity of climatic and geographical conditions in Indonesia and it is believed that Nusantara Architecture is basically Indonesian Classical Architecture [3,4,5]. In this article, the term Nusantara Architecture will refer to the wooden construction architecture which is the main structure. In order to revive the Nusantara Architecture, Indonesian people must compose it, not duplicate it [1,2]. As Indonesian architecture evolves to adapt to modern standards, architects may look for various ways to incorporate uniqueness into their designs. Can architecture recreate tradition, a common ground upon which the criteria for authenticity and quality are based?

[3]Corresponding author: didit.novianto@its.ac.id

Nusantara architecture, which is rich in local wisdom, the result of the development of tens of generations, might be the answer. Meanwhile, contemporary architecture is a design style that is based on everything that exists today and a combination of architecture at some time. Thus, contemporary architecture can always keep up with the times and is dynamic [6].

1.2. Sustainable Architecture

Sustainable Architecture, defined as a concept that supports a sustainable environment, is a concept of preserving natural resources nature in order to be able to stand longer, which fills with the age of potential resources nature and human ecological environment, such as planetary climate system, agricultural system, industry, forestry, and of course the architecture. There are two main objectives of sustainable architecture, such as: Sustainable buildings must minimize the impact on the environment; Buildings must be able to make a positive contribution to the social environment in it, by addressing community needs while improving environmental quality [7].

2. Research method

This research and study use a descriptive-qualitative method which is expected to be able to produce data in accordance with the conditions in the field. The data obtained comes from literature studies conducted, both from expert opinions, journals, and drawing documents, considering that there is still very little published literature on Nusantara Architecture. The case studies that are appointed as research objects are not to be compared, but to be studied and compared based on each characteristic of each region where the architecture is located. The discussion focuses on principles related to sustainability and resilience, both in terms of meaning and updating, examples of buildings/architecture which, according to the author, adopt the concepts of Nusantara Architecture.

3. Case study

3.1. The Tongkonan House

The Toraja ethnic is one of the ethnic groups in Indonesia. The Toraja ethnic lives in the mountains to the north of South Sulawesi Province, with a population estimated at more than one million people. The Toraja ethnic is scattered in several areas, including Mamasa, North Toraja, and Tana Toraja. One of the cultural treasures of the Toraja ethnic is the Tongkonan House, which is their traditional house. Tongkonan comes from "Tongkon" which means "to sit". Sitting is defined as a place to gather with their family. Even though it's categorized as a traditional house, the Tongkonan House has been able to implement a sustainable design and has good environmental resistance. The sustainable design of the Tongkonan house is a form of architectural response to natural conditions, climate and geography in the area. Natural conditions have a major influence on Nusantara architecture. In terms of structure and construction, Tongkonan House also implements a sustainable construction system and has good resistance. Due to its sustainable and resilient architectural design, Tongkonan Houses are able to survive until now, even adapted, and developed nowadays. From the analysis table it is known that the lighting in Tongkonan houses depends on natural lighting. There are several responses from this building to maximize the potential for lighting. Such as the direction of the orientation of the building, the use of gaps in wood material, to the placement of windows and openings in the eastern and western parts. The intensity of irradiation and long exposure duration (12 hours) in a day are used as the main source of lighting during the day.

Also, from the analysis of climatic conditions, it is known that the average temperature in South Sulawesi ranges from 22°C-32°C. The air conditions tend to be warm and hot. According to data released by the Passive House standard, it is stated that a comfortable temperature for humans is 25°C. From these data, it is known that there is a gap between the thermal conditions of South Sulawesi and the comfortable air temperature. Forms of adaptation to this situation by maximizing natural ventilation from the wind. In the Tongkonan House, there are many wall openings and gaps that allow air to enter.

3.2. The Betawi Stage House

This case study discusses the Betawi traditional house, where the Betawi traditional house has two types of houses, namely land houses, and stilt houses. In Asia Pacific Region, there were some basic similarities in traditional architecture that the structure consists of bottom-primary-top structure [4]. What will be discussed this time is about the traditional house on stages from Betawi, which was built on land from the river coast to the coast. The form of the building of this Betawi stage traditional house has a building mass that stands on its feet above the ground. Its function is to defend against puddles and waves on rivers or beaches, especially on the Ciliwung river.

The condition of the community requires sufficient resources to meet the needs of life, there are many waterways in the city of Jakarta. Many people build their houses on the banks of the river, which serves to prepare for floods that occur due to overflowing rivers and to avoid wild animals at that time. The dominant material used in Betawi stilt houses is dominated by wood originating from coastal resources, making it easier to find and search for building materials [10]. With various considerations, there are natural situations and conditions that affect the shape of the structure and construction of this Betawi traditional house on stilts, explained in figure 2.

Table 2. Structural anatomy and characteristics of Betawi Stage House [11].

Element	Characteristic
Head	Wooden roof truss with response elbows, a combination of saddle and shield; Ceiling from woven bamboo (*bilik*); Roof from red brick tile
Body	House foundation frame placed on *umpak*; Floor from lined up bamboo; Walls from boards that are lined up tightly with nails; Rooms framed by 4 main columns
Base	Ground pavement with coral fragments; The building supports are made of jackfruit wood which is built on *umpak*

3.3. The Gadang House

Gadang House is a Minangkabau traditional house that has a long rectangular plan, with several gable roofs forming the ends like buffalo horns. Gadang House is usually built close to the previous family house. So that, in an area, usually there is still a kinship relationship [12]. The process of building a Gadang House includes deliberation and notification of the construction of Gadang House in a village [13].

Figure 3. Gadang House floorplan and façade
(left: floorplan; middle-right: typical faced).

In the Minangkabau custom, one of the requirements for an area is having *Babalai Bamusajik* or the existence of a meeting place and a mosque for religious and scientific activities and providing public facilities in it with a consideration of natural resource utilization as we know that Minangkabau tribe very concern with nature. The Grand Mosque of West Sumatra or Masjid Mahligai Minang by Rizal Muslimin, which is located in Padang, is one of the unique places of worship for Muslims. Every Minangkabau society competes to build and prosper mosques. Thus, the mosque becomes a center for social activities. The mosque is the main building of Mahligai Minang which took and re-wrote the art and architecture of the building "Minangkabau in the early civilization era".

Table 3. Reinterpretation of Traditional Minangkabau Value in Masjid Mahligai Minang.

Reinterpreting tradition	Minangkabau 's values	Meaning	Principles	Application to objects	Images [2]
Reinterpreting the values contained in the construction of the Gadang with a new form without losing the original meaning.	Deliberation.	The existence of a deliberation system, namely *saiyo sakato* to reach a consensus at every traditional event such as establishing a Gadang House in a village.	Reinterpretation.	Space that allows occupants interact with one another (wide room without partition)	
	Cooperation.	Togetherness in finding solutions in a village to achieve a goal that is not detrimental to various parties.	Combination.	The roof's form indicates a combination of Gadang's roof and the value of Prophet's turban when placing *hajar aswad* to emphasize the concept of togetherness	
	Respect for nature.	Nature is the source of all life. From nature back to nature.	Transformation.	Natural energy utilization and interpreting the nature values in Minangkabau vernacular architecture by transforming it into a new form	
Reinterpreting the values contained in the construction of the Gadang with a new form without losing the original meaning.	Kinship.	Uphold a harmonious kinship	Defamiliarization	Interior space without partition increases family value.	
	Religion.	Adat basandi syara', syara' basandi Kitabullah, which means tradition and syara' to achieve perfection	Combination	The application of togetherness, integration, deliberation, faith, and nature.	

3.4. The Lamin House

Lamin House is the Nusantara architecture of Dayak Kenyah tribe located in the Long Nawang Subdistrict, Bulungan Regency, East Kalimantan Province, to be precise around Apo Kayan or in the upper reaches of the Mahakam and Kayan Rivers, where the river flows across different districts and reaches into the Sulawesi Sea. The tribe originated from the beaches of Kalimantan but then resided upstream due to conflict with an invasive Malay tribe [14]. From a geophysical perspective, in general, the East Kalimantan region is dominated by undulating topography, from gentle to steep slopes, with a height from sea level ranging from 0-1,500-meter s, with a slope of 0-60%, which is more separating the slope class of more than 40%.

The climate experienced by the Kenyah tribe on a daily basis is tropical, with two seasons that alternate every year, which are the rainy season and the dry season, and the air tends to be hot and humid, with the lowest rainfall of 74 and the highest of 306. The soil found on their territory is mainly peat soil which contains high organic matter, and its ability to hold large amounts of water so that it is fertile and usually overgrown by rainforests. The types of trees available near the habitation of the Kenyah tribe include ironwood and meranti trees, whose wood is known to be strong and durable when used as a house construction material, so it is often used in the construction of Lamin Houses [15].

The Lamin House is a structure with an approximate length of 300m, 15m width, and 3-4m height, which can hold 25-30 family groups in one house. They are usually built near the riverbank, along the river with the entrance facing it. The roof is unusually a gable roof, which tiles are constructed from ironwood planks, with dragon ornaments attached to its ends. The walls are made from ironwood bark, and are supported by columns of ironwood up to 6m in length, and decorated with Dayak ornamentations. The structure is elevated from the ground, accessible using stairs, with the space under the house used for keeping livestock [16].

Figure 4. Lamin House structure and exterior.

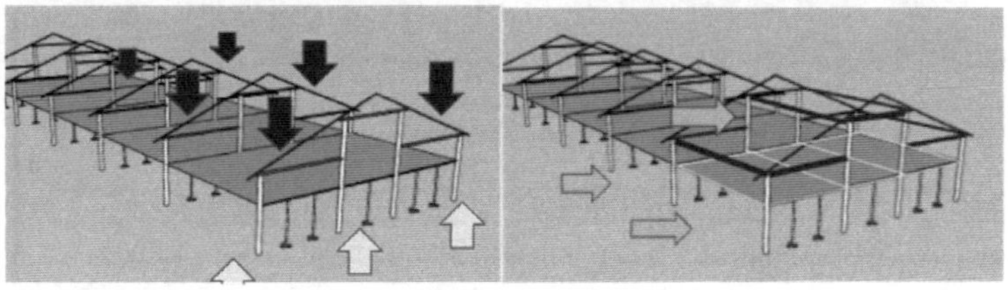

Figure 5. Load distribution analysis.

4. Conclusions

Nowadays, sustainable architecture is one of the things that is intensified in future development. Sustainable architecture is important which can support the concept of sustainability, which is preserving natural resources so that they can be maintained and have a long-term lifetime. However, without realizing it, the concept of sustainable architecture has in fact been applied for a long time, one of which is in buildings spread across the archipelago, Indonesia [17]. In this paper, the author tries to criticize the principle of sustainable architecture in Nusantara Architecture buildings with several case studies, namely the Tongkonan House, Betawi Stage House, Gadang House, and Lamin House. From the case studies that have been criticized, it can be concluded that the principles of sustainable architecture can be applied in various aspects, both structure, material, style, and meaning in the building to respond to the surrounding climate and society in order to achieve human comfort. The complex design knowledge that reflects the logic of genius, technological civilization, identity, lifestyle, and aesthetic value, developed by past builders is priceless and can be considered as a precedent for future architecture. Where the principle of sustainable architecture is maintained and adapted to today's modern buildings. As for the sustainable architecture development in Indonesia, the basic physical design characteristic is almost the same as Nusantara architecture, but it is visually more simple, modern, and uses nowadays material technology.

5. References

[1] Prijotomo J 2020 *Webinar Kebhinekaan Arsitektur Nusantara#6*. Accessed 12 January 2021. Video available at: https://www.youtube.com/watch?v=Q4QnRF8KTG0&t=2780s

[2] Abdel H 2020 *Umah Hati Villa/ Studio Jencquel*. ArchDaily. Accessed 12 January 2021. available at: https://www.archdaily.com/954050/umah-hati-villa-studio-jencquel

[3] Octavia L and Prijotomo J 2018 Arsitektur Nusantara bukan Arsitektur Tradisional maupun Arsitektur Vernakular. *Jurnal Lingkungan Binaan Indonesia,* **7** (4) pp 249-253. Accessed 12 January 2021. available at: https://jlbi.iplbi.or.id/arsitektur-nusantara-bukan-arsitektur-tradisionalmaupun-arsitektur-vernakular/

[4] Bakhtiar, Waani J O and Rengkung J 2014 Tipe Teori Pada Arsitektur Nusantara Menurut Josef Prijotomo. *Jurnal Media Matrasain,* Vol. **11** No 2 (Manado: Media Matrasain)

[5] Koesmartadi C 2018 *Mengenal Arsitektur Nusantara Lebih Mendalam.* Portal Berita Unika Soegijapranata. Accessed 12 January 2021. available at: http://news.unika.ac.id/2018/05/mengenal-arsitektur-nusantara-lebihmendalam/

[6] Lim W and Tan H 1998 *Contemporary Vernacular: Evoking Traditions in Asian Architecture* (Singapore: Select Books)

[7] Kurniawan R A and Pamungkas L S 2020 Penerapan Arsitektur Berkelanjutan (Sustainable Architecture) Pada Perancangan Taman Budaya Di Kabupaten Sleman. *Jurnal Arsitektur,* Vol. **2** No 1 (Yogyakarta: Universitas Teknologi Yogyakarta Press) p 35-39

[8] Sir M M 2015 *Pengetahuan Tektonika Arsitektur Tongkonan.* (Makassar: Lembaga Sejarah Arsitektur Indonesia)

[9] Sir M M, Wunas S, Herman P and Parung H and Patandu J 2011 *Model Tektonika Arsitektur Tongkonan Toraja* (Makassar: Universitas Hasanuddin Makassar Press)

[10] Lim W and Bang T 1998 *Contemporary Vernacular*. Microlibrary Warak Kayu by SHAU Indonesia (Singapore: Tien Wah Press) available at: https://www.archdaily.com/936421/microlibrary-warak-kayu-shauindonesia?ad_source=search&ad_medium=search_result_all [Accessed 11 Januri2021] DA 184505 - Q - Nusantara Architecture Autumn 2020-2021

[11] Wardiningsih S 2015 Arsitektur Nusantara Mempengaruhi Bentuk Bangunan yang Berkembang di Indonesia. *Physics of Semiconductor Devices,* **2** (2) pp 274-283 Sze S M 1969 (New York: Wiley–Interscience)

[12] Nurdiah E A 2011 *Studi Struktur dan Konstruksi Rumah Tradisional Suku Batak Toba, Minangkabau dan Toraja.* Master Thesis. Universitas Kristen Petra. Surabaya. 09/Pen.Arsitektur/UKP/2011 http://repository.petra.ac.id/id/eprint/15677

[13] Prijotomo J 1995 *Analisa Formal Bagi Identifikasi Jatidiri* Proceeding for Nusantarian Architecture, Change and Continuity 1995. Surabaya

[14] Prasetyo Y and Astuti S 2017 Ekspresi Bentuk Klimatik Tropis Arsitektur Tradisional Nusantara dalam Regionalisme. Jurnal Koesmartadi C and Prijotomo J 2020 Ciri-ciri Struktural-Konstruksional Arsitektur Kayu Nusantara. *Prosiding Seminar Struktur dalam Arsitektur*

Pemukiman, **12** (2). Accessed 12 January 2021. available at: http://jurnalpermukiman.pu.go.id/index.php/JP/article/view/149

[15] Purnamaria E, Azizah S and Ramadhani S 2019 *Pendekatan Arsitektur Vernakular Rumah Lamin Pada Desain Kompleks Studio Photography Etnik Kalimantan Timur di Samarinda*. Accessed 12 January 2021. available at: https://ejurnal.itats.ac.id/sntekpan/article/view/615

[16] Noviana M 2013 Konsep Arsitektur Berkelanjutan Arsitektur Vernakular Rumah Lamin Suku Dayak Kenyah. *Jurnal Kreatif,* **1** (1). Accessed 12 January 2021. available at: http://ejournal.polnes.ac.id/index.php/kreatif/article/view/70

[17] Sato K 1991 Menghuni Lumbung: Beberapa Pertimbangan Mengenai Asal-usul Konstruksi Rumah Panggung di Kepulauan Pasifik. *Journal Universitas Indonesia Antropologi,* **49** Pp 31-47. http://journal.ui.ac.id/index.php/jai/article/viewFile/3291/2578

Acknowledgment

This research is part of the Final Project on Nusantara Architecture Class, Department of Architecture, Sepuluh Nopember Institute of Technology 2021. The authors gratefully acknowledge financial support from the Institut Teknologi Sepuluh Nopember for this work, under project scheme of the Publication Writing and IPR Incentive Program (PPHKI). Also, we would like to acknowledge and express our deepest gratitude to all members of our class, as well as our family for their amazing support and understanding throughout our research and writing project.

Rehabilitation of substandard housing program's benefits: lesson learned from Plandi Village, Jombang Regency

N Pusporini[1,2], W R Palupi[1], H R Santosa[1], D Septanti[1], and S Cahyadini[1]

[1]Architecture Department, Faculty of Civil, Planning, and Geo-Engineering, Institut Teknologi Sepuluh Nopember. Kampus ITS, Sukolilo, Surabaya 60111, Indonesia

Abstract. The rehabilitation of substandard housing (Rumah Tidak Layak Huni or RTLH) program is an effort of both central and local governments to provide decent housing for the low-income community. Over the past four years, Plandi Village has received substandard housing rehabilitation grants in the form of building materials and cash. This research will describe the physical and non-physical benefits felt by the beneficiaries using a qualitative descriptive method based on interviews, literature studies, and related documents. The results show that the beneficiaries can experience physical benefits in the form of a house that is suitable for habitation which is sturdy and healthy in providing enough light and sanitation as well as a standardized building floor area. Meanwhile, the non-physical benefits obtained are improved and more functional space to carry out daily activities in the house, even increasing their social status.

1. Introduction

The Indonesian Government, through Law Number 1 of 2011, is responsible for protecting the entire nation in the provision of adequate and affordable housing. Both the central and local governments are obliged to provide facilities, especially for low-income communities, in obtaining decent houses. One of the efforts to realize this responsibility is through the rehabilitation of substandard housing programs. The program is intended for low-income substandard self-help housing owners so that their houses become standardized based on government regulations.

The local government of Jombang Regency, through its vision "*JOMBANG BERKADANG*" (Jombang with character and competitiveness) program, helps the village through financial assistance in the form of physical development, one of them is the rehabilitation of substandard housing. The Housing and Settlement Service is responsible to carry out the program alongside the Construction of Individual Toilets [1]. Plandi Village is one of the targetted villages.

In 2017 - 2020, Plandi Village has carried out the rehabilitation program. The source of funds for the aid program comes from the central government through the Village Fund (Dana Desa) and the local government through the Village Intention Budget (Pagu Intensi Desa). Each village is obliged to propose as many as 4 (four) beneficiaries each year.

In "Housing as a Verb", Turner defined housing as a process [2]. A house cannot be seen as a once-completed physical result but as a continuing process related to social, economic, and occupant mobility in a period. The community must better regulate the housing procurement process to create a better residential environment. In the process of building a house, a distinction must be made between the physical form of the house and the non-physical activities it can accommodate. The government and

[2]Corresponding author: npusporini83@gmail.com

institutions will not be able to provide affordable housing if they only look at the house from a physical perspective, because the more limited resources, the more expensive the house will be. On the other hand, the poor can build their own houses. Even though they are economically deficient, they can build individually or collectively and have more free time to develop their houses.

Rehabilitation of substandard housing programs is in line with the view of housing as a process (see Figure 1). Whereas in the context of self-help housing owned by low-income communities, the government and the community (and the owner) are actors of the housing rehabilitation program (activities) and standardized housing as the achievement. The expected feedback is the fulfilment of the physical and non-physical needs of the house.

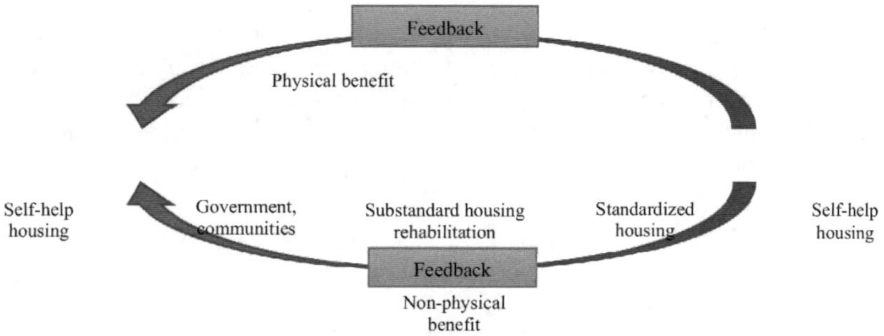

Figure 1. Definition of housing as a process in substandard rehabilitation housing program [2].

Based on the Indonesian statute, a self-help house is a house built on the initiative and efforts of the community. Self-help houses consist of independent houses, government-assisted and institutionalized self-help houses [3]. In this research, a low-income community built a self-help house, then received a rehabilitation program so that it became a government-assisted house.

Through the supervision mechanism, the implementation of the program has been technically and administratively successful. However, can it also be called successful if it is viewed from the perspective of the beneficiaries? What benefits can beneficiaries feel from the programs? Or does it impose additional financial burdens for beneficiaries, as happened in India? Loan programs for home improvement have added to the debt burden for the poorest communities [4]. Therefore, it is necessary to study the benefits of the program based on the experiences of the beneficiaries.

2. Methodology

The method used in this research is descriptive qualitative. This method seeks to explain a phenomenon in general and in-depth. The technique used is to dig up information based on observations, interviews, literature, and related documents. They are then processed and analysed. Interviews were conducted with village officials and four beneficiaries from Plandi Village. The sample taken in this study is 1 of 4 houses that are renovated every year, intending to know the progress of renovations every year.

Aspects and variables studied include:

1. Program implementation stages, including requirements for beneficiaries and the financing process;
2. Physical benefits, including three aspects (based on government regulations):
   a. Meeting building safety requirements: building structural components, and quality of the roof, floor, wall covering material;
   b. Meeting health requirements: adequacy of light (natural lighting), ventilation (air circulation), provision of sanitation (indoor bathroom);

c. Having a minimum area: 7,2 m^2 per person.

The physical components are assessed qualitatively to obtain the most beneficial component for improvement using the Social Network Analysis (SNA) method. SNA can be interpreted as a study of human/ community relations with graph theory [5]. An application that is a tool for creating SNA structures is Gephi 0.9.2, to map small-world networks. SNA focuses on patterns of relationships between actors and describes the network relationships [6]. In this case, the actors used are physical components that have a role as objects of recommendations for improvement in the coming year.

The use of the SNA will be carried out in 2 stages. First, assess perceptions of the building safety components to obtain recommendations for the most important components to be repaired. Perceptions of building safety components are assessed using a scale of 1 (Not Important) - 10 (Very Important).

Second, studying social structure helps us to understand how actors cluster or group in social space [7]. From the selected components, the Inter-Component Relations Adjacency Matrix is formed. The relationship in a graph is expressed in the form of an Adjunct Matrix. The relationship value between components in the matrix is processed to obtain the Eigen Centrality value, which determines the cluster nucleus.

3. Non-physical benefits, according to functions and activities in the house. Functions can be determined by knowing the number of rooms, and activities can be classified as basic as food preparation, eating, and bathing [8]. The need for house space according to Indonesian standards, regarding Space Matra, consists of a sitting room, dining room, bedroom, kitchen, bathroom, toilet, ironing room, and warehouse. While, Sahid states that the types of space in a house type consist of six groups of spaces, namely 1) living room, family room, dining room; 2) the bedroom is a closed and separate space; 3) kitchen; 4) bathrooms and latrines which are closed spaces 5) storage rooms; 6) garage. The basic activities that must be accommodated in a simple house are sleeping, sitting, eating, cooking and toilet activities [9].

Next, Visibility Graph Analysis (VGA) is carried out. It is used to test how efficient and effective a spatial arrangement is [10]. The analysis is done using depthmapX. It provides 3 methods, namely visual analysis, agent analysis, and axial and segment analysis. However, this study uses only agent analysis.

Agent analysis allows depthmapX for simulating different types of pedestrian behavior by providing various ways for an agent to choose where to walk, such as towards larger spaces, long lines of sight, or parts of their view that are occluded. The movement simulation is carried out in the interior of the house with 5,000 steps. The number of agents is adjusted to the number of residents in the house.

The analysis results are shown through colour gradations (see Figure 2). The low-value parameter will be indicated by the blue colour gradation towards the green, where the green colour gradation towards yellow is the intermediate value, and the yellow colour towards the red colour gradation is the highest value [11]. The higher the value indicates the most accessible space.

Figure 2. Colour gradation and value parameter [12].

3. Results

3.1. Plandi Village Overview

Plandi Village is one of the villages in the Jombang District. Administratively, Plandi Village covers an area of 1,115 km^2 and populates 7,717 people in 2019. Geographically, it is located on the urban border of Jombang so that the settlement facilities and infrastructure are complete. Facilities provided are health facilities such as community health centres and pharmacies; educational facilities from kindergarten to higher education; two mosques in each hamlet, one of which is a historical mosque; two funeral sites; village government offices; and several security posts. Meanwhile, basic infrastructures and utilities that have been fulfilled are electricity, drinking water system, nice condition of the asphalt roads except for the rocky agricultural roads, also, a working technical irrigation system that supports the people's economy for agriculture. Regarding individual sanitation, majorities of the residents already have private toilets at home. It is because, during these 4 years, alongside the housing rehabilitation program, there was also an allocation of funds for the construction of 87 units of private toilets.

3.2. The Substandard Housing Rehabilitation Program and Financial Support

The substandard housing rehabilitation program will be given to the community with the following criteria:

1. The condition of the house is unfit for habitation. Priority for houses with non-permanent conditions, such as bamboo walls and soil floors;
2. The land is privately owned;
3. Are underprivileged people/ low-income communities.

The house selection is made by the village government which will be reported to the central government. If it is approved, there will be a disbursement of funds for rehabilitation.

From 2016 to 2019, the program received funds from the central government (Village Fund). Funds are handed over directly to the village government. The village government was responsible to choose the beneficiaries and the budget given by looking at the physical conditions of the house. Then it is reported to the central government. However, the central government provides a budget limit for each house between IDR 10,000,000 (US$ 706.9) and IDR 20,000,000 (US$ 1,413.81). The amount of budget obtained by each house is optimized for the cost of house rehabilitation. The funds obtained are used to buy construction materials and labour costs by using individual workers or non-governmental organizations who receive the grant.

In 2020, the source of funding comes from the local government through the Village Intention Budget. A budget of IDR 16,000,000 (US$ 1,131.05) for each house (IDR 2,500,000 (US$ 176.73) for workers' wages and IDR 13,500,000 (US$ 954.32) for building materials. The central government has provided references and standards for standardized house repairs. So, the village government is responsible for selecting houses that match the standards given by the central government.

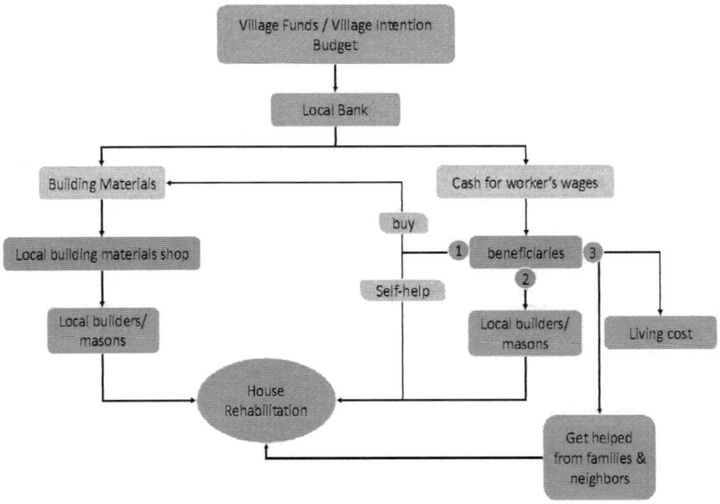

Figure 3. Grant distribution and house rehabilitation process.

Figure 3 shows the process of grant distribution and house rehabilitation. Grants from the Village Fund or Village Intention Budget are channelled through local banks. Then it is divided into a building material that is directly distributed to the nearest building materials shop, and cash for workers' wages. Cash is given to beneficiaries which can be used in three ways. First, beneficiaries have the expertise to build their own houses and use the money to buy additional building materials so that the finishing of their houses is better. Second, the beneficiary was unable to build their own houses, so the grant is used to hire local builders. Third, beneficiaries are not able to build on their own but receive assistance from neighbours and families. In this case, they can use the cash for living expenses.

The grant, although it may appear small when compared to the cost of commercial housing, it is significant for beneficiaries. Raw materials for the building are given from the nearest building shop. The cash for labourers is mostly converted into building materials. It is because the beneficiaries can use their energy to build. While the elder beneficiaries are received helps from neighbours and relatives.

The selection of building materials is based on availability at the nearest building material shops. It is following the principle of using local resources for building a more affordable house [13]. In addition, if beneficiaries are unable to build their own houses, they employ local workers. Channelling aid through this method can help the local economy because money is circulating at the district level.

The provision of this kind of grant is very suitable for the recipient's economic condition and does not result in additional debt or conditions that are choking for their financial condition. However, one of the beneficiaries suggested that the amount of the grant should be adjusted to the initial conditions of the house. And it had to be increased every year based on the rising inflation of building material prices.

3.3. House Conditions Before and During Rehabilitation

The majority of houses that receive the housing rehabilitation program are in non-permanent building conditions. Figure 4 is a picture of the house's conditions before receiving the grant.

2017

2018

2019

2020

Figure 4. House conditions before rehabilitation 2017, 2018, 2019, 2020 [14].

From Figure 4, it can be seen that the conditions of the houses are non-permanent. In 2017, the house was made from bamboo walls and GRC (Glass Reinforced Fibre Concrete) roofs. The structure was not strong enough, so it is prone to collapse during a disaster and cannot protect against crime. In 2018, the house was already using brick walls and tile roofs. But, there are some hollow parts due to incomplete construction. So, it is classified as a substandard house. In 2019, the house was made from bamboo walls, tiled roofs, and plaster floors. In 2020, the house was made from brick walls and tiled roofs. But, it is in bad condition and has a lot of damage.

The beneficiary's house conditions are different each year. In 2017, the house's conditions were the worst and slightly better with only minor damages. The village officers said that each year, it is getting hard for them to select the houses for the program because the number of non-permanent houses is decreasing.

During the building process, it is carried out by the beneficiary's community independently. A group of neighbours and local builders worked together to maximize the existing budget. Some of the workers were not paid, just given food and water while working.

Figure 5 shows images of the building process in Plandi Village. The walls were replaced using bricks and the roof using a new GRC with a wooden truss. The vents are added, as well as the windows. The door and window frames are made of wood.

Figure 5. House conditions during rehabilitation 2017, 2018, 2019, 2020 [14].

3.4. Physical Benefits

Physical feedback concerns material aspects that can be measured and the specific needs of the house [13]. The following is a physical description of the house and the home owner's interview regarding the perceived physical benefits:

1. The house resulting from the rehabilitation has met the building safety requirements. The substructure (foundation) uses concrete with a composition of cement, river stone, and sand. The floor is cemented. The walls are made from bricks and cement, but not painted. The windows and doors are using wood. The roof frame is made of wood, and the roof itself is made of GRC.
2. The house that has been rehabilitated has guaranteed health. Each house is given two windows and vents, to provide air circulation and natural light. If the minimum house area of 30,89 m^2 is combined with a ceiling height of at least 2,8 m, it still meets the requirements for fresh indoor air for two adults and two children [9]. For houses that do not have sanitation or that are in bad conditions, there is a sanitation development with an indoor place for bathing and washing.
3. The house that has been rehabilitated has met the minimum area of 7.2 m^2 per person. The land areas after rehabilitation are 30 to 56 m^2. With a maximum of four family members, the minimum area has been fulfilled. Moreover, the minimum area of the rehabilitated house is still following the reference for the minimum area of a simple house. The area of the house is determined by the old house foundation or by the width of the owned land.

The rehabilitated house provides security and fulfils the main function of the house as a shelter. The beneficiaries feel more comfortable and secure in the solid conditions of their house. The existence of solid walls makes them more comfortable and does not allow outside air to enter the house. During the rainy season, there are almost no house leaks found. Before rehabilitation, they often could not sleep because of the large amount of water that entered their house, and sometimes even their property was damaged due to wetness. Some people who previously did not have toilets were very happy with the rehabilitation of this house because it also provided toilets for houses that did not have private ones. They find it very easy because they no longer need to go to public toilets, which are also quite far away.

Moreover, to examine the components that give the most physical benefits, the social network analysis method is used. The aim is to describe the perceptions of the beneficiaries of which components provide the most benefits to them and their relationship to other components. Analysis of the relationship between components can also produce clusters of physical aspects that can be used as aid packages in the future.

The quality of the roof, the floor, the wall covering material, and the air circulation are the components that provide the most physical benefits (see Figure 6). These results indicate that after rehabilitation, the owners experienced improvements in the roof, walls, and floors when compared to their previous house. The addition of windows and vents has provided benefits in terms of adequate air circulation.

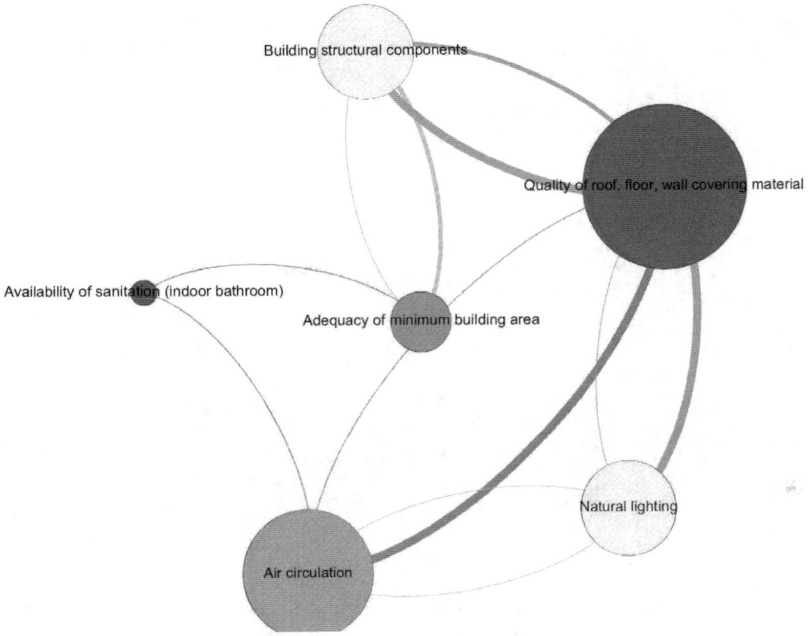

Figure 6. Network of physical benefit components.

Based on the perception of the relationship between components, two clusters were formed (see Figure 7). The SNA structure created with Gephi Software can describe the pattern of relationships in a network where nodes can be people, keywords, which in this study are building safety components.

The number of nodes indicates the most preference. The width of the line shows how big the relationship between the nodes is. The colour indicates modularity, meaning similarity (clusters/groups). The first cluster consists of quality of covering material, air circulation, building structure, and natural lighting. The second cluster consists of air circulation, quality of covering material, natural lighting, and indoor bathroom. The formation of clusters can be used as a basis for emphasizing future physical improvements.

The initial condition of the house affects the type of repair. The two clusters can be used as the basis for prioritizing budget allocations. If the initial housing conditions are non-permanent, then the first cluster is more appropriate. If the initial housing conditions are semi-permanent, then the second cluster can be used. So that the final results of repairs can be more targeted and satisfy the homeowner.

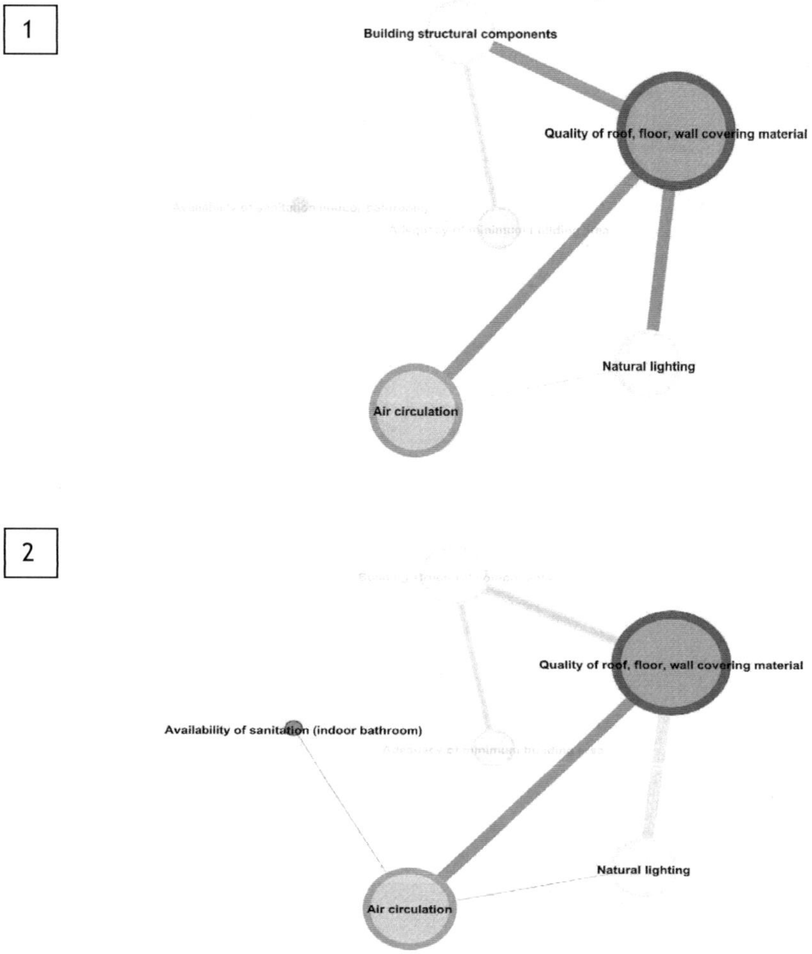

Figure 7 Clusters formed.

Housing is a process allowing people to help themselves in providing their own house [15]. In the rehabilitation program, the beneficiaries have control over the housing construction process. Their involvement in the construction process and the physical benefits of the house have created a sense of pride and a sense of belonging to their home. Although there are cases where the grant is not sufficient to perfect the covering material, the beneficiary tries to solve it with his money.

3.5. Non-physical Benefits

Non-physical feedback in the definition of housing as a process is a vital or non-material need beyond the physical needs of housing. In this study, the vital feedback is translated as non-physical benefits of the house, which are manifested in activities or use of space after rehabilitation. Furthermore, it will also explore the psychological and economic benefits that can be felt by the beneficiaries.

House 1: 1 occupant. House 2: 2 occupants.

House 3: 3 occupants. House 4: 4 occupants.

Figure 8. Actual use of space (left) and simulation by agent (right).

Based on the interview, the rehabilitated house consists of a bedroom, living room, bathroom, and kitchen. Four of the six divisions of the house have been fulfilled. There is no dedicated storage room and garage. The storage function is integrated into other rooms. While the garage function, usually only for motorcycles, uses a terrace or the living room.

The use of space in the house is indicated by activities carried out in the house shown in Figure 8. Basic activities in a simple home are sleeping, sitting, eating, cooking, and toilet activities. The results of the interviews showed that residents were able to do more than basic activities. Analysis of space usage will be carried out through comparisons between the actual use of rooms and the simulation use of rooms by the agent(s). The number on the left plan shows the activities of the occupants: (1) sleeping, (2) napping, (3) resting, (4) chatting, (5) cooking, (6) washing dishes, (7) washing clothes, (8) eating, (9) drying clothes (10) taking a bath, (11) watching TV, (12) studying, (13) receiving guests, and (14) parking. The colours on the plan on the right indicate agent movement and space accessibility.

The actual use of space in all houses shows that most activities are in the kitchen. While the simulation results show the high movement in houses 1 and 3 are in the living room, houses 2 and 4 are in the bedroom. The room with the most movement in the room that has the greatest access among the other rooms. A comparison of space usage shows that there is no synchronization between the activity load and space position. This condition is a problem that needs to be solved when planning the floor plan in providing rehabilitation assistance.

However, the problem of spatial position was not viewed as a negative thing by the beneficiaries. They are grateful because, after rehabilitation, there was a clear division of rooms. They feel that rehabilitation improves their standard of living and the function of their house. Previously, some activities could not be done inside the house. Now with an indoor bathroom and functional kitchen, they can do bathing and washing clothes in the bathroom, even drying their laundry in the kitchen. They are also very comfortable receiving guests who come to their house without shame after they have proper flooring and living room. It makes them feel that their social status increase. Psychologically, beneficiaries can feel the benefit of the program. However, economically, there is no immediate benefit. There is no increase in income due to home improvements. The effect is more on comfort and diversion of consumption from purchasing building materials to home contents or furniture.

4. Conclusion

The government's effort to provide adequate housing for all, especially low-income people, is through the home rehabilitation program among others. The implementation of the program can be evaluated in terms of the principle of housing as a process. Housing as a process in this research starts from a substandard self-help house built by low-income people, then receives aid from the government through a rehabilitation program. The rehabilitation process involves beneficiaries and nearby community groups to help the building process. The results of the program are expected to provide physical and non-physical benefits to beneficiaries. These benefits are described as feedback from program implementation, obtained through interviews with beneficiaries. The physical benefits successfully fulfilled the function of the house as a shelter. The rehabilitated houses now have a secure building structure, meet health requirements and standard minimum floor area. The non-physical benefits obtained are a clear division of space and an increase in activities that can be done inside the house.

The rehabilitation program has succeeded in helping low-income people to have a decent home. However, two things need to be improved for the program in the future. First, the allocation of funds should be allocated more on the quality of roofs, floors, and wall covering materials. Second, house plan development needs to consider the activities that will be carried out by residents.

5. References

[1] Regent of Jombang 2019 Regulation of the Regent of Jombang Number 10 of 2019 concerning the Jombang Program with Character and Competitiveness

[2] Turner J F C 1972 Housing as a Verb. *Freedom to Build, Dweller Control of Housing Process* (New York: The Macmillan Company) p 148–75

[3] Lejone J N 2011 *Self-Help Housing in South Africa : Paradigms, Policy and Practice.* Doctor Thesis. University of the Free State. Bloemfontein

[4] Datta K and Jones G A 2012 Housing and Finance in Developing Countries. *Housing and Finance in Developing Countries.* https://doi.org/10.4324/9780203217689

[5] Tsvetovat M and Kouznetsov A 2011 *Social Network Analysis for Startups.* Edited by Shawn Wallace and Mike Hendrickson. Zhurnal Eksperimental'noi i Teoreticheskoi Fiziki (Sebastopol: O'Reilly Media, Inc) http://scholar.google.com/scholar?hl=en&btnG=Search&q=intitle:No+Title#0

[6] Oktora R and Alamsyah A 2017 Interaction Patterns and Actors Who Have Most Role in the 2013 JGTC Event Through Twitter Social Media (Study Using Social Network Analysis Method). *Jurnal Manajemen Indonesia*, Vol. **14**, Issue 3, p 201. https://doi.org/10.25124/jmi.v14i3.370

[7] Nugroho Y and Saritas O 2009 Incorporating network perspectives in foresight: A methodological proposal. *Foresight,* **11** (6), 21–41. https://doi.org/10.1108/14636680911004948

[8] Belcher J C and Vazquez-Calcerrada P B 1972 A Cross-Cultural Approach to the Social Functions of Housing. *Journal of Marriage and Family,* **34** (4): 750–61

[9] Suryo M S 2017 Analysis of Minimum Area Requirements in Simple Landed Houses in Indonesia. *Jurnal Permukiman,* **12** (2): 116–23. http://jurnalpermukiman.pu.go.id/index.php/JP/article/viewFile/62/pdf_1

[10] Wahyuningrum T and Prabawati I 2014 *Evaluation of the Rehabilitation Program for Unfit for Living (RTLH) in Depok City*. Magister Thesis. Universitas Indonesia. 1304067402
[11] Turner A, Doxa M, O'Sullivan D and Penn A 2001 From isovists to visibility graphs: A methodology for the analysis of architectural space. *Environment and Planning B: Planning and Design,* **28** (1), 103–121. https://doi.org/10.1068/b2684
[12] Pinelo J and Turner A 2010 *Environmental Modeling: Using Space Syntax in Spatial Cognition Research* UCL Depthmap 10. Portland. Retrived January 10th, 2021. https://www.academia.edu/15667098/Introduction_to_Depthmap10_SC2010
[13] Turner J F C 1977 Housing by People: Towards Autonomy in Building Environments (New York: Pantheon Books) p 1-205
[14] Beneficiaries' collection 2020
[15] Cohen M 2015 John F.C. Turner and Housing as a Verb. *Built Environment,* **41** (3): 412–18. https://doi.org/10.2148/benv.41.3.412
[16] Indonesian National Standard Agency 1990 SNI **03- 1979-1990** Specifications for Space Matrix for Houses and Buildings
[17] Imran M 2016 Environmentally Friendly Construction Materials. *Procedia Engineering Journal,* **6** (2): 146–57
[18] Law of the Republic of Indonesia Number 1 of 2011 concerning *Housing and Settlement Areas*. Vol. **11**. https://doi.org/10.16194/j.cnki.31-1059/g4.2011.07.016
[19] Mangeswuri D R 2016 Policy on Housing Loan through Housing Loan Liquidity Facility (FLPP). *Jurnal Ekonomi & Kebijakan Publik,* **7** (1): 83–95
[20] Marsh A and Gibb K 2014 *Housing Economics* (SAGE Publications Ltd) https://doi.org/10.4135/9781446261958
[21] McDonald J F 2015 Affordable Housing: An Economic Perspective. *Architecture_MPS,* Vol.**7**(1). https://doi.org/10.14324/111.444.amps.2015v7i3.001
[22] Metzel S 2009 *Turner's 3 Laws and Housing Microfinance*. Retrieved February 8th, 2021. http://housingandmicrofinance.blogspot.com/2009/07/turners-3-laws-and-housing-microfinance.html
[23] Pinelli A and Favalli L 1973 Allylisopropylacetamide-induced porpyria-protective effect of 3,5-dimethylisooxazole. *Biochemical Pharmacology*, Vol. **22**, Issue 7. https://doi.org/10.1016/0006-2952(73)90358-4
[24] Pramitha P P 2016 Performance Evaluation of Substandard House Rehabilitation Program in Purbalingga Regency. *Journal of Chemical Information and Modeling*, **53** (9): 1689–99
[25] Rizka R K 2010 *Evaluation of the Implementation of the Home Improvement Program Unfit for Living in Surakarta City* 1–148
[26] Taryono 2012 Introduction to Economic Development Theory. *Pustaka.Ut,* 1.2-1.51. http://repository.ut.ac.id/4234/1/MMPI5204-M1.pdf
[27] Winarno B 2018 Evaluation of the Program for Provision of Liveable Houses for the Poor in Belitung Regency. *Jurnal Pengembangan Kota*, **6** (1): 66. https://doi.org/10.14710/jpk.6.1.66-74

Housing challenges in sinking coastal city: rethinking urban housing in subsidence area for a more resilient community

R S Hamdani[1,3], S P Hadi[1], and I Rudiarto[1,2]

[1] Graduate Study Program of Environmental Science, School of Postgraduate, Diponegoro University
[2] Department of Urban and regional planning, Faculty of Engineering, Diponegoro University

Abstract. Housing in coastal cities is facing complex environmental hazards. The risk of being permanently submerged caused by sea level rise-related flooding is aggravated by the localized hazards, which is land subsidence. We explored the land subsidence conditions in the housing zone of Semarang City using overlay spatial analysis, and we expand the discussion to the regulatory aspect and local adaptation through content analysis. We found that challenges in housing zone are the severe subsidence that is occurring on dense housing zone, regulation that still less sensitive towards incorporating the slow onset disaster and not properly provided housing hydrological consideration nor the water provision, and poor drainage system as well as limited adaptation capacity that makes people lives in a false harmony of being resilient. Affordable alternative housing such as stilt house and drainage infrastructure improvement as well as detailing and reframing building regulation that is adequate to address both subsidence and flooding is undeniably important to ensure the adequate housing is accessible for all coastal city residents.

1. Introduction

Coastal area, where most populated cities are located, is in the front line in facing uncertainties and fatalities caused by changing climate. In Indonesia, the largest archipelago country in the world, 150 cities where 60% of the national population resides are on the coasts [1,2]. Impact of global hazard exposure such as sea level rise that threatens coastal cities had also been recently raised by Joe Biden, the USA President, which brought this issue to be discussed in various national-level debates [3]. Most of the important cities in Indonesia are located on the northern part of Java Island and are susceptible to this environmental challenge. Such as Jakarta, the national capital in the western part of this island and Semarang City in the centre, which is chosen to be the study location, are experiencing local land degradation that is making the situation become more severe.

The local irreversible geo-environmental degradation, land subsidence, is a vertical displacement of the upper soil layer [4–6], which is mainly caused by unsustainable groundwater extraction and building loads that exceed the bearing capacity of the soil [6–8]. This resulted in higher coastal flooding risk [9], forcing houses and land to be submerged and abandoned [10–12], and the most severe: bringing the coastal areas to be completely permanently flooded [13].

Coastal housing is one of the most vulnerable zones in facing this challenge. Rapid urbanization in Semarang had resulted in increasing residential and industrial zones [14]. Coastal communities that are severely affected by land subsidence and tidal floods are dominated by low-income households [15]. If the industrial areas had the financial capability to develop alternative plans, residents in sinking coastal areas are on the other edge. Coastal housing is not equipped by adequate drainage and sanitation

[3] Corresponding author: rizkiana.sidqi@gmail.com

infrastructure, as well as limited capability of the people to do such costly adaptation measures. In addition, people's reliance on coastal natural resources makes their income become vulnerable because it is highly affected by the climate condition [16].

The global mission on creating safe and resilient housing, indicates the importance of this issue. The Sustainable Development Goals Target 11.1 explicitly stated that "we shall ensure adequate housing is available for all" [17]. Housing itself is one of basic human rights [18]. While on the other side, as the population increases, the housing needs also increase. Consequently, within the urban context, we commonly found the dense housing increment. Density might be perceived as a land efficiency measure, but speaking of environmentally, higher density increases stresses to resources [19].

Adapting to flooding is the most viable way to provide proper housing to coastal communities. Regarding our study case, there were lots of households choosing to rather adapt and live in harmony with the floods and other coastal hazards than to move to a safer place [20]. To academically criticize this phenomenon, the discussion in this paper started by exploring theoretical understandings on how ideal resilient housing is. According to Shelter Cluster's 8 Build Back Safer Key Messages [21], preventing houses from sinking showed how the houses shall be built on strong foundations. They also mentioned the importance of giving more weight to the house, so it could hardly be attacked by strong wind and storm, flooding, or ground motion such as earthquake. That framework also emphasizes to consider the housing location and avoid building housing in floodplains. People's awareness is also mentioned in the framework as the basis in determining the most appropriate measures.

Through the lens of settlement resilience, infrastructural aspects shall also be considered in upgrading the coastal housing. Sen *et al.,* [18] stated that drainage infrastructure plays a crucial role in making resilient housing as it is able to reduce the severity of floods, lowering the potential loss and risk, as well as longer the infrastructure lifespan (if only be followed by proper maintenance). They later added that this drainage infrastructure system is more important than the structural housing measures itself. Considering infrastructure in assessing coastal communities' vulnerability had previously been done by Rudiarto and Pamungkas [16]. They argue that the higher the availability of drainage infrastructure, would reduce the climate-related flood risk. In addition, they said that access to sanitation infrastructure also contributes to forming the adaptive capacity of the people since its adequacy may help people get rid of any pollutants surrounding people's houses.

Speaking in the context of policy, Aquino *et al.,* [22] discussed that the Sendai Framework Disaster Risk Reduction is mentioning the importance of building codes regulation, a part of land utilization control in spatial planning, to be clearly mentioned in the policy. It could be used to address the poor housing structure and construction. They also synthesized that there are two important rules that shall be included in the policy, they are: (i) implementing house improvement measures by respecting, educating, and gaining citizen knowledge; and (ii) the importance of stipulating disaster-prone area zones.

Prior studies within the academic field of coastal housing are mostly done through micro perspective approaches. Ameir *et al.,* [23] qualitatively explored cultural-based architectural adaptation in Bandarharjo, a village in Semarang City where the most severe subsidence occurs. They focused on understanding the local genus. Local housing strategy had also been explored by Saharom *et al.,* [24], which resulted in recommendation of building more stilt houses as it was regarded as the most relevant alternatives. Both studies rarely discuss housing-related policies. Research that included discussion about governmental programs and policies had been done by Dewi *et al.,* [15]. They quantitatively assessed the physical quality of the settlement using a set of resilience indicators and found that housing in those areas is categorized as less resilient which they mentioned as slums. Regarding the policies, they descriptively explain about programs that had been implemented and factors that affected its success or failure. Although this research had gone further from micro views on housing to more macro views, they still overlooked the spatial planning policy which actually is the core of every governmental program as said by Handayani *et al.,* [14] that it *"include various infrastructure provisions under the local government authority".* On the other hand, the urge to do integrated management is rising as the scholars and policy makers started to realize the complexity of coastal hazard that threatens the coastal community's future [25], as well as the importance of incorporating appropriate resilience housing strategy into the policy [26]. Accordingly, intertwining micro-level adaptation and policy in understanding critical coastal housing issues in coastal cities is needed.

This research is, therefore, aiming to portray a comprehensive understanding of coastal housing challenges through discussing the micro aspect as well as looking from the policy perspective. The intention of this research is to understand the macro context by assessing the land subsidence rates and condition in the residential zones, identify the current policies (zoning, building codes, etc), and criticize current adaptation measures taken by the people.

2. Research method

This research is descriptive research which deployed a mixed methodology using both quantitative and qualitative. This approach enables enrichment and complement between one result and another [27], which is aligned with our aim in doing this research. We combined the result of spatial analysis, field observation, and also enriched the discussion through regulatory and literature reviews. The mixed method research was previously done by Sen *et al.,* [18] in evaluating the implementability of housing infrastructure resilience framework. The mixed method that is being dominated by quantitative studies was also used to assess current land subsidence management in Semarang city.

The quantitative method is done to achieve the first research objective which would be discussed in point 3.1 by doing overlay analysis using an open-source spatial analysis software, the QGIS. We used the land subsidence data year 2007-2010 and 2015-2017 from Yastika *et al.,* [28], and we overlayed it with the housing zone stipulated in Semarang City's Spatial Planning as extracted from the government's official One Map site <https://distaru.semarangkota.go.id/semarang/index.php?webgis=tataruang>. The information about increasing or decreasing rates per specific housing zones characteristics is gained from through extracting the Attribute Table. The analysis continued by qualitative descriptive analysis using content analysis techniques in understanding the building codes and other relevant regulations within the residential zones that would be elaborated with the previous analysis. The regulation that is considered in this research is the Semarang City Local Regulation Number 14/2011 about Semarang City's Spatial Planning year 2011-2031. Phrases like "spatial planning regulation" or "spatial planning" mentioned in this paper are referring to this regulation. The content analysis is also used to review related literature about housing in the subsidence area. We use the computer aided qualitative data analysis software (CAQDAS), the Atlas.Ti 9.1., which is able to help us read through, coding, and highlighting relevant sentences/paragraphs, before we finally could illustrate our synthesis result through a code-network graph. Lastly, we also used the classic triangulation technique to intertwine between one finding and another. Our study took place in Semarang City. A city that is located at the north-centre of Java Island as seen in Fig. 1 below. This city is directly adjacent to the Java Sea. Although this city located more than 400 km to the west away from Jakarta, they both had the same anthropogenic challenges which is the groundwater overexploitation induced land subsidence, that increase their vulnerability towards sea level rise and coastal flooding.

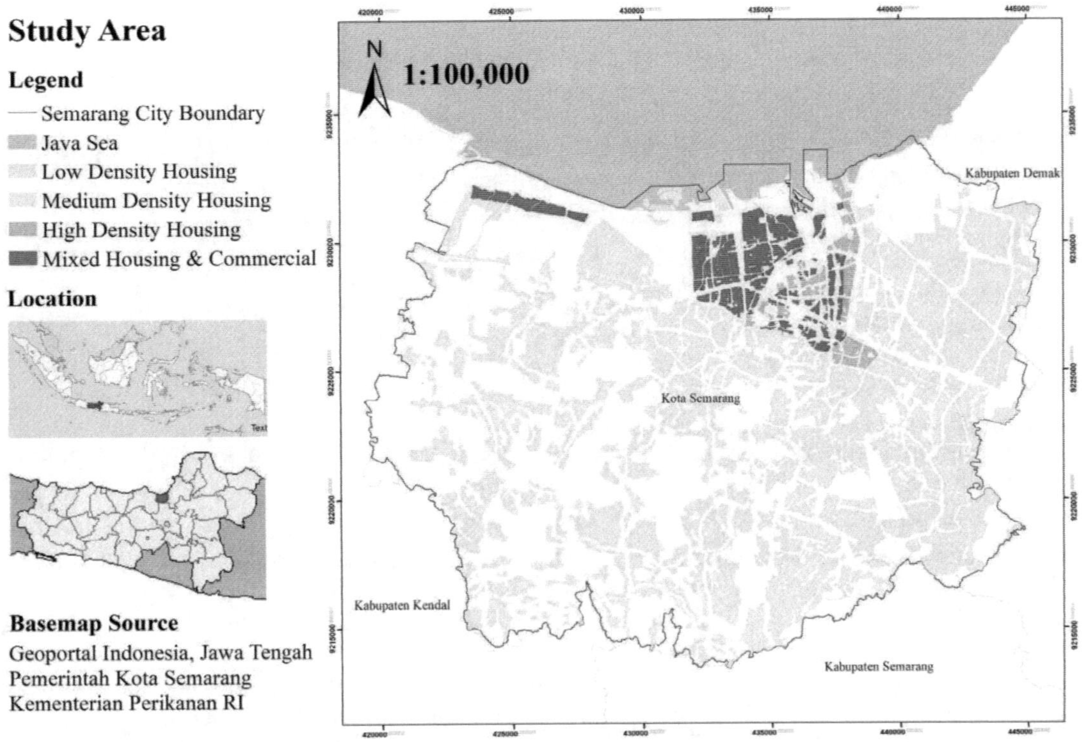

Figure 1. Study area map with the spatial pattern of housing zones as stipulated in the spatial planning

3. Result and discussion

Challenges on housing in the sinking areas are discussed under three main points, they are: (1) the denser the housing the higher the land subsidence rates; (2) the conventional building codes regulation needed to be tightened up; and (3) current adaptation measures are needed to be improved. The following section would discuss more about these points.

3.1. Rethinking dense housing

Housing in Semarang City spatial planning is categorized into four zones as mentioned in the Table 1 below, they are: (i) low-density housing (LDH); (ii) medium density housing (MDH); (iii) high density housing (HDH); and (iv) mixed housing and commercial (MHC). Spatially, Semarang City housing zone is dominated by the LDH and MDH. The rest zones are mainly located in the city centre (see Fig. 1). Based on our overlay analysis, the rates in all four zones are increasing which means that the latest rate is higher than the previous period. The highest increase showed in the MDH zones. Although the MDH had the highest increase, its rate is lower than HDH and MHC. The MHC zones facing high rates increment. The lowest subsidence rates change, which indicates the most stable subsiding zones, is in the low-density housing. In summary, the highest subsidence rates occur in the highest housing density zones. According to the spatial pattern as shown in Fig. 1., the lowest density is sparsely located in the Mijen and Gunung Pati district, which geologically is composed of a stable layer of rocks, meanwhile the housing zone that is dense and mixed, are in the north-eastern part of the city, which is consists of young alluvium sediments soil layers that is vulnerable to subsidence due to its natural compaction and unsustainable groundwater utilization.

Table 1. Land subsidence rates change in housing zone

Zone	Median Land Subsidence Rates*		Changes*	Note**
	2007-2011	2015-2017		
Low Density Housing (LDH)/ *Perumahan Kepadatan Rendah*	0,12	-1,67	-1,79	Increasing
Medium Density Housing (MDH)/ *Perumahan Kepadatan Sedang*	-35,75	-55,73	-19,98	Increasing
High Density Housing (HDH)/ *Perumahan Kepadatan Tinggi*	-49,62	-68,27	-18,65	Increasing
Mixed Housing and Commercial (MHC)/ *Perumahan Perdagangan dan Jasa*	-44,78	-59,84	-15,06	Increasing

*All rates in this table mentioned in mm year^{-1} unit
**Increasing rates means the rate in the second period is higher than the previous period

As we explored in the literature, it could be understood why higher density could lead to higher subsidence rates. We graphed the synthesis process as shown in Figure 2 below. Mulyana *et al.,* [29] explained that the dense housing zones have a high population. Consequently, the water demand is high. Unfortunately, the people shall fulfil it through extracting water from the ground as the local water supplier is inadequate to provide the piped water [30]. Spatial correlation research by Prasetyo *et al.,* [31] showed that the shallow aquifer change is highly correlated with the high-density housing areas. It strengthens our findings where there is a negative interaction between current dense housing and the land subsidence rates. Besides, the more houses built on a land, it consequently increases the building loads [8,30]. Where then, the suggestion to increase the house weight in order to make the housing more resilient based on the 8 Build Back Messages [21], is not properly suitable to be implemented here in the subsidence vulnerable areas.

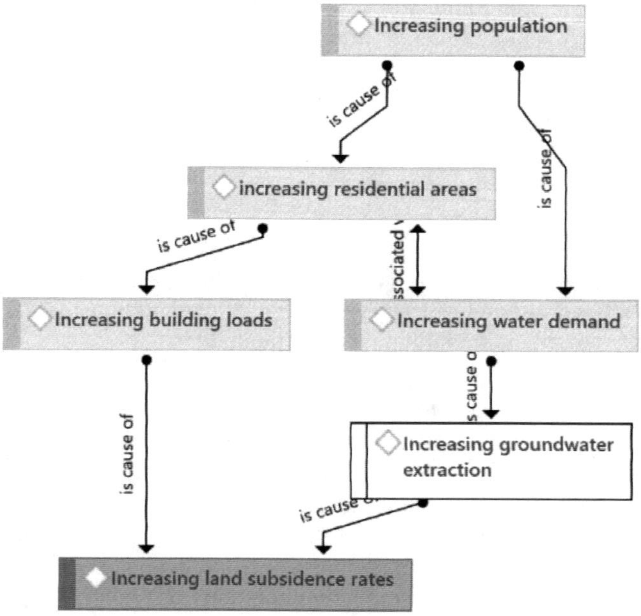

Figure 2. The conceptual framework in understanding how housing might affects land subsidence rates synthesized from [10] [30] [32] [33] [34] [35] [36] [37]

Urbanization in the means of increasing population and increasing urbanized areas also influencing the land subsidence rates. More housing is developed as a response to more people coming and living in Semarang. Historically, population increase has happened since the 1990s and people continued to occupy the north-eastern part of the city, the coastal area [38]. Increasing urbanized area is explained through the increasing density in the existing housing zones and the land use change from the green area to the built environment [39]. Assessed by Indrawati *et al.,* [39] that the urban expansion itself is increasing in the year 2011-2015. This could be the reason behind land subsidence rates increasing within the observed period. The impact is that dense housing brings the community closer to high vulnerability. As the ground is lowered and the sea level keeps rising, it leads to higher risk of flooding occurrence [29]. Vulnerability assessment was also proved that high vulnerability is found at the dense housing area [16,40,41].

3.2. Revisiting spatial planning policy

Our prior study Hamdani *et al.,* [42] found that land subsidence is currently not incorporated in spatial planning. In this continued study, we found that in City Official's One Map website, we could easily find the land subsidence map. This map states the severity of land subsidence in the whole city. But unfortunately, this map is not officially mentioned as the disaster-prone area that is stipulated in the spatial planning regulation. Consequently, the detailed requirement of appropriate housing in subsiding areas is not yet regulated. This empirical evidence is in contrast with what has been said in the literature.

This incorporating regulation is also portrayed in the building codes mentioned in spatial planning, which is loose yet comprehensive. The building coverage ratio (BCR) and floor area ratio (FAR) is mentioned generally based on the sub-city region which is defined by district administrative delineation. The maximum BCR for HDH, MDH, LDH are mentioned as 80%, 60%, and 40% respectively. The maximum FAR itself is 1.8 for all housing zones, with a maximum three floors. With no further requirement on how the building shall be built regarding the local challenges on every different part of the city. We assumed this general law is caused by the level of the regulation itself that a macro-level that still needed more detailed regulation.

Environmental consideration in developing housing areas has been mentioned in this general law. The development of a single house "shall consider the environmental carrying capacity". This mandate is also mentioned in the article that regulates the housing developers. Beside also mentioning about the minimum zone for infrastructure (40%) and green area (20%), the developers shall also consider the hydrological aspect. Where they should make a water retention pond or well and make sure that the water runoff exempted from the developed housing area is minimised. This hydrological consideration is not mandated to personal house owners that are living by their own right to land and to the building.

The housing zone that is experiencing highest subsidence rates, is also having the highest BCR, vice versa. This finding may help us to rethink about developing resilience and sustainable cities that is often considered a high-density area is an important aspect. Moreover, the latest spatial planning we reviewed, was previously reported that it intended to increase the building density [43]. The scrutiny on how the dense housing is implemented sustainably is also crucial, at least, on how it utilized the groundwater or other subsurface resources. Besides, it is also shown that the regulation aimed to create resilience housing by ensuring space for water recharge in every parcel, might also be further evaluated. Exploring between detailing more the regulation or strengthening the implementation and control shall also be done to achieve the ideal aim of creating sustainable dense cities.

3.3. Reframing the perceived harmony

People in coastal areas had a perceived harmony to live with the hazards. They accepted their local condition. This belief brought them to be willing to take any affordable adaptation measures. As we explored through previous literature and field observation, we see that the perceived harmony is needed to be reframed. Where the measures taken by the people and the government are leading nowhere but to deteriorate the housing's environmental condition.

Bringing back the discussion to the literature we had mentioned in the Introduction, it could preliminary concluded that there is an indication of not yet resilience housing. As the dense housing subsides, it indicates that the house structure and surrounding infrastructure could not be able to keep the houses safe neither from being flooded nor sunk. This is the contrary of what should have been done based on the 8 Build Back Messages [21]. Adaptation measures taken by the local people are referred to as false harmony in the literature, because as they elevate their houses and surroundings the load given to the land is increased. This would cause the subsidence and flooding to become more severe.

Poor drainage system makes this condition more severe. Uneven land subsidence might cracks the infrastructure. Correlated with the social and financial unjust ability to cope with this hazard, it caused uneven housing's ground level as seen in figure 3. Drainage system in this kind of environment is challenging. Where now this challenge still could not be solved yet by the respected stakeholder, resulting in trapped water that is easily found through the field observation. Prior flood water that could not be exempted back into the sea or the drainage system caused by the different ground level is the reason why the water is trapped. This is commonly found in a subsiding house parcel that is being abandoned by the owner, as shown in figure 4. This static water could be the source of environmental diseases [44]. This finding showed the importance of drainage systems in increasing the resilience and improving the quality of living in the subsiding housing areas by Sen *et al.,* [18].

Figure 3. The uneven ground level of housing caused by land subsidence and different ability to raise the house foundation

Figure 4. The trapped water in a subsided and abandoned house

Still in the context of drainage failure, housing challenges are also coming from river overflooding. Our field observation that was done right after major flooding in Semarang coastal area on February earlier this year [45], found that the people in flooded neighbourhood is aware that the cause of flooding is caused by their land that is subsiding along the years, but they are more concern towards the river flooding and extreme rainfall rather than the sea level rise exposure or the land subsidence itself. Their concern might be caused by their awareness about their housing area's elevation that is already lower than the river embankment (showed in the lower left of Fig. 4) that makes them become more prone to river flooding, and their housing area is already surrounded by mangrove barrier, so they hardly bothered by the sea level dynamics (showed in the lower right of Fig. 4). This housing area is in Mangkang Wetan village, at the western part of Semarang. The perceived resiliency caused by community-developed mangrove barriers is also mentioned by Hadi [46]. This mangrove barrier is not only supported by the community's initiative but also the seed funding from the government and international assistance [47]. This housing is near the industrial area, but when we look back to the housing zone plan in Fig. 1, we could see that this area might not be explicitly stated as a housing area. This zone was flooded for more than two days at that time. As what had been said by one of the inhabitants there, the situation we have seen at the observation period is already not as severe as the first day of the flooding. So, we assumed that the flooding conditions were more devastating than these pictures shown below. The trapped water also found in this area (upper middle part of Fig. 4), where showing no sign to be able to recede. The uneven ground level which impacted the flooded environment was also found in the field as shown in the upper right of Fig. 5. Prior studies had also mentioned these drainage challenges. Marfai *et al.,* [44] stated that this poor drainage system is found in the dense housing area that is experiencing land subsidence. Rudiarto and Pamungkas [16] had also found that the dense housing without adequate drainage system is the place where inundation occurs.

Figure 5. Flooded housing area in Mangkang Wetan

People's awareness has the potential to be further developed. Aforementioned field observation result explains how the people are already aware of hazards that threaten their life. This corresponds with their understanding about the consequence and measure that they should take. On the other hand, they also needed financial support to be able to do effective measures well as achieving economic justice to prevent the uneven ground level caused by the unjust financial capacity. As we came into the flooded area, the people there were expecting us to give them something. This is another issue that needs to be more focused on.

Proper knowledge is important in facing these challenges. Sen *et al.,* [18] had explained about how people's understanding could help the recovery phase become faster. It also contributes to preparedness for the future disaster. Besides, the people hardly implement the resilient housing structure ideas, such as a stilt house, because it isn't affordable [18].

4. Conclusion

We concluded that the challenges on housing issues in the sinking coastal city are found at least in three aspects: the unsustainable density, the regulation that is neither compliant with the local condition nor being implemented as how it was planned, as well as the limited housing alternative implementation due to financial restriction. Density is one of the most important characteristics of a city to be sustainable and resilient. But when it is not accompanied by adequate infrastructure or proper regulation and people action, it could become a disaster as what we have found in our quantitative analysis. The regulations needed to be tightened up based on the local challenges, where in this case we emphasized shall be considering more the land subsidence issue. On the local level, people had the awareness but were not fully able to do the proper measures as they had limited financial capabilities. Despite all the challenges, local people's willingness to do whatever it takes to save their environment means that they are not abandoning the ship at all. Any assistance, financially, academically, and socially, could be given to this initiative society.

This research expands the discussion on housing issues in the sinking city, as it is able to comprehensively discuss the macro to micro context. Nevertheless, finding in this study may somewhat be limited as it started from the lens of spatial planning regulations, which often hardly portray the actual land use zoning delineation. Further study is needed to dig more in any kind of housing related planning policies, analysing adaptation patterns for every different housing zone might be useful to help develop better building codes regulation. Above all else, financial constraint is also needed to be researched more in order to be able better resilience housing implementation. The consideration of environmental carrying capacity in private housing also needed more discussion in the future. And as the environmental condition is continuing to deteriorate (indicated by the increasing land subsidence rates), the implementation of the building codes regulation is questionable and consequently, also need further research.

5. Conflict of interest

The authors declare that they have no known competing financial interests or personal relationships that could have appeared to influence the work reported in this paper.

6. References

[1] Rahmat A, Syadiah N, Subur B 2016 *Smart Coastal City: Sea Pollution Awareness for People in Surabaya Waterfront City* Procedia - Soc Behav Sci. **227** November 2015 pp 770–7
[2] Rudiarto I, Handayani W, Setyono JS 2018 *A regional perspective on urbanization and climate-related disasters in the northern coastal region of central Java, Indonesia* Land. **7** 1
[3] Muthiariny DE, Afifa L 2021 Aug 4 *Joe Biden's Remark on Jakarta Sinking; Megawati Asks Jokowi to Seek Solutions* Tempo.
[4] Marfai MA, King L 2007 Oct 25 *Monitoring land subsidence in Semarang, Indonesia* Environ Geol. **53** 3 pp 651–9
[5] Pipkin B, Trent D, Hazlett R, Bierman P 2008 *Subsidence and Collapse* In: Geology and The Environment. 7th ed. Cengage Learnings; 2008.
[6] Widada S, Zainuri M, Yulianto G, Satriadi A, Wijaya Y 2020 Nov 1 *Estimation of Land Subsidence Using Sentinel Image Analysis and Its Relation to Subsurface Lithology Based on Resistivity Data in the Coastal Area of Semarang City, Indonesia* J Ecol Eng. **21** 8 pp 47–56
[7] Chaussard E, Amelung F, Abidin HZ 2012 *Sinking cities in Indonesia: space-geodetic evidence of the rates and spatial distribution of land subsidence* Proc 'Fringe 2011 Work Frascati, Italy, 19–23 Sept 2011 (ESA SP-697, January 2012). September 2011 pp 19–23
[8] Abidin HZ, Andreas H, Gumilar I, Sidiq TPP, Fukuda Y 2013 Sep *Land subsidence in coastal city of Semarang (Indonesia): characteristics, impacts and causes* Geomatics, Nat Hazards Risk. **4** 3 pp 226–40
[9] Andreas H, Z. Abidin H, Gumilar I, P. Sidiq T, A. Sarsito D, Pradipta D 2018 Dec 12 *Insight into the Correlation between Land Subsidence and the Floods in Regions of Indonesia* In: Natural Hazards - Risk Assessment and Vulnerability Reduction. IntechOpen; 2018.
[10] Bott L-M, Schöne T, Illigner J, Haghshenas Haghighi M, Gisevius K, Braun B 2021 Oct *Land subsidence in Jakarta and Semarang Bay – The relationship between physical processes, risk perception, and household adaptation* Ocean Coast Manag. **211** pp 105775
[11] Abidin HZ, Andreas H, Gumilar I, Sidiq TP, Fukuda Y 2013 *On the Roles of Geospatial Information for Risk Assessment of Land Subsidence in Urban Areas of Indonesia* In: Lecture Notes in Geoinformation and Cartography. 2013. p. 277–88
[12] Wahyudi SI, Adi HP, Schultz B 2017 *Revitalizing and Preparing Drainage Operation and Maintenance to Anticipate Climate Change in Semarang Heritage City* J Environ Sci Eng B. **6** 1 pp 17–26
[13] Masvika H, Adi AD, Faris F 2019 Jan 5 *Evaluasi penurunan konsolidasi tanah di Semarang Utara berdasarkan korelasi N-SPT dengan mv* Rekayasa Sipil. **7** 1 pp 1
[14] Handayani W, Chigbu UE, Rudiarto I, Putri IHS 2020 Sep 23 *Urbanization and Increasing Flood Risk in the Northern Coast of Central Java—Indonesia: An Assessment towards Better Land Use Policy and Flood Management* Land. **9** 10 pp 343
[15] Dewi SP, Ristianti NS, Kurniati R 2021 Jan 9 *Coastal settlement resilience to water-related disasters in Semarang City* IOP Conf Ser Earth Environ Sci. **623** 1 pp 012067

[16] Rudiarto I, Pamungkas D 2020 Jul 15 *Spatial Exposure and Livelihood Vulnerability to Climate-Related Disasters in The North Coast of Tegal City, Indonesia* Int Rev Spat Plan Sustain Dev. **8** 3 pp 34–53

[17] United Nations. Dept. of Economic and Social Affairs 2015 *Sustainable Development Goals 11* [Internet] Sustainable Development Goals. 2015.

[18] Sen MK, Dutta S, Kabir G 2021 Mar *Development of flood resilience framework for housing infrastructure system: Integration of best-worst method with evidence theory* J Clean Prod. **290** pp 125197

[19] Sajjad M, Chan JCL, Chopra SS 2021 Jun *Rethinking disaster resilience in high-density cities: Towards an urban resilience knowledge system* Sustain Cities Soc. **69** pp 102850

[20] Amin C S R 2019 May 19 *Staying Decision Making Process in Disaster Prone Area: A Grounded Theory Method Study on Fishermen Community in Semarang Coast, Indonesia* Humanit Soc Sci Rev. **7** 2 pp 322–30

[21] Global Shelter Cluster 2014 *8 Build Back Safer Key Messages* [Internet] Geneva; 2014

[22] Aquino DH, Wilkinson SJ, Raftery G, Potangaroa R, Chang-Richards A 2018 *Challenges to building housing resilience: the case of Fiji post-cyclone Winston* Procedia Eng. **212** pp 475–80

[23] Ameir MJ, Ekomadyo AS, Hutomo CS 2020 Jan 15 *Cultural Capital of a community to adapt in prone areas with floods, land subsidence and its Architectural representation. Case Study in Bandarharjo Semarang* Local Wisdom J Ilm Kaji Kearifan Lokal. **12** 1 pp 61–70

[24] Saharom NS, Diana SC, Kusyala D 2018 May *Alternative Housing System & Materials Criteria For Land Subsidence Area (Case Study: Bandarharjo, Semarang)* IOP Conf Ser Earth Environ Sci. **152** 1 pp 012015

[25] Hadi SP, Anggoro S, Purnaweni H, Yuliastuti N, Ekopriyono A, Hamdani RS 2020 *Assessing the giant sea wall for sustainable coastal development: Case study of Semarang City, Indonesia* AACL Bioflux. **13** 6 pp 3674–82

[26] Wijaya N 2015 *Climate Change Adaptation Measures in the Coastal City of Semarang, Indonesia: Current Practices and Performance* J Perenc Wil dan Kota. **26** 1 pp 28–42

[27] Bamberger M, Rao V, Woolcock M 2015 *Using Mixed Methods in Monitoring and Evaluation: Experiences from International Development* In: SAGE Handbook of Mixed Methods in Social & Behavioral Research. 2455 Teller Road, Thousand Oaks California 91320 United States: SAGE Publications, Inc.; 2015. p. 613–42

[28] Yastika PE, Shimizu N, Abidin HZ 2019 *Monitoring of long-term land subsidence from 2003 to 2017 in coastal area of Semarang, Indonesia by SBAS DInSAR analyses using Envisat-ASAR , ALOS-PALSAR , and Sentinel-1A SAR data* Adv Sp Res. **63** pp 1719–36

[29] Mulyana W, Daniel DD, Zhang S, Schensul D 2013 *Climate vulnerability and adaptation in the Semarang Metropolitan Area: a spatial and demographic analysis* [Internet] Vol. October 20, Urbanization and Emerging Population Issues. 2013

[30] Lubis AM, Sato T, Tomiyama N, Isezaki N, Yamanokuchi T 2011 Mar *Ground subsidence in Semarang-Indonesia investigated by ALOS–PALSAR satellite SAR interferometry* J Asian Earth Sci. **40** 5 pp 1079–88

[31] Prasetyo Y, Firdaus HS, Diyanah 2019 Aug 27 *Land Subsidence of Semarang City Using Permanent Scatterer Interferometric Synthetic Aperture Radar (Ps-Insar) Method In Sentinel 1a Between 2014-2017* IOP Conf Ser Earth Environ Sci. **313** 1 pp 012044

[32] Abidin HZ, Andreas H, Gumilar I, Sidiq TP, Gamal M, Murdohardono D 2010 *Studying Land Subsidence in Semarang (Indonesia) using Geodetic Methods* FIG Congr. April 2010 pp 11–6

[33] Arimoto M, Fukushima Y, Hashimoto M, Takada Y 2013 *Land Subsidence in Semarang, Indonesia, Observed by InSAR Time-Series Analysis Using ALOS/PALSAR Data* J Geod Soc Japan. **59** 2 pp 45–66

[34] Andreas H, Abidin HZ, Gumilar I, Sidiq TP, Yuwono B 2017 *Adaptation and mitigation of land subsidence in Semarang* In: AIP Conference Proceeding. AIP Publishing; 2017. p. 060005

[35] Putranto TT, Widiarso DA, Susanto N 2017 Jul *Assessment of Groundwater Quality to Achieve Sustainable Development in Semarang Coastal Areas* IOP Conf Ser Earth Environ Sci. **79** 1 pp 012001

[36] Prasetyo Y, Bashit N, Simarsoit Y 2019 Aug 12 *Study of Correlation of Residential and Industrial Growth Pattern in Semarang City to the Aquifer Capacity Changes in the Year 2014-2017* IOP Conf Ser Earth Environ Sci. **280** 1 pp 012001

[37] Husnayaen, Rimba AB, Osawa T, Parwata INS, As-syakur AR, Kasim F, et al. 2018 *Physical assessment of coastal vulnerability under enhanced land subsidence in Semarang, Indonesia, using multi-sensor satellite data* Adv Sp Res. **61** 8 pp 2159–79

[38] Maimunah S, Rosli N, Rafanoharana S, Sari K, Higashi O 2011 *Strengthening Community To Prevent Flood Using Participatory Approach (a Case of the Semarang City)* J Int Dev Coop. **18** 2 pp 19–28

[39] Indrawati L, Sigit Heru Murti BS, Rachmawati R, Aji DS 2020 Mar 31 *Effect of Urban Expansion Intensity on Urban Ecological Status Utilizing Remote Sensing and GIS: A Study of Semarang-Indonesia* IOP Conf Ser Earth Environ Sci. **451** 1 pp 012018

[40] Sejati AW, Buchori I 2010 *A GIS Model for Predicting Disaster Prone Areas Affected by Global Sea-Level Rise : a Case Study of Semarang City Vulnerable Regions and Disaster Management* In: ICRD proceeding. 2010. p. 5–12

[41] Wiratuningsih D, Setyowati DL, Suhandini P 2018 *The Adaptation of The Society in Coping with Tidal Flood in Kemijen Village Semarang City* J Educ Soc Stud. **7** 29 pp 146–53

[42] Hamdani RS, Hadi SP, Rudiarto I, Purnaweni H 2020 Nov 10 *Do we care enough? revisiting land subsidence and coastal spatial planning policy in Semarang, Indonesia* Warsito B, Sudarno, Triadi Putranto T, editors. E3S Web Conf. **202** August pp 06005

[43] Buchori I, Sugiri A, Mussadun M, Wadley D, Liu Y, Pramitasari A, et al. 2018 Mar *A predictive model to assess spatial planning in addressing hydro-meteorological hazards: A case study of Semarang City, Indonesia* Int J Disaster Risk Reduct. **27** pp 415–26

[44] Marfai MA, King L 2008 Oct 8 *Coastal flood management in Semarang, Indonesia* Environ Geol. **55** 7 pp 1507–18

[45] Afifa L 2021 Feb 6 *Flood Inundates Semarang City* [Internet] Tempo. 2021.

[46] Hadi SP 2018 Feb 1 *Integrated Community Based Coastal Management: Lesson From The Field* IOP Conf Ser Earth Environ Sci. **116** 1 pp 012064

[47] Hadi SP 2017 Feb *In Search for Sustainable Coastal Management: A Case Study of Semarang, Indonesia* IOP Conf Ser Earth Environ Sci. **55** pp 012054

Acknowledgement

The authors send their greatest gratitude to Dr Putu Edi Yastika and Prof. Norikazu Shimizu for allowing us to access their data and use it for our project. We are also thankful to the Directorate of Research and Community Service of the Ministry of Research and Technology/National Research and Innovation Agency of the Republic of Indonesia that has given funding to this land subsidence research with Contract Number: 206/SP2H/PMDSU/DRPM/2020.

Development of post-disaster shelters (case study: Gampong Pande post-rehabilitation and reconstruction of the earthquake and tsunami disaster)

W Setyawan[1], P Setijanti[1,2], and A Hayati[1]

[1] Architecture Department, Faculty of Civil, Planning, and Geo-Engineering, Institut Teknologi Sepuluh Nopember. Kampus ITS, Sukolilo, Surabaya 60111, Indonesia

Abstract: Indonesia has a high risk of earthquake and tsunami disasters, especially in areas around the long and relatively sloping coastline. Nanggroe Aceh Darussalam as an area that has been hit by an earthquake and tsunami, in its development of settlements it is more directed to disaster-prone areas in coastal areas. This study aims to identify the development of housing after rehabilitation and reconstruction due to the earthquake and tsunami in the *Gampong Pande* area, in the context of mitigating earthquake and tsunami disasters in the future. This research is related to ethnographic research that uses an emic approach in field surveys, emphasizes the natural setting context, focuses on interpretation and meaning, involves the community deeply, and focuses on contemporary phenomena as a whole. Post-disaster mitigation actions carried out in residential development are structural modifications referring to and in line with the Building Code. Post-disaster mitigation actions carried out in the development of residential areas are land grants for road widening as an evacuation route during disasters and the construction of sea dikes to reduce tidal waves of sea water entering settlements and planting mangroves along the sea wall as a green belt that doubles as a green belt and source of fishermen's economy.

1. Introduction

As an archipelagic country, Indonesia, which is located in the ring of fire area and the confluence of 3 tectonic plates, has a risk of earthquake and tsunami disasters [1]. With the character of the area around the long and relatively sloping coastline, resulting in the growth of settlements in coastal areas, both rural and as development areas in urban areas that are prone to earthquake and tsunami risks.

Post-earthquake and tsunami disaster management in Aceh [2] and Nias [3] over a period of about 5 years (2005 – 2009) which has become one of the role models for disaster rehabilitation and reconstruction throughout the world, and should become a valuable lesson for Indonesia in the context of disaster mitigation [4]. Nanggroe Aceh Darussalam as an area that has a building code and spatial planning based on earthquake and tsunami disaster mitigation, in its development of settlements it is more directed to areas prone to tsunami disasters [5], in this case the coastal areas. Likewise, settlements that have been reconstructed after the earthquake and tsunami in Aceh have developed towards new conditions, the new normal [6].

This study aims to identify the development of housing after rehabilitation and reconstruction due to the earthquake and tsunami in the Aceh region, in particular the development of housing and the environment in the context of mitigating the earthquake and tsunami disaster in the future. This research was conducted based on the phenomenon of the development of coastal settlements in earthquake and

[2] Corresponding author: psetijanti@arch.its.ac.id

tsunami prone areas. This phenomenon is important to study considering that many settlements in coastal areas of Indonesia are categorized as earthquake and tsunami prone areas.

An understanding of mitigation (preparedness, response, recovery) for earthquake and tsunami disasters as a multi-hazard disaster is closely related to understanding areas affected by disasters, in this case coastal settlements, especially in the context of development typology (folk, transitional, industrial, post-industrial). The development of settlements in coastal areas needs to be seen from the aspect of public understanding of the dangers of earthquake and tsunami disasters, especially those dominated by self-help type houses [7]. With the character of a self-supporting house that prioritizes the participation of residents in determining the form of their dwelling, the cultural elements of the occupants and the surrounding community will greatly affect the physical character of the dwelling and the residential environment. Community understanding of the threat of earthquake and tsunami disaster in a settlement is also influenced by collective memory of the disasters that have occurred, and ultimately affects community attitudes and actions towards disaster threats (coping strategy) [8,9]. Collective memory and understanding of community disasters can be traced by looking at the development of the physical condition of housing and settlements (built environment) associated with cultural aspects/elements (worldview, values, ideals, norms, lifestyle, activity system, kinship) [10].

2. Methodology

In accordance with the stated objective that this research focuses on investigating the pattern of coastal settlements in earthquake and tsunami prone areas, the factors that influence the development of coastal settlements in earthquake, the research this is categorized as naturalistic/constructivism research [11] or qualitative research. Naturalistic/constructivism research according to Groat and Wang [11] emphasizes the natural setting context, focuses on interpretation and meaning, involves the community deeply, and focuses on intact contemporary phenomena. According to Niezabitowska [12], qualitative research in the context of architecture is not only used to look at building and examine criteria for the quality of a design, but also to find the relationship between architecture and other factors behind its presence. This research is also closely related to ethnographic research [13] which uses an emic approach in field surveys.

2.1. Case study

The criteria set out in the research strategy, namely coastal residential areas that have experienced the earthquake and tsunami disaster and have been rehabilitated and reconstructed so that they have developed until the time this research is carried out, then another important criterion is the ease of access to data in the field. Because the researcher was involved in the emergency response, rehabilitation and reconstruction period after the earthquake and tsunami disaster in Aceh 2004, the case study selected was a coastal settlement area in the city of Banda Aceh.

The area selected as a case study is *Gampong Pande*, Kuta Raja sub-district, Banda Aceh City. *Gampong Pande* has an area of 260 ha which is located in the northern part of Banda Aceh City, with a population of 690 people and 199 households [14]. As an area located on the north coast of Banda Aceh city, *Gampong Pande* was one of the areas most affected by the 2004 earthquake and tsunami. Almost all buildings in this area were heavily damaged with large loss of life and significant changes in the landscape [4]. *Gampung Pande* is also one of the areas where the rehabilitation and reconstruction program was the first to receive assistance from the Asian Development Bank (ADB). *Gampong Pande* is one of the oldest villages in the city as the forerunner of the city where Banda Aceh was founded [15].

During the post-disaster rehabilitation and reconstruction period, researchers were assigned by ADB to prepare plans for the repair and rebuilding of *Gampong Pande* after the earthquake and tsunami. Planning preparation activities include identifying survivors and then reorganizing them for the rehabilitation and reconstruction of the area. Furthermore, together with the organized community, they collect data on assets and boundaries of each family as the basic data for the assistance program. The researcher is also responsible for determining the residential design that will be built according to the character of the land with a community participation approach in the design process. Another important

activity carried out by researchers is determining the pattern of rehabilitation and reconstruction of the area with the community, with a combination of top down and bottom up strategies.
Another involvement of the researcher during the rehabilitation and reconstruction phase was as an architectural expert in the preparation of the Building Code for the Province of Nanggroe Aceh Darussalam after the disaster. The Building Code generated from these activities became the basis for rehabilitation and reconstruction activities by all parties involved at that time.

With the considerable involvement of researcher during the rehabilitation and reconstruction period in *Gampong Pande*, researcher has broad access to information and data considering that there are still local residents who are known by researchers and can be used as resource persons for searching data in the field.

2.2. Key person
Ethnographic research according to Lucas [13] was conducted by conducting in-depth interviews with people (participants) who were known, and the researchers had a strong relationship with the research location. The first participant interviewed was a local community leader who became a key figure and played a major role in the rehabilitation and reconstruction of *Gampong Pande* after the earthquake and tsunami. This key figure is Mr. Irdus, a 58-year-old resident of *Gampong Pande* and a retired employee of the Department of Food, Agriculture, Marine and Fisheries of Banda Aceh City.

Mr. Irdus naturally became an important figure in the rebuilding of *Gampong Pande* after the earthquake and tsunami, through a long process, a journey full of challenges both from within (the residents) and with external parties such as donor agencies and the city government as well as other interrelated parties. His persistence in fighting for assistance for rehabilitation and reconstruction programs from donor agencies often puts him in a difficult position as a leader as well as a bridge between residents and aid providers. The struggle as a citizen leader with the provision of sincerity without financial rewards from the government has forged Mr. Irdus as an informal community leader who was born from a multi-dimensional post-disaster hardship.

Key participant interviews were conducted using the WhatsApp application communication media on cell phones. The initial interview was aimed at digging up information about the events that occurred during the earthquake and tsunami, and important events that followed until *Gampong Pande* was finally able to be rebuilt. Considering that key participants have been blind due to glaucoma since 2018, it is necessary to be accompanied by his wife (Mrs. Tengku Raimah) at the time of the interview to help answer written questions if needed.

3. Results and discussion
Before the ADB house in *Gampong Pande* was built in 2006, Mr Irdus and other residents also received assistance from the International Federation of Red Cross (IFRC) in the form of an emergency house built with a light steel structure (figure 1). The concept of this emergency house is a house that can be built on land with various conditions because it uses a screw system foundation that can adjust to the contours of the land. The main structural material is mild steel which can be assembled and disassembled quickly (knock down), while the roofing material is a corrugated plate, and the floor and wall materials can use local materials such as boards and bamboo strands.

When Mr Irdus received the IRFC aid emergency house, he decided to place those houses at the back of the land, with the hope that a more permanent generic ADB aid house (figure 2) will be built in the front area (figure 3). Mr. Irdus planned when the ADB house was built, it would be developed backwards. Because he really likes plants, the front area of the house has a large yard as a beautiful green open space.

Figure 1. IFRC relief house by ADB ETESP PPC

Figure 2. ADB ETESP PPC generic house

Figure 3. Mr. Irdus's house was assisted by ADB ETESP PPC when construction was completed in 2006

ADB's generic house is the standard ADB-assisted house for general land conditions. For land conditions where it is not possible to build a generic house design, ADB also provides different design alternatives according to the area or shape of the land, but still meets the requirements of the Government of Indonesia, namely an area of no more than 36m2.

At the time the ADB-assisted house was built, Mr. Irdus had submitted a request for modifications to the house design, by changing the door and window designs and adding a terrace area with additional structures. ADB's housing assistance scheme allows those modification, but any excess costs resulting from modifications are the responsibility of the beneficiary. The size of the main building area of the house must follows the standard of the Indonesian government at that time, which was 36 m2 plus a kitchen and bathroom.

In 2007, Mr. Irdus gradually expanded the house to the back area. To meet the needs of the family, the addition of several rooms and rooms. Considering the limited land in the rear area, it is necessary to construct a new 2-storey building with the consequence of dismantling the emergency housing building donated by IFRC and moving it to the 2nd floor of the new building (figure 4.). During the construction of the additional house, Mr. Irdus explicitly asked the builders to make a solid structure, because the building "must be able to last a long time". According to Mr. Irdus, because the builders used were builders who build ADB-assisted houses, they used the typology of ADB's house structures as a reference.

In an additional building in the front porch of the main house, the visual structure of the column and concrete beams may be considered redundant because of their large dimensions. According to Mr. Irdus, this is influenced by the design of the ADB house which also uses a column larger than that used by residents in general before the disaster (figure 5.)

Figure 4. Mr. Irdus/ participant A house in 2021

Figure 5. The front porch of the main house which is a modification of the standard ADB-assisted house

According to Mr Irdus, the ADB-assisted houses have good quality, judging from their physical condition for more than 14 years, they are still in good condition, with no cracks in the walls and main structure (figure 6.). He added, the supervision factor at the time of construction is very important in maintaining the quality of the building as it has been determined.

At the back, the development of the house was carried out separately from the main house assisted by ADB, with a design typology similar to the ADB house but made 2 storeys according to needs but with limited land area conditions (figure 7.). At the top of the 2nd floor, a clapboard building which is a modification of the IFRC-assisted house which was originally on the ground (figure 8.). The IFRC relief house which is an emergency house with a knock down construction system of light steel can easily be dismantled and reassembled in a different place. In the back area there is also a green open space that is quite spacious and beautiful, as a form of Mr. Irdus' hobby of gardening.

Figure 6. The interior of the main house of ADB assistance belonging to Mr. Irdus of participant A

Figure 7. Rear residential development with a typology similar to ADB's house of participant A

At the below of the development building, there is a structure of concrete columns and beams with quite large dimensions (figure 9.), which is a manifestation of efforts to further strengthen the building structure from the threat of earthquake hazards. The use of concrete construction was to improve the structural physical quality of his house development by referring to ADB's house design. Likewise, in the first-floor interior of the development house, a concrete block structure with fairly large dimensions can be seen (figure 10.)

Figure 8. IFRC house moved to the 2nd floor of the development building in the rear area of participant A

Figure 9. The atmosphere of the first-floor area of the development house of participant A

Figure 10. The atmosphere of the space on the first floor of the development house of participant A

In general, it can be identified that the dominant earthquake hazard mitigation effort carried out by Mr. Irdus in the development of post-disaster housing is by strengthening the building structure, especially with reinforced concrete construction. The thin form of the building is one of the principles in the design of buildings that are responsive to the threat of earthquake hazards. The availability of an open space that is wide enough and evenly distributed is also an opportunity for mitigation efforts from earthquake hazards as a gathering point area.

4. Conclusion

Regarding the participants' views on natural disasters, especially earthquakes and tsunamis, there are two things related to the views of *Gampong Pande* residents towards natural disasters (1) "steps, fortunes, meetings, and death are decided by God", this principle makes Acehnese people in general so quickly rise from natural disasters, (2) "boils never grow in the same place", this principle is believed by the residents of disaster victims so that they continue to rebuild their homes in their original places. The participants' initial knowledge about disaster mitigation was by lying down when an earthquake occurred and seeing the overflowing cloudy river water carrying building debris as a sign that "our village has disappeared".

Post-disaster mitigation actions carried out in residential development are housing development with a structural typology referring to ADB's house buildings and Building Code. Post-disaster mitigation actions carried out in the settlement context are land grants for road widening as an evacuation route during disasters, and construction of sea dikes to reduce tidal waves of sea water flood entering village. Another mitigation action is planting of mangroves along the sea embankments as a green belt, as a natural wall retaining waves and sea breezes, and as a habitat for the growth of sea crabs which are one of the economic sources of fishermen.

The results of this initial interview opened up opportunities for interviews with other key figures to get a more complete picture of the architectural mitigation carried out by the *Gampong Pande* community in dealing with earthquake and tsunami disasters in the future. It is hoped that the identification of architectural mitigation carried out in housing and settlements will be useful as a reference for disaster mitigation in other areas with a similar context.

5. References

[1] Pusat Studi Gempa Nasional 2017 *Peta Sumber dan Bahaya Gempa Indonesia* Tahun 2017 (Bandung: Pusat Penelitian dan Pengembangan Perumahan dan Permukiman, Badan Penelitian dan Pengembangan , Kementrian Pekerjaan Umum dan Perumahan Rakyat)

[2] Blunt A and Silas J 2010 *Planning the Earthquake and Tsunami Emergency Support Project (ETESP) Housing Program Rebuilding Lives and Homes in Aceh and Nias Indonesia*, ed F Steinberg and P Smidt (Mandaluyong City: Asian Development Bank) pp 20-47

[3] Lang H 2010 *Rehabilitation and reconstruction in South Nias Heritage villages Rebuilding Lives and Homes in Aceh and Nias Indonesia*, ed sF Steinberg and P Smidt (Mandaluyong City: Asian Development Bank) pp 143-168

[4] Steinberg F 2010 *Rebuilding lives and homes in Aceh and Nias: A retrospective, Rebuilding Lives and Homes in Aceh and Nias Indonesia*, ed F Steinberg and P Smidt (Mandaluyong City: Asian Development Bank) pp 1-19

[5] Akbar A and Ma'rif S 2014 Arah Perkembangan Kawasan Perumahan Pasca Bencana Tsunami di Kota Banda Aceh *J. Teknik PWK* **3** 274-284

[6] Vale L, Shamsuddin S, Goh K 2014 Tsunami + 10: Housing Banda Aceh After Disaster *Places J* https://doi.org/10.22269/141215

[7] Rapoport A 1969 *Housing Form and Culture* (Englewood Cliffs: Prentice Hall-Inc)

[8] Sorensen J H and White G F 1980 *Natural Hazard – A Cross Cultural Perspective Human Behavior and Environment, Advances in Theory and Research*, ed I Altman , A Rapoport , J F Wohlwill (New York: Springer Science+Bussiness Media)

[9] Kates R W 1976 *Experiencing the Environment as Hazard Experiencing the Environment*, ed S Wapner, S B Cohen, B Kaplan (New York: Plenum Press)

[10] Turner J F C 1977 *Housing by People: Towards Autonomy in Building and Environments* (New York: Pantheon Books)

[11] Groat L and Wang D 2013 *Architectural Research Methods* (New Jersey: John Wiley & Sons Inc)

[12] Niezabitowska E D 2018 *Research Methods and Techniques in Architecture* (New York: Routledge)

[13] Lucas R 2016 *Research Methods for Architecture* (London: Laurence King)

[14] Sistim Informasi Kawasan Permukiman, http://simkap.bandaacehkota.info/page.php?page=profile&view=kecamatan&id=5 (accessed 26 April 2021)

[15] Yanti I D, Izziah, Isya M 2018 Konsep Zona Kawasan Situs Sejarah Gampong Pande Banda Aceh *J Arsip Rekayasa Sipil dan Perencanaan* **1** 63-71 DOI: 10.24815/jarsp.v1i1.10354.

Discovering pattern of space layouts from low-income housing: A discussion of the New Normal preparedness after the Covid-19 pandemic

A S Yahya[1,3], M H Ahmad[1], A A G Hassan[2]

[1]Faculty of Built Environment and Surveying, Universiti Teknologi Malaysia, Johore, Malaysia
[2]Azman Hashim International Business School, Universiti Teknologi Malaysia, Kuala Lumpur, Malaysia

Abstract. By the end of 2021, it is expected that the coronavirus (COVID-19) will shift from pandemic to endemic phase as the number of vaccinated people is increasing over time. It gives us hope as the endemic phase will allow us to live together with the virus. During the Covid-19 situation, people have become more creative and innovative for familiarising themselves with the New Normal. The next question is how is the existing housing for the low-income group ready to be suited with this situation? Therefore, this study will examine the concept of space design in housing units for the lower income group. The study adopted a case study approach based on housing layout from Kuala Lumpur, Bandung and Buckinghamshire. These housing layouts were analysed using the Space Syntax method for describing the meaning of space layout from a low-income perspective. This study found two consideration points, (1) the willingness of space to accommodate small business activities and to avoid the spread of epidemics in the domestic space, and (2) the willingness of the government to educate the public to cultivate small home entrepreneurship with the existing low-income routine and limited domestic spaces.

1. Introduction

The Covid-19 Pandemic has dramatically changed the employability and social system in countries around the world [1]. These changes force people to increase the functioning of their home to accommodate the work-from-home activity. In many countries, individual freedom is limited with regards to 'new norms' rules for curbing the Covid-19 spread [3]. The New Norm rules seem to change the public awareness on life experiences which are related to movement control order, social distancing and travelling outdoors alone. These changes have made home an important component for virtual social gatherings, working, attending virtual classes, playing sports and shopping [10].

This new norm phenomenon has been seen in people who have been laid off from work. Due to this, they started adapting home as a small-scale business platform to gain income and indirectly benefit the local community. At the same time, people should be prepared to change the function of their home as a quarantine place if one member of the household is infected. They also need to be prepared to face conflict due to stress of working at home [2]. In this regard, a house is important for people towards accommodating complex survival activities coherently during COVID-19 virus era.

[3] Corresponding author: ashuib2@live.utm.my

This paper will explore three types of house design for low-income groups in search for its social logic in space. The underlying meaning of social logic in space may be useful to inform spatial awareness of the whole configuration as a feature key for shaping work-from-home policy. The research will focus on how homes were designed before the covid-19 phase. This will accommodate more complex functions in the next era of endemic. The study intended to understand further meaning displayed in the layout pattern of space for low-income groups. Space syntax analysis method solves the absence of an interview session from the architect who is responsible for the house design. This analysis also revealed the perspective meaning from social logic of space. The finding found that the housing design for low-income groups are homogeneous. However, it acknowledged that experiences in housing studies may vary based on different contexts such as social logic, technology, policy, and social and demographic dynamics. People have different families and may work either full time or part-time. People live in various locations such as urban, suburban, and rural. We live in a house that witnessed how Covid-19 phenomenon had changed our way of life from normal to new norms.

2. Literature review and methodology

'Space Syntax' analysis is a popular configurational analysis that has been widely used by researchers to explore the cultural pattern of traditional and vernacular houses [13]. The configuration analysis of the space and layout plan displays its morphological values where the social logic of the household can be conveyed. The household social logic is also described in the form of housing units. The main recommendation of this method is based on social relationships and self-expressing events with regards to spatial configuration. Configuration is the relationship between two spaces by considering all the other spaces in the building. Hillier, Hanson and Graham (1987) explain:

> *"Therefore, spatial configuration is a more complex idea than spatial relationships, which do not just require a pair of related spaces."*

In this study, several configurational analysis techniques were imported as part of research activity. The research activity aimed to investigate the meaning of social logic contained in the housing spaces of the lower income group. The main goal of this study is to extract the meaning of social logic from the house configuration that has been offered to dwellers. Site visits have been made to social housing in Bandung, Indonesia (Appendix 2), and Kuala Lumpur, Malaysia (Appendix 3) in 2019. Based on the initial view, both housing spaces have different design concepts and design goals. The Cigugur Housing Prototype in Bandung adopted the open-plan concept with shared common spaces among urban dwellers temporarily before successfully owning the house. Meanwhile, PPR Seri Semarak Housing in Kuala Lumpur adopted the concept of redesigning three-bedroom units for city dwellers to choose either to purchase or rent the residential unit. Housing design samples from Bean Hill's 1973 by the world-renowned architect Norman Foster were used as a cross-reference to describe designer's intuition.

The history of architectural spatial analysis began when Hillier and Hanson (1984) proposed a scientific research method for the purpose of understanding space and its arrangement in architecture. According to them, the layout plan has difficulty in revealing its configurational characteristics of residential units even by simply comparing it to others. The argument was based on their understanding that social structure is naturally a space. Therefore, the configuration of space has a basic form of social logic. Hillier and Hanson (1984) proposed space mapping analysis that could be generated into a Justified Permeability Graph. The value of integration of space could be numerically identified through the graph (figure 1). Hence, the characteristics of space arrangement can be objectively justified. According to Maina (2014), the calculation process must first start with the value of MD (mean depth) which is the value given by d / k-1, where d is the amount of depth (d) of all space and k is the number of rooms put together in configuration form. The value of integration value (IV) space for space (1) is shown as in figure 1.

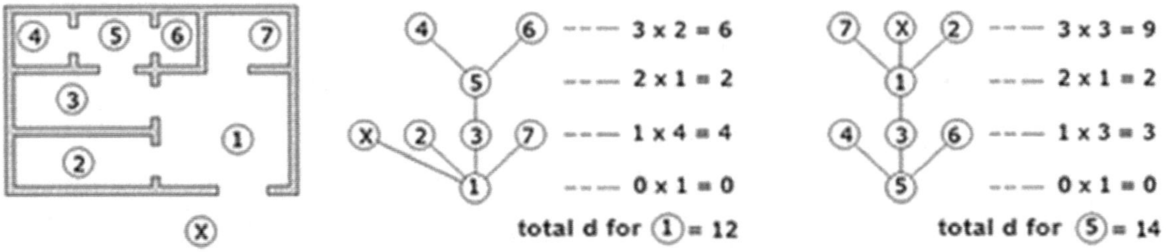

Figure 1. Examples of a layout plan (a) defined in the form of Justified Permeability Graph from ①as a carrier (b), and from ⑤as a carrier (c).

The second step is to obtain RA (relative asymmetry) via the formula; RA = 2 (MD – 1) / (k – 2) where, d = 12, k = 8, MD = 12/8 - 1 = 1.714., RA is the MD specified as the maximum fraction against possible depth ranges for any spaces with the same number assembled as spatial system [9]. For the space ① in figure 1b,

$$\begin{aligned} RA &= 2\,(MD - 1) / (k - 2) \\ &= 2\,(1.714 - 1) / (8 - 2) \\ &= 1.4285 / 6 \\ &= 0.238 \end{aligned}$$

For the third step, the value of RRA (Real Relative Asymmetry) which the value for the ratio of RA for space provided by the spatial system with the same number of spaces where RRA = RA / D value for the ideal k space (Hillier & Hanson, 1984) is obtained. RRA means that the value of a space contained in a spatial system has a centrality value and is related to all spaces [12]. The value of RRA explains how highly permeable the space is. A low RRA value means higher integration, while a high RRA value means high segregation [4]. For figure 1b, the D value for 8 spaces = 0.328 (Appendix 1)

$$\begin{aligned} RRA &= 0.238 / 0.328 \\ &= 0.7259. \end{aligned}$$

The final step is to obtain the IV (Integration Value) via formula 1/RRA. The following equation is shown:

$$\begin{aligned} IV &= 1 / 0.7259 \\ &= 1.3776. \end{aligned}$$

These are numerical figures that count for each space as a reflection of the extent to which space integrates or organises its access and movement. Furthermore, the integration has its implications towards social networks and activities inside the complex space [6][8][9][14]. It is the inverse of the RRA. The higher this value, the more integrated the space will be on the floor plan or the layout, from normal to new norms.

3. Findings

3.1 Sample 1: Cigugur social housing, Bandung, 2005

Sample 1 in figure 2 is a social housing prototype only for the people of Bandung, Indonesia. The house is designed for rental demand, while the number of occupants is limited to a couple with only a child. Therefore, the residential units share the Lobby (figure 2) with other units to cater for the need to entertain their guests. The unit layout is arranged in one-way configuration where there is no branch of linkage to the other spaces. The finding shows that the layout pattern of Sample 1 has low integration value (IV) from the Exterior (IV=0.46) due to high segregation of space (RRA=2.16). What is interesting in this arrangement of space. It can be seen when the integration value is measured from the Living/Dining. This space becomes the centrality point in controlling social movements in configuration. Through the Justified Permeability Graph, the Living/Dining is in a strategic position to control the social movement of two types of space groups namely, Group of space 1 (Lobby, Passage 1 and Passage 2) and Group of space 2 (Bedroom 1, Kitchen, Toilet).

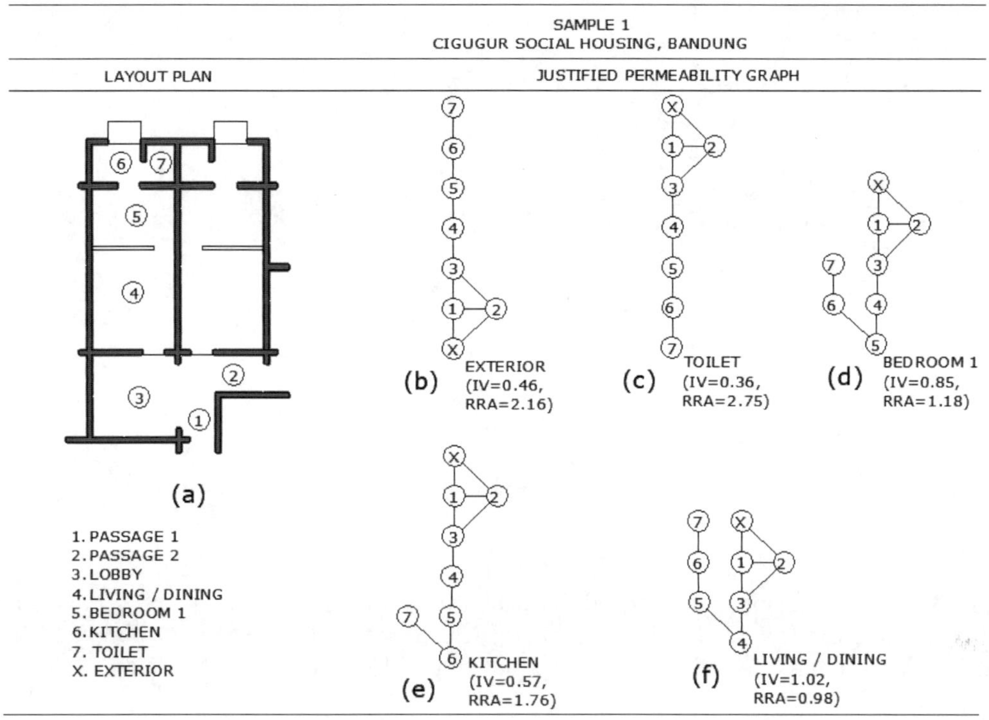

Figure 2 (a) The floor plan & Justified Permeability Graph (JPG) of the Cigugur Public Housing Prototype from space (b) exterior, (c) toilet, (d) bedroom 1, (e) kitchen, and (f) living/Dining. The number in the JPG refers to the name of the space function in the layout plan of the house.

3.2 Sample 2: PPR Seri Semarak, Kuala Lumpur, 2005

Sample 2 is one of housing estates that was completed in 2005 to replace squatter villages in Government's efforts to achieve the Zero Squatter mission in Kuala Lumpur city. The housing is designed to be purchased or rented to accommodate a household of five people in a single house unit. In accordance with the standard of residential space at the time, the house provides three bedrooms, two Toilet, a Yard, a Kitchen, and a Dining. Figure 3 shows the space arrangement of Sample 2 that focuses to develop from the exterior where space is on passage1 area with the main junction that goes to other five space branches. Group of space 1 also has a separate Passage area to go to both Toilet 1 and Toilet 2, while the second branch along with the Passage area gives dwellers the option to choose a branch to the Bedroom, Group of space 2 (Kitchen and Yard) and Group of space 3 (Living / Dining, Bedroom 2 area and Bedroom 3). Therefore, the integration value (IV) of the Exterior is at moderate level with configuration of (IV=0.92) median as a result of segregated space (RRA=1.09, the median configuration value). The interesting part of the space arrangement is that the Living / Dining is set to become the central point (RRA=0.73, the lowest configuration value) for social configuration movement to be highly permeable to all spaces. The Justified Permeability Graph shows the Living / Dining has the position to potentially create social gathering (IV=1.38, the highest configuration value) through the informal encounter meeting resulting in Bedroom 2 and Bedroom 3 area that (segregation value of RRA=1.45) easily accessible from Exterior.

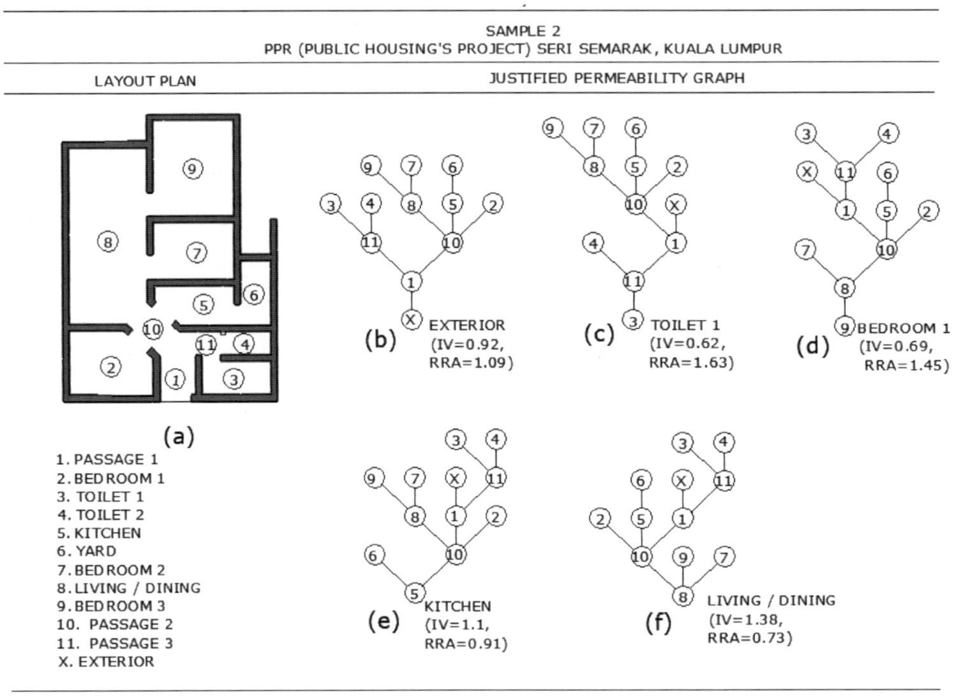

Figure 3 (a) The floor plan & Justified Permeability Graph (JPG) of the PPR Seri Semarak Housing, Kuala Lumpur from (b) the exterior, (c) toilet, (d) bedroom 1, (e) kitchen / dining, and (f) living. The number in the graph refers to the name of the function of the space in the floor plan.

3.3 Sample 3: Bean Hill Housing, England, 1973

Sample 3 is housing completed in 1973 by Architect Sir Norman Foster. This design is considered a thoughtful interpretation of modern life in architecture [11]. This housing design is an attempt to offer experiences from the occupancy and employment culture. Housing is also could be considered as an attempt to a conducive living within the industrial and commercial environments with design that reflects the economic urge. While this housing has suffered a shock in society's acceptance of minimal physical appearance (iron frames, steel walls, leaking roofs and damaged floors according to economic roads [5], but criticism of the layout of the spaces remains unspoken. Figure 4 shows layout patterns in sample 3 that evolve from the Exterior area where the Lobby area acts as the main junction to go to all segregated spaces.

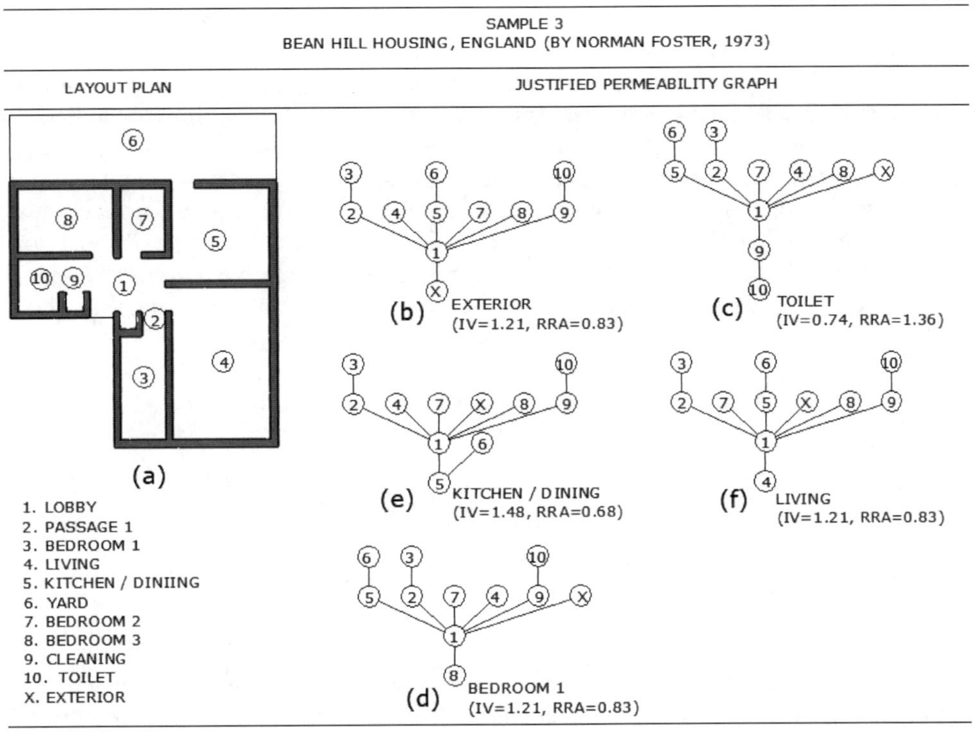

Figure 4 (a) The floor plan & Justified Permeability Graph (JPG) of the BeanHill Housing estate from (b) the exterior, (c) toilet, (d) bedroom 1, (e) kitchen / dining, and (f) living. The number in the graph refers to the name of the function of the space in the floor plan.

Living, Bedroom 2, Bedroom 3, were segregated without any shared space. Meanwhile, the Cleaning is shared with Toilet, Bedroom 1 shared with Passage 1 and Kitchen / Dining shared with Yard. Therefore, the integration value of the Exterior area is at a moderate level of configuration (IV =1.21, the median value in the configuration) as a result of the segregation of space (RRA = 0.83, the median value in the configuration). What is interesting in this layout is that when the integration value is measured from the Living, this space has a centrality point (RRA=0.83, median value in configuration) that is the same as the Exterior and Bedroom 1 subsequently with social movement. This spatial configuration has the potential to control movement to all spaces even with segregation possibilities such as a single entrance to the Living. Justified Permeability Graph shows that the Living is in a position of potential to gain more private experience as Living is not directly related to other spaces. Social gathering through formal encounters is seen as a social system in the house. It can be observed by seeing the Lobby area as a medium for movement in the entire space accessibility. The configuration also works for the dweller and visitor who visit the internal space from the Exterior (figure 4b). This configuration pattern of housing layout is based on segregation of space for creating gathering spots through formal encounters as all spaces are separately arranged from the Exterior area. The lobby area is a system that directs people to reach the intended space.

4. Discussion

Despite the varied shapes of space arrangements from these samples, this study explored the designer's nature of the space configuration that is introduced to occupants. The Cigugur prototype is a one-way way in inducing potential formal meetings where the Living /Dining area acts as a controlling centre for social movement. Meanwhile, PPR Seri Semarak has five growing branches of access to the three groups of spaces. These branches can potentially generate informal gatherings where the Exterior has high access to the whole space. Bean Hill housing also expands from the Exterior by collecting six access branches into the Lobby area to entire spaces for generating formal meetings. This type of housing will invite dwellers into any intended space. Space arrangements (Table 1) descriptively indicate that each house has similarities and differences in organising social encounters. All samples show that the Toilet should be placed at the most segregated space. As for the easiest space to arrive from the Exterior, both Cigugur Prototype and PPR Seri Semarak agreed to position a Living / Dining for accommodating all functions compared to Bean Hill Housing which proposes a Kitchen / Dining. All three houses have different opinions for placing the space with a balanced value towards integration and segregation of space housing layouts. Basically, Cigugur prototype proposes Kitchen, PPR Seri Semarak suggests Bedroom 1 area and Bean Hill Housing proposes Bedroom 1 area and living.

Table 1 Basic syntactic data in three housing samples

House	k	MD	RRA		
			Min	Median	Max
Cigugur	7	2.8	0.98 (Living / Dining)	1.76 (Kitchen)	2.75 (Toilet)
PPR Seri Semarak	10	2.9	0.73 (Living / Dining)	1.45 (Bedroom 1)	1.63 (Toilet 1)
Bean Hill	11	2.3	0.68 (Kitchen / Dining)	0.83 (Bedroom1, Living)	1.36 (Toilet / Cleaning)

Based on the understanding of all permeability levels, two new questions have arisen in relation to the control of outbreaks spread. The considerations of domestic spaces control which based on the new normal practice for work-from-home consideration are:

1. *Is the existing housing layout design able to accommodate small business activities and at the same time avoid the spread of virus in the domestic space?*
2. *How to educate the public to think and cultivate entrepreneurship that starts from small business in limited space with existing daily home routines?*

Bean Hill housing can generate formal encounters for households to reach the intended space. However, the design will also face the urgency of outbreak in the Lobby. The lobby of Bean Hill design is the most humane and accessible space ever. Similarly, Cigugur housing has differences in encounters for creating social gathering as all spaces are segregated from the exterior. The urgency of the spread of the outbreak remains only to one-way movement over time and it is difficult to control. Meanwhile, Living/Dining spaces in PPR Housing Seri Semarak are vulnerable to the outbreak because it was the central point for accessibility. Social movement in central configuration is highly permeable to all spaces. Therefore, these three housing samples showed a vulnerability of covid-19 spread and faced the complication from work-from-home practice. This situation might have a socio-spatial impact on households in managing built form activities.

5. Conclusion

Each house shows its arrangement of a social space between men and women, young and old, hosts and guests in its building. Designers should think about the implication of these dimensions towards its space configuration. In addition, designers should not design merely with only self-intuition or visual evaluation. Architecture can enrich human experiences in various social aspects. This can be achieved by placing the architect in the experience of the dweller through space and form configuration during design practice. The nature of this research assessment concluded that the Cigugur prototype has exhibited the uniqueness of composition in providing an adequate space with clear occupancy level. Meanwhile at PPR Seri Semarak, the space composition is distorted with uncontrolled social movement configuration. Bean Hill Housing also shows the lobby as a transition space and acts as a space management system for a specific visitation purpose.

Throughout this study, it is understood that the design of these houses has size limitation and high space integration value except for the house design produced by Architect Norman Foster. Preparing the endemic phase of Covid-19 requires undefined home spaces to rest after all day's work. However, a house is a microhabitat and all exterior social spaces need to be integrated with its domestic space. Therefore, home functions need to be flexible in accommodating new routines. A new perspective on the relationship between architecture and livability can be realised through design determination that encourages social logic thinking in mathematics and as well as design intuition in art. The research provides architectural possibilities for realising a suitable condition with regards to the endemic phase of Covid-19. It is suggested that mathematical evaluation should be combined with an intuitive compositional thinking in art to obtain truly an innovative architectural housing design. Therefore, it can accommodate complex and various social activities that could result from the pandemic.

6. Appendices

Appendix 1. Table of D values. Source: Hillier & Hanson (1984), p. 112

k	D value	k	D value	k	D value	k	D value	k	D value
1		8	0.328	15	0.259	22	0.214	29	0.184
2		9	0.317	16	0.251	23	0.209	30	0.181
3		10	0.306	17	0.244	24	0.205	31	0.178
4		11	0.295	18	0.237	25	0.2	32	0.174
5	0.352	12	0.285	19	0.231	26	0.196	33	0.171
6	0.349	13	0.276	20	0.225	27	0.192	34	0.168
7	0.34	14	0.267	21	0.22	28	0.188	35	0.166

Appendix 2. Fieldwork, Cigugur Housing, Bandung Indonesia

Appendix 3. Fieldwork, Projek Perumahan Rakyar (PPR) Seri Semarak, Kuala Lumpur, Malaysia

7. References

[1] Couch D L, O'Sullivan B, Malatzky C 2020 *What COVID-19 could mean for the future.* Gender Work Organ., 267-274.

[2] Dockery M, Brings S 2020 *Working from Home in the COVID-19 Lockdown*. Bankwest Curtin Economics Centre Research Brief COVID-19.
[3] Dubey A D, Tripathi S 2020 Analysing the Sentiments towards Work-From-Home Experience during COVID-19 Pandemic. *Journal of Innovation Management*, **8** (1) 13-19.
[4] Faris Ali M, Ahmad Sanusi H 2010 Using Space Syntax Analysis in Determining Level of Functional Efficiency: A Comparative Study of Traditional and Modern House Layout in Erbil City, Iraq. *International Seminar on Tropical Eco-Settlements* pp 131- 144 Sanur Denpasar, Indonesia.
[5] Ortolano G 2019 *Thatcher's Progress: from social democracy to market liberalism through an English Town* Cambridge University Press.
[6] Hillier B, Hanson J 1984 *Social Logic of Space*. London: Cambridge University Press.
[7] Kramer A, Kramer K Z 2020 The potential impact of the Covid-19 pandemic on occupational status, work from home, and occupational mobility **119** *Journal of Vocational Behavior*, 1-4.
[8] Maina J J 2014 *Discovering hidden patterns: An overview of space syntax methods in architecture and housing research*. Ahmadu Bello University.
[9] Ostwald M J 2011 The Mathematics of Spatial Configuration: Revisiting, Revising and Critiquing Justified Plan Graph Theory. *Nexus Network Journal*, **13** 445-470.
[10] Rubin O, Nikolaeva A, Nello-Deakin S, Brömmelstroet M T 2020 *What can we learn from the COVID-19 pandemic about how people experience working from home and commuting?*.Centre for Urban Studies. https://urbanstudies.uva.nl/content/blog-series/covid-19-pandemic-working-from-home-and-commuting.html
[11] MacCormac R 1979 Housing and The Dilemma of Style. *Architectural Review*
[12] Seo K W 2012 DNA of the House. *Home Culture*, **9** 77-98.
[13] Hanson J 1994 Deconstructing Architects' Houses. *Environment and Planning B: Urban Analytics and City Science* **21**(6) https://doi.org/10.1068/b210675
[14] B. Hillier, J. Hanson, H. Graham 1986 Ideas are things: an application of the space syntax method to discovering house genotypes. *Environmental Planning B: Planning and Design* **14** pp 363-385

Acknowledgment
This study is funded by the LRGS Research Grant program; New Housing Development Concept for B40 Group in Malaysia based on Social-Spatial and Economic Integrations. Ministry of Higher Education. LRGS/1/2016/UKM/02/1/3.

IOP Conf. Series: Earth and Environmental Science 1007 (2022) 012020 doi:10.1088/1755-1315/1007/1/012020

Imputation of missing values for generating typical meteorological year (TMY) with data decomposition and recurrent neural networks

K Chreng[1,2], H S Lee[1,5], R P Pradana[1,3], T Q Trong[1], I D G Arya Putra[4] and H Nimiya[4]

[1] Transdisciplinary Science and Engineering Program, Graduate School of Advanced Science and Engineering, Hiroshima University, 1-5-1 Kagamiyama, Higashi-Hiroshima, Hiroshima 739-8529, Japan
[2] Division of Planning and Technique, Electricité du Cambodge, Phnom Penh, Cambodia
[3] Center for Research and Development, Indonesian Agency for Meteorology Climatology and Geophysics (BMKG), Jakarta, Indonesia
[4] Graduate School of Science and Engineering, Kagoshima University, Kagoshima, Japan

Abstract. Typical meteorological year (TMY) for a specific location is critical information when designing low-carbon and energy-saving buildings. However, in developing countries, long-term observations of weather are now readily available and even mixed with missing values. In this study, a nonlinear autoregressive (NAR) recurrent neural network model in combination with improved complete ensemble empirical mode decomposition with adaptive noise (ICEEMDAN) method is demonstrated for treating the missing values in observed daily average air temperature at Bogor City in Indonesia. The prediction results for missing values indicate that the ICEEMDAN-NAR hybrid model performs very well with high accuracy when compared with the observed in the validation and model comparison phases. Particularly, the relatively lower frequency oscillation modes in observed data can be predicted well. Thus, this method can be used for relatively medium- and long-term prediction of missing values with respect to the given data/input period. critical to consider cities and buildings from the eyes of older people in society.

1. Introduction

Typical meteorological year (TMY) is a collation of selected weather data for a specific location, containing one year of hourly data of solar radiation and meteorological elements that best represents median weather conditions over a multiyear period, for example 10 years. TMY data is frequently used in building simulation, in order to assess the expected heating and cooling costs for the design of the building. It is also utilised by building designers and others for modelling renewable energy conversion systems such as solar energy systems including solar domestic hot water systems and large-scale solar thermal power plants. Since it represents typical rather than extreme conditions, it is not suited for designing systems to meet the extreme conditions occurring at a location [1, 2].

[5] Corresponding author: leehs@hiroshima-u.ac.jp

Low-carbon is strongly required in the building sector even in developing countries to achieve the targets of the Paris Agreement. In the Global South, the growing energy consumption caused by air-conditioning is a major concern in its hot and humid climates. However, TMY data for a specific site in developing countries is not usually available, and even the multiyear meteorological observations are not readily available with frequently presenting missing values. Therefore, this study demonstrates the treatment of missing values in multiyear meteorological records from the major cities in Indonesia with improved complete ensemble empirical mode decomposition (EMD) with adaptive noise and recurrent neural networks for the generation of TMY towards the design of affordable low-carbon buildings with passive cooling techniques.

2. Material and method

2.1 Data

Daily weather element observations, such as daily minimum, maximum, and average temperature, daily average relative humidity, rainfall, duration of solar radiation, maximum wind speed, wind direction at maximum speed, daily average wind speed, and average wind direction, are collected from Indonesian Agency for Meteorology Climatology and Geophysics (BMKG) for 19 major cities in Indonesia. Table 1 shows the summary of the daily weather records from the major cities.

The observation period varies from the longest data, 1972-2020, with 17.52% missing values at Medan City to the shortest data, 2010-2020, with 7.45% of missing values at Semarang City. The percent of missing values ranges from 2.86% at Tangerang Selatan to 32.32% at Kupang City. In this study, the daily weather observation at Bogor City from 1984 to 2020 with 7.1% missing values is used to demonstrate the treatment of missing values with improved complete ensemble EMD with adaptive noise (ICEEMDAN) and recurrent neural networks.

Table 1. Summary of daily weather elements observations at 19 major cities in Indonesia.

Station ID	Coordinate		Station Name	Regency/City	City	Elevation	Period		Years	Missing data	
	Lat.	Long.				(m) ASL	From	To		Period	%
96017	5.40	95.46	Stasiun Klimatologi Aceh Besar	Kab. Aceh Besar	Banda Aceh	40	1994	2020	26	1994-2020	11.02
96031	3.62	98.71	Stasiun Klimatologi Deli Serdang	Kab. Deli Serdang	Medan	25	1972	2020	48	1972-2020	17.52
96167	-0.55	100.30	Stasiun Klimatologi Padang Pariaman	Kab. Padang Pariaman	Padang	128	1985	2020	35	1985-2020	18.72
96255	-3.87	102.31	Stasiun Klimatologi Bengkulu	Kota Bengkulu	Bengkulu	12	1985	2020	35	1985-2020	14.73
96191	-1.60	103.48	Stasiun Klimatologi Muaro Jambi	Kab. Muaro Jambi	Jambi	24	1998	2020	22	1998-2020	4.46
96223	-2.93	104.77	Stasiun Klimatologi Palembang	Kota Palembang	Palembang	11	1976	2020	44	1976-2020	11.58
96753	-6.5	106.75	Stasiun Klimatologi Bogor	Kota Bogor	Bogor	207	1984	2020	36	1984-2020	7.1
96733	-6.26	106.75	Stasiun Klimatologi Tangerang Selatan	Kota Tangerang Selatan	Jakarta	27	1979	2020	41	1979-2020	2.86
96783	-6.88	107.60	Stasiun Geofisika Bandung	Kota Bandung	Bandung	791	1975	2020	45	1975-2020	6.53
96835	-6.98	110.38	Stasiun Klimatologi Semarang	Kota Semarang	Semarang	6	2010	2020	10	2010-2020	7.45
96943	-7.90	112.60	Stasiun Klimatologi Malang	Kab. Malang	Malang	590	1989	2020	31	1989-2020	4.08
96933	-7.22	112.72	Stasiun Meteorologi Perak I	Kota Surabaya	Surabaya	3	2010	2020	10	2010-2020	10.77
96583	0.08	109.19	Stasiun Klimatologi Mempawah	Kab. Pontianak	Pontianak	2	1985	2020	35	1985-2020	1.75
96687	-3.46	114.84	Stasiun Klimatologi BanjarBaru	Kota Banjarbaru	Banjarbaru	55	2010	2020	10	2010-2020	18.51
97236	-8.34	114.62	Stasiun Klimatologi Jembrana	Kab. Jembrana	Jembrana	25	1999	2020	21	1999-2020	13.05
97242	-8.64	116.17	Stasiun Klimatologi Lombok Barat	Kab. Lombok Barat	Lombok	55	1998	2020	22	1998-2020	10.04
97184	-4.99	119.57	Stasiun Klimatologi Maros	Kab. Maros	Maros	4	1999	2020	21	1999-2020	7.74
97012	1.55	124.92	Stasiun Klimatologi Minahasa Utara	Kab. Minahasa Utara	Minahasa	84	1973	2020	47	1973-2020	10.71
97374	-10.14	123.67	Stasiun Klimatologi Kupang	Kota Kupang	Kupang	19	2010	2020	10	2010-2020	32.32

2.2 Empirical mode decomposition (EMD) and improved complete ensemble EMD with adaptive noise (ICEEMDAN)

The EMD is an adaptive method to decompose a signal $x(t)$ into a number of intrinsic mode functions (IMFs), which become the basis representing the signal. The algorithm can be described as follows [3, 4]:

Step 1. Set the IMF index $k = 0$ and find all extrema of the *0*th residue $r_0 = x$.
Step 2. Interpolate between minima (maxima) of r_k to obtain the lower (upper) envelope e_{min} (e_{max}).
Step 3. Compute the mean envelope $m = (e_{min} + e_{max})/2$.
Step 4. Compute the IMF candidate $d_{k+1} = r_k - m$.
Step 5. Is d_{k+1} an IMF?
Yes. Save d_{k+1}, compute the residue $r_{k+1} = x - \sum_{i=1}^{k} d_i$, do $k = k + 1$, and treat r_k as input data in step 2.
No. Treat dk+1 as input data in step 2.
Step 6. Continue until the final residue r_k satisfies some predefined stopping criterion.

The refinement process (steps 2 to 5) is needed to extract every mode by a certain number of iterations and is named as *sifting* process. EMD is ideally suited for analysing data from non-stationary and nonlinear processes. However, EMD still cannot resolve the most complicated cases, when the processes are nonlinear, and the noises also have the same time-scale as the signal; their separation becomes impossible (mode mixing). Ensemble EMD (EEMD) defines the "true" modes (here noted as $\overline{IMF} = \overline{d}$ in what follows) as the average of the corresponding IMFs obtained from an ensemble of the original signal, x, plus different realisations of finite variance white noise [5]. However, EEMD has still disadvantages: (i) the decomposition is not complete and (ii) different realisations of signal plus noise might produce different number of modes particularly in low frequency [6].

To improve these drawbacks, a new ensemble method called complete EEMD with adaptive noise (CEEMDAN) was proposed [7]. However, CEEMDAN still has some aspects that have to be improved further: (i) its modes contain some residual noise; and (ii) the signal information appears "later" than in EEMD with some "spurious" modes in the early stages of the decomposition. Taking into account the two drawbacks of CEEMDAN, an improved algorithm for CEEMDAN (ICEEMDAN) is proposed [3]. In order to obtain noise realisations with smaller amplitudes for the late stages of the decomposition, in the rest of the modes the noise will be used resulting from its pre-processing by EMD, i.e., without normalising them by its standard deviation. In this study, ε=0.02, I=500, a few hundred of realisation, and the same signal noise ratio (SNR) are used for all the stages in all analysis. The explanations on EMD and its variations are greatly reduced in this paper. Therefore, refer to Lee [6] for the full descriptions on EMD, EEMD, CEEMDAN, ICEEMDAN, and their algorithms.

2.3 Recurrent neural networks – NAR and NARX

The artificial neural network model consists of many neural network models. A nonlinear autoregressive model (NAR) and a nonlinear autoregressive model with exogenous inputs (NARX) are among the groups of recurrent neural networks (RNNs). Also, NARX and NAR are nonlinear dynamic models that fit the time series prediction. NAR is suitable for time series prediction. NAR neural network model is basically used the historical dataset only as the input data to predict itself for future values [8]. The output function of the NAR model can be mathematically given in Eq. (1):

$$y(t) = f\left(y(t-1), y(t-2), \cdots, y(t-d_y)\right) \quad (1)$$

where y is historical power demand dataset over (t) time, d_y is the feedback delay or lagged feedback output, and f is the activation function in the neural network model. The feedback delay impacts the simulated closed-loop output as well as multistep predictive outcome, and it was defined by determining the auto-correlation function of the input dataset (weather elements). Consequently, sufficient feedback delay is crucially important, which permits the training process could visibly understand the characteristic of the historical data.

The NARX (nonlinear autoregressive neural network with exogenous inputs) has the same structure as the NAR model, but the NARX model has external inputs. The equation of the NARX model is expressed as Eq. (2).

$$y(t) = f\left(y(t-1), y(t-2), \ldots, y(t-dy), x(t-1), x(t-2), \ldots, x(t-dx)\right) \quad (2)$$

Where x is the exogenous input dataset, and d_x is the input delay (ID). The NARX's configurations is based on the NAR's configuration.

2.4 ICEEMDAN-Recurrent neural network hybrid model

In this study, the hybrid modelling framework is the combination of NAR model and the ICEEMDAN decomposition procedure. The original dataset was used to decompose into IMF signals, and each IMF was then inputted into NAR model. After the prediction outcome of each IMF is obtained, the composition process is applied to estimate the original amplitude of the dataset. The final stage is sought to the best hybrid model between ICEEMDAN-NAR hybrid and single NAR approaches. Figure 1 depicts the hybrid model framework. If the decomposition of the input data is not considered in figure 1, then it becomes identical with the single NAR model.

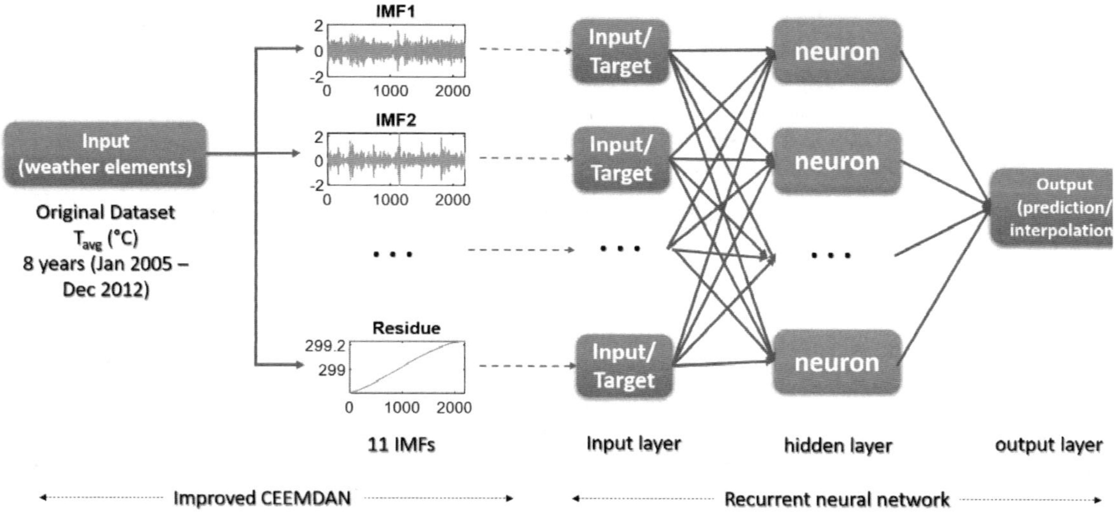

Figure 1. Improved CEEMDAN – NAR hybrid Model Framework

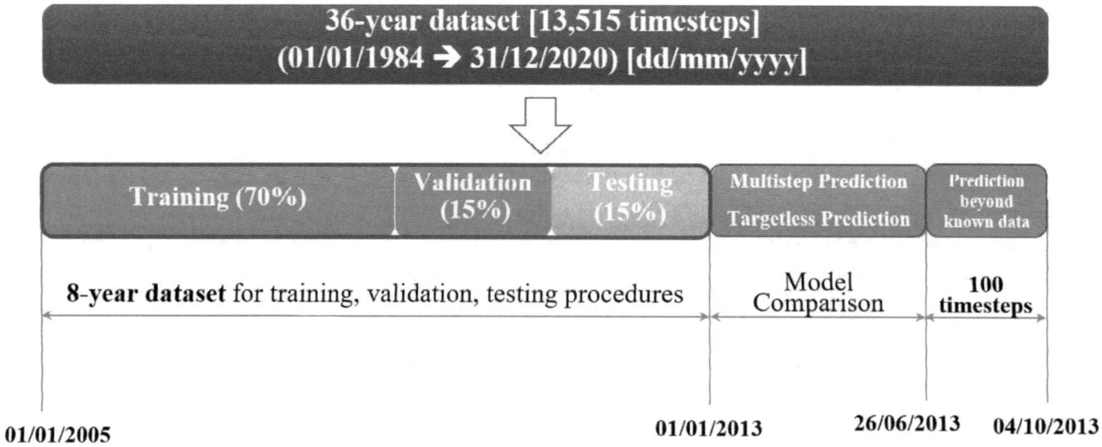

Figure 2. Application and prediction phases of neural network models for T_{avg} data at Kota Bogor.

2.5 Data processing and experiment

As described, the daily weather observation at Kota Bogor is used for the treatment of missing values for the period from Jan 2005 to Dec 2013. First, the missing values within the data period are set to *zero* based on the assumption that it would not affect the local characteristics of data in the decomposition process since the EMD and its variations are adaptive for nonlinear and non-stationary data. Then, the daily average temperature (T_{avg}) is applied to ICEEMDAN to decompose the original time-series into

IMFs. Statistical significance test is then carried out for the resulting IMFs to prove the significances of the modes.

After the data decomposition is completed with ICEEMDAN, experiments were conducted to predict the missing values with neural network model. The statistically significant modes are used as input in the prediction of missing values using the NAR model. Figure 2 shows the phases of training, validation and testing of neural networks from Jan 2005 to Dec 2012, model comparison from 1st Jan to 26th June 2013, and the prediction phase for the actual missing values of T_{avg} from 27th June to 4th October in 2013.

2.6 Performance evaluation

The performance of the models was evaluated by using error indices: the normalised root-mean-square error (NMSE) [9], coefficient of determination (R^2), mean absolute error ($MAE([MW])$, mean absolute percent error ($MAPE([\%])$), and root-mean-square error ($RMSE([MW])$, which were expressed as given in Eq. (3) to Eq. (7).

$$NMSE = \frac{\sum_{i=1}^{N}\left(P_{predicted}^{i} - P_{actual}^{i}\right)^2}{\sum_{i=1}^{N}\left(P_{actual}^{i} - \overline{P}_{actual}\right)^2} \quad (3)$$

$$R^2 = 1 - \frac{\sum_{i=1}^{N}\left(P_{predicted}^{i} - P_{actual}^{i}\right)^2}{\sum_{i=1}^{N}\left(P_{actual}^{i} - \overline{P}_{actual}\right)^2} \quad (4)$$

$$MAE = \frac{1}{N}\sum_{i=1}^{N}\left|P_{predicted}^{i} - P_{actual}^{i}\right| \quad (5)$$

$$MAPE = \frac{100}{N}\sum_{i=1}^{N}\left|\frac{P_{actual}^{i} - P_{predicted}^{i}}{P_{actual}^{i}}\right| \quad (6)$$

$$RMSE = \sqrt{\frac{1}{N}\sum_{i=1}^{N}\left(P_{predicted}^{i} - P_{actual}^{i}\right)^{2}} \quad (7)$$

where $P_{actual}^{i}, P_{predicted}^{i}, \overline{P}_{actual}, \overline{P}_{predicted}$ are the i^{th}actual value, the i^{th}predicted value, the mean of the actual values in period $N\left(\frac{1}{n}\sum_{i=1}^{N} P_{actual}^{i}\overline{P}_{actual}\right)$, and the mean of the predicted values in period $N\left(\frac{1}{n}\sum_{i=1}^{N} P_{predicted}^{i}\overline{P}_{predicted}\right)$ for the power demand, respectively, and N is the total value during the testing period. The R^2 demonstrates the goodness of fit, where a high value (close to 1) expresses a linear relationship between the actual and predicted values [10-12]. The MAPE represents the best model's behaviour when the value is close to zero. The MAE and RMSE are used as indicators of the deviation between the actual and predicted values in MW units [13]. According to Guiamel and Lee [14, 15], goodness of fit can be identified by the R^2 value, as described in Table 2.

Table 2. The statistical performance evaluation for determining goodness of fit

Rank	Description	Performance Ratting
1	$R^2 \geq 0.80$	Excellent
2	$0.70 < R^2 < 0.60$	Good
3	$0.60 < R^2 < 0.50$	Satisfactory
4	$R^2 \leq 0.50$	Not satisfactory

3. Results

Figure 3 illustrates the results from the model comparison phase from 1st January to 26th June in 2013. As notified in figure 3, the predicted output of T_{avg} agrees very well with the target (the observed) values, in particular, in the relatively lower frequency IMFs from IMF3 even to the residual. In the higher frequency IMFs for IMF1 and IMF2, the differences between the target (the observed) and the output (the predicted) values are remained very slight and acceptable. Then, the results of real prediction for the missing period for 100 timesteps starting from 27th June are presented also in Figure 3.

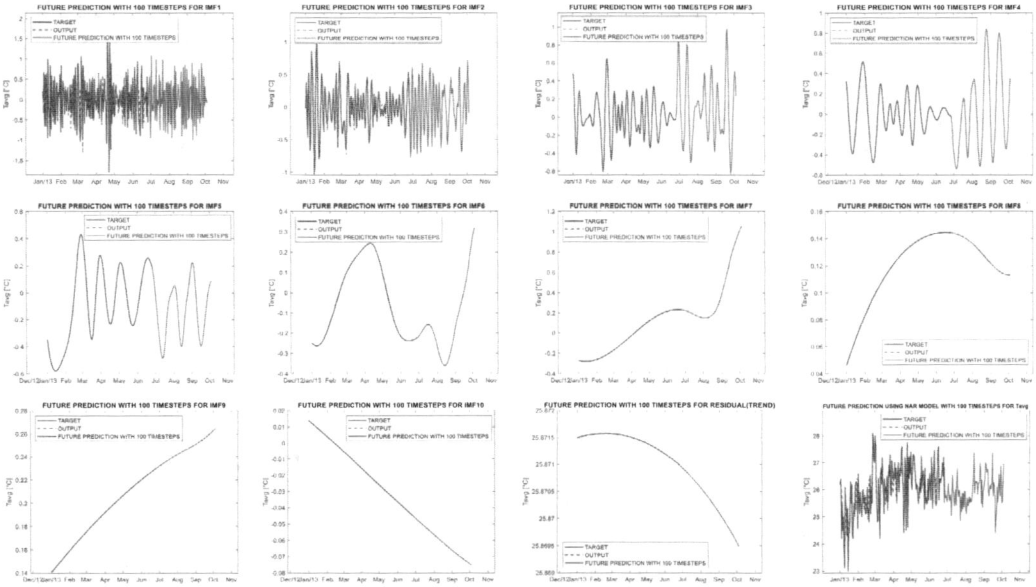

Figure 3. Results of model comparison (from 1st January to 26th June 2013) and 100-timestep future prediction (from 27th June to 4th October, 2013) phase in Fig. 2 of each IMF and NAR model using original T_{avg} data.

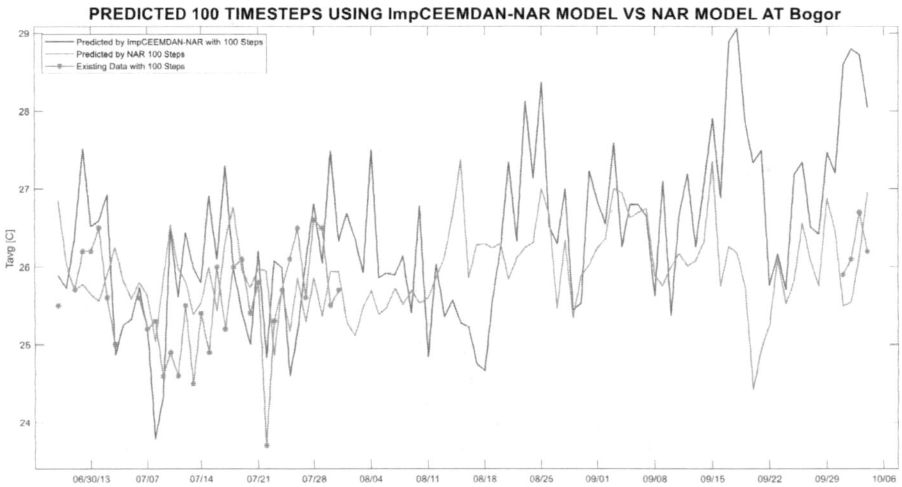

Figure 4. Comparison of prediction results of T_{avg} by NAR model (blue), Improved CEEMDAN-NAR hybrid model (red), and the observed data with missing values (green)

To assess the prediction performance of the single and the hybrid model, the statistical indices, such as NMSE, R^2, MAE, MAPE, and RMSE as defined in the previous section, are calculated and presented in Table 3. It shows that the model performance increases considerably by combining the two methods from 0.728 to 0.923 and from 0.4760 to 0.2530 for R^2 and RMSE, respectively. As mentioned, the prediction errors are dominant in the higher frequency IMFs, such as 0.1369 of MAE and 0.0239 of MAE for IMF1 and IMF2, respectively, while their MAPE are 18.380% for IMF1 and 1.285% for IMF2. However, their absolute values still remained small. The accuracies of the relatively lower frequency IMFs are very high such that relatively longer oscillations in the data can be predicted with high accuracy by considering those low frequency modes. Figure 4 exhibits the prediction results of T_{avg} by the single NAR model and the hybrid model in the comparison with the original T_{avg} data with a missing period.

Table 3. Summary of statistical indices showing the performance of various predictive models. Single model indicates the NAR model without the combination with decomposition method. Model for IMFx indicates the prediction of single IMF in various frequency levels. Hybrid models consider all IMFs in the prediction.

Model	NAR				
	NMSE	R^2	MAE [°C]	RMSE [°C]	MAPE [%]
Single model	0.272	0.728	0.2538	0.4760	0.1171
IMF1	0.2100	0.790	0.1369	0.2459	18.380
IMF2	0.0171	0.983	0.0239	0.0431	1.2850
IMF3	0.0003	0.999	0.0021	0.0039	0.1009
IMF4	5.3e-05	0.999	0.0012	0.0015	0.0012
IMF5	1.5e-05	0.999	0.0009	0.0011	0.0106
IMF6	9.2e-07	0.999	0.0002	0.0002	0.0024
IMF7	1.7e-07	0.999	6.3e-05	7.8e-05	0.0013
IMF8	4.0e-07	0.999	1.5e-05	1.8e-05	0.0003
IMF9	2.9e-07	0.999	1.2e-05	1.4e-05	0.0001
IMF10	1.2e-07	0.999	5.0e-06	5.9e-06	0.0009
Residue	0.0007	0.999	3.8e-06	4.6e-06	1.8e-07
Hybrid model	0.0765	0.923	0.1415	0.2530	0.0046

4. Conclusions

This study illustrates how to treat the missing values in the observed weather elements by using a data decomposition method and artificial neural network model. The decomposition method used are the empirical mode decomposition (EMD) and its variation, ICEEMDAN, and the neural network model used is a recurrent neural network model, NAR. In the demonstration, the long-term weather observations with frequent missing values within the observation period at Kota Bogor in Indonesia is applied to the ICEEMDAN-NAR hybrid model for the prediction of missing values.

The prediction results indicate that the hybrid model performs very well with high accuracy when comparing with the observed in the validation and model comparison phases. Particularly, the relatively lower frequency oscillation modes in observed data can be predicted well. Thus, this method can be used for relatively medium- and long-term prediction of missing values with respects to the given data/input period. Also, more careful investigations and applications are necessary in future for the limitations and improvements of the proposed method in this study.

5. References

[1] Wikipedia Typical meteorological year
[2] Wilcox S and Marion W 2008 *Users manual for TMY3 data sets*. National Renewable Energy Laboratory (NREL) p 58
[3] Colominas M A, Schlotthauer G, Torres M E 2014 *Biomedical Signal Processing and Control* **14** 19-29
[4] Huang N E, Shen Z, Long S R, Wu M C, Shih H H, Zheng Q, et al. 1998 *Proc R Soc London* **454** 903-95

[5] Wu Z, Huang N E 2009 *Advances in Adaptive Data Analysis* **1** 1-41
[6] Lee H S 2018 *Journal of Coastal Research* **SI85** 526-30
[7] Colominas M A, Schlotthauer G, Torres ME and Flandrin P 2012 *Advances in Adaptive Data Analysis* **04** 1250025
[8] Zhang G, Zhou H, Wang C, Xue H, Wang J, Wan H 2020 *IEEE Transactions on Geoscience and Remote Sensing* **58** 3544-57
[9] Li Q, Liang S Y, Yang J, Li B 2016 *Entropy* **18** 23
[10] Hussainzada W, Lee H S, Vinayak B, Khpalwak G F 2021 *Journal of Hydrology: Regional Studies* **36** 100835
[11] Hussainzada W, Lee H S 2021 *Hydrology* **8** 16
[12] Mohammadi K, Shamshirband S, Tong C W, Arif M, Petković D, Ch S 2015 *Energy Conversion and Management* **92** 162-71
[13] Prasad R, Deo RC, Li Y, Maraseni T 2017 *Atmospheric Research* **197** 42-63
[14] Guiamel I A, Lee H S 2020 *Civil Eng Journal* **6** 626-48
[15] Guiamel I A, Lee H S 2020 *Energy Reports* **6** 1010-28

Acknowledgments
This research was supported by the Science and Technology Research Partnership for Sustainable Development (SATREPS) in collaboration between Japan Science and Technology Agency (JST, JPMJSA1904) and Japan International Cooperation Agency (JICA)

Impregnation analysis of phenol-formaldehyde resin on sengon wood to explore the potential of sengon wood as construction material

A A Birrnaqiy[1,3], Basyaruddin[1] and J Malik[2]

[1] Kalimantan Institute of Technology (ITK), Karang Joang, Balikpapan 76127, Indonesia
[2] Forest Products Research and Development Centre, Pasir Jaya, Bogor 16610, Indonesia

Abstract. The current climate emergency is a growing concern, while construction activities are one of the biggest causes. Therefore, research is needed that leads to the application of construction with environmentally friendly materials such as wood. In this study, the wood used is sengon wood (*Falcataria moluccana*) because it has very abundant availability. However, sengon wood has low strength and durability, so treatment is needed to improve the quality of sengon wood so that its potential can be utilized. There were seven test variables, namely untreated sengon sawn wood, vacuum pressure impregnation with 5%, 10%, 15% phenol-formaldehyde (PF) resin solution, cold immersion impregnation with 5%, 10%, and 15% PF. The results of this study showed that treatment of cold immersion concentrate PF 10% increases tensile strength parallel to fibre up to 91.5%, the biggest increase in value against one of the mechanical properties, but to improve the mechanical properties of the highest average is variable B with vacuum pressure treatment of concentrates 5% with the increase up to 9.1%. Regardless of the comparison of these variables, that the value of the mechanical properties of sengon wood as sawn wood without treatment already has the potential to be used as a construction material compared to the value of the lowest strength class of construction sawn wood material. According to this research, the average density of sengon wood is about 300 kg/m^3 which tends to be categorized as a lightweight construction material, where the concept of a lightweight structure can trigger increased innovation in creating more varied architectural and structural designs.

1. Introduction

To create an ideal living place, it needs to be based on the concept of sustainable development, especially with the current climate emergency. The main principle in sustainable development is the defence of the quality of life for all human beings in the present and in the future in a sustainable manner based on economic prosperity, social justice, and environmental preservation. Healthy ecosystems and environments are necessary for the survival of humans and other organisms [1]. The current climate emergency due to excessive carbon emissions continues to be the focus of the world in dealing with climate change, as has been an amendment to the United Nations Framework Convention on Climate

[3] Corresponding author: alibirrnaqiy@gmail.com

Change (UNFCCC), Paris Agreement 2015 [2]. Carbon emission has been a serious issue in the construction sector for the past few years, attracting the world's attention. The construction industry accounts for around 36% of global emissions [3]. One way to reduce the problem of the climate emergency is to use materials from renewable resources and minimize the carbon footprint [4]. Wood is a material that can be directly taken from nature in the form of trees. Wood requires less energy than concrete and steel in its production and recycling process. Its natural character makes wood not produce waste and can be reused before being used as fuel. Wood is an inexhaustible source of natural wealth if it is managed/managed in good ways. This means that when trees are cut down (in the forest) for wood, the forest land must be replanted immediately so that the source of wood is not exhausted. Thus, wood can be said to be a renewable resource (a source of natural wealth that can be renewed/held again) [5]. The use of wood as the main material for house construction is one solution to minimize the current climate emergency. Efforts to improve the efficiency of the use of wood raw materials can be made by utilizing fast-growing, and low-quality wood species, one of which is sengon wood (*Falcataria moluccana*) [6]. Sengon wood is a lightweight material, where the concept of a lightweight structure can trigger increased innovation in creating more varied architectural and structural designs [7] [8]. Sengon as forest vegetation is quite prospective in an effort to increase the income of rural communities, and it also has environmental benefits in terms of reducing CO2 emissions [9]. Although sengon wood has low strength and durable qualities that cannot last long, it is not impossible to develop it as a construction material. For example, sengon sea wood (*Paraserianthes falcataria*), which has been preserved in a house building made in 1963 by the Centre for Research and Development of Forest Engineering and Forest Product Processing (Pustekolah), is still standing for more than 50 years [10]. In this study, to improve the quality of sengon wood, the vacuum pressure method and cold immersion were used because it has been generally proven as a way of impregnation of a solution into wood [11]. The solution used for impregnation is a solution of phenol-formaldehyde (PF) resin because it can increase the dimensional stability of the wood [12]. Thus, the purpose of this study is to explore the potential of sengon wood as construction material that has been treated with impregnation.

2. Methods

There were seven treatment variables for the specimens in this study, namely wood that was not given any treatment as a comparison variable (A), vacuum pressure impregnation with 5% phenol-formaldehyde (PF) resin solution (B), vacuum pressure impregnation with 10% phenol-formaldehyde (PF) resin solution (C), vacuum pressure impregnation with 15% phenol-formaldehyde (PF) resin solution (D), cold immersion impregnation with 5% phenol-formaldehyde (PF) resin solution (E), cold immersion impregnation with 10% phenol-formaldehyde (PF) resin solution (F), cold immersion impregnation with 15% phenol-formaldehyde (PF) resin solution (G). The total number of test objects is 252 pieces.

2.1. Procedure for preparation of materials

Preparing wood as a test object with dimensions that have been determined in SNI and Australian Standard for physical and mechanical tests, with a total of 210 specimens for mechanical testing and 42 for physical testing as follows (table 1):

Table 1. Variable number of test objects.

Variable	Fc	$Fc\perp$	Ft	Fb	Fv	ρ
A	A1	A7	A13	A19	A25	A31
	A2	A8	A14	A20	A26	A32
	A3	A9	A15	A21	A27	A33
	A4	A10	A16	A22	A28	A34
	A5	A11	A17	A23	A29	A35
	A6	A12	A18	A24	A30	A36

B	B1	B7	B13	B19	B25	B31
	B2	B8	B14	B20	B26	B32
	B3	B9	B15	B21	B27	B33
	B4	B10	B16	B22	B28	B34
	B5	B11	B17	B23	B29	B35
	B6	B12	B18	B24	B30	B36
C	C1	C7	C13	C19	C25	C31
	C2	C8	C14	C20	C26	C32
	C3	C9	C15	C21	C27	C33
	C4	C10	C16	C22	C28	C34
	C5	C11	C17	C23	C29	C35
	C6	C12	C18	C24	C30	C36
D	D1	D7	D13	D19	D25	D31
	D2	D8	D14	D20	D26	D32
	D3	D9	D15	D21	D27	D33
	D4	D10	D16	D22	D28	D34
	D5	D11	D17	D23	D29	D35
	D6	D12	D18	D24	D30	D36
E	E1	E7	E13	E19	E25	E31
	E2	E8	E14	E20	E26	E32
	E3	E9	E15	E21	E27	E33
	E4	E10	E16	E22	E28	E34
	E5	E11	E17	E23	E29	E35
	E6	E12	E18	E24	E30	E36
F	F1	F7	F13	F19	F25	F31
	F2	F8	F14	F20	F26	F32
	F3	F9	F15	F21	F27	F33
	F4	F10	F16	F22	F28	F34
	F5	F11	F17	F23	F29	F35
	F6	F12	F18	F24	F30	F36
G	G1	G7	G13	G19	G25	G31
	G2	G8	G14	G20	G26	G32
	G3	G9	G15	G21	G27	G33
	G4	G10	G16	G22	G28	G34
	G5	G11	G17	G23	G29	G35
	G6	G12	G18	G24	G30	G36

2.2. Wood impregnation procedure

For the immersion impregnation method, the test object used has a moisture content of 8%. Because the density of wood is lighter than water, the wood is held back by stones so that it does not stick out to the surface of the water, and the test object is left in a solution of phenol-formaldehyde (PF) resin for 168 hours (7 days) [13]. For vacuum pressure impregnation, the specimen must have a moisture content of 8%. In the vacuum pump process, it was maintained for 30 minutes, then drained the phenol-formaldehyde (PF) resin solution from the supply container into the vacuum pressure tank while maintaining the initial vacuum, then switched to applying pressure of 15 kg/cm^2 for 60 minutes [14].

2.3. Wood physical test procedure

The physical test of the wood is carried out in the form of measuring dimensions, moisture content, and density of the test object. Physical tests were carried out before and after the impregnation treatment with the condition that the specimens were in the condition of water content 8%. Physical test objects amounted to 42 pieces.

2.4. Wood mechanical test procedure

There are five wood mechanical tests that were carried out which are testing the compressive strength parallel to the fibre, compressive strength perpendicular to the fibre, tensile strength, flexural strength, and shear strength. These five values are the mechanical properties that are needed in the planning of the wooden building structure [15]. These five wood mechanical tests that were carried out could be seen as follows (figure 1):

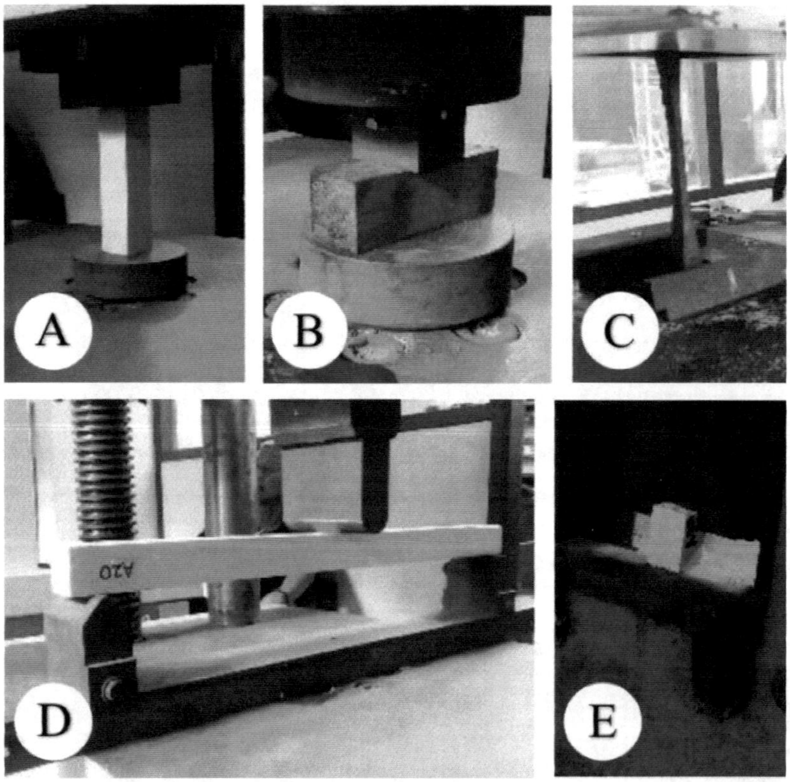

Figure 1. Wood mechanical test: compressive strength parallel to the fibre (A); compressive strength perpendicular to the fibre (B); tensile strength (C); flexural strength (D); shear strength (E).

Mechanical tests were carried out after impregnation treatment with the condition that the specimens were in a condition of moisture content 8% and had been left for 7 days from the time the specimens were removed from the oven with the aim that the specimens underwent conditioning prior to mechanical testing. Mechanical test objects totalled 210 pieces consisting of 6 pieces for each mechanical property, but only 5 pieces were taken in the mechanical strength analysis for each mechanical property.

3. Result and discussion

According to the physical test of sengon sawn wood that has been carried out, the average density is about 0.3 gram/cm^3 (300 kg/m^3). To easily understand the physical changes that occur after the test object is impregnated, the density difference between before and after the impregnation treatment should be known. Because, with this value, we can find out whether the impregnation treatment increases the density and mass or even decreases it. The value of the difference in density between before and after the impregnation treatment is as follows (table 2):

Table 2. Difference in density between before and after the impregnation treatment.

Variable	Density Change (gram/cm^3)
A	0.001
B	0.001
C	0.018
D	0.071
E	0.002
F	0.006
G	0.013

Thus, it can be seen that the higher the concentration of the solution, the more the wood density increases.

The following is a recapitulation of the five strength values reviewed, which are taken from the average value of 5 samples for each treatment variable (table 3):

Table 3. Recapitulation of mechanical properties values.

Variable	*Fc* (kg/cm^2)	*Fc*⊥ (kg/cm^2)	*Ft* (kg/cm^2)	*Fb* (kg/cm^2)	*Fv* (kg/cm^2)
A	253.927	52.828	250.808	418.704	53.984
B	233.209	45.159	431.079	469.442	55.363
C	297.757	62.347	391.691	333.855	48.853
D	237.034	47.528	250.870	359.743	65.549
E	230.249	41.365	366.885	369.725	50.859
F	229.886	42.749	480.286	387.528	56.595
G	235.237	53.262	373.679	413.905	51.362

Comparisons cannot be made directly with values taken directly from the table 3 because the values between these strength capacities cannot be compared and are not comparable values. So that first, the data must be processed into a percentage change in the strength value based on treatment A (test object without treatment) as a reference for changes as follows (table 4):

Table 4. Percentage of strength capacity comparison between treatments.

Variable	*Fc* (%) - 1	*Fc*⊥ (%) - 2	*Ft* (%) - 3	*Fb* (%) - 4	*Fv* (%) - 5
A	0.0	0.0	0.0	0.0	0.0
B	-8.2	-14.5	71.9	12.1	2.6
C	17.3	18.0	56.2	-20.3	-9.5
D	-6.7	-10.0	0.0	-14.1	21.4
E	-9.3	-21.7	46.3	-11.7	-5.8
F	-9.5	-19.1	91.5	-7.4	4.8
G	-7.4	0.8	49.0	-1.1	-4.9

Visualization of the comparison of the strength capacity values in the table 4 can be seen as follows (figure 2):

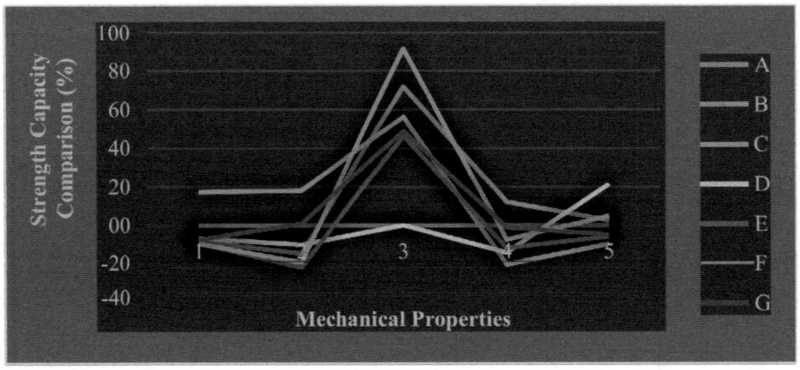

Figure 2. Graph of comparison of strength values between variables.

Shown that there is an increase of up to 91.5% of the tensile strength on the variable F. So, if averaged, the rating of the mechanical properties is obtained as follows (table 5):

Table 5. Inter-treatment strength capacity rating.

Variable	Total (%)	Mean (%)	Rank
A	0.0	0.0	5
B	63.9	9.1	1
C	61.7	8.8	2
D	-9.3	-1.3	7
E	-2.2	-0.3	6
F	60.3	8.6	3
G	36.4	5.2	4

Based on the average value of the increase in mechanical strength, the highest value obtained is the variable B, with a value of 9.1%.

Regardless of the comparison of these variables, that the value of the mechanical properties of sengon wood as sawn wood without treatment already has the potential to be used as a construction material compared to the value of the lowest strength class of construction sawn wood material as seen below (table 6) [16]:

Table 6. Lowest strength class of construction sawn wood material.

Fc (kg/cm^2)	$Fc\perp$ (kg/cm^2)	Ft (kg/cm^2)	Fb (kg/cm^2)	Fv (kg/cm^2)
163.155	20.394	81.577	142.760	17.335

Finally, focus on the aspect of the weight of sengon wood which tends to be categorized as a lightweight construction material. Based on this research, sengon wood has a density of about 300 kg/m^3, which is relatively lighter than normal concrete, which has a density of about 2400 kg/m^3, which means that sengon wood is 8 times lighter than normal concrete [17]. The uses of the devices and configurations which produce efficient and therefore lightweight structures – the complex cross-section, the circular 'lightening' hole, triangulation of elements and profiling to conform to bending moment diagrams – are not always appropriate from the technical viewpoint in the context of architecture where they are justified technically only in situations in which an efficient, lightweight structure is required [7].

Lightweight structures, built responsibly and disciplined, may contribute heavily to an enriched architecture. Light, filigree and soft evokes more pleasant sensations than heavy, bulky and hard. In the typical lightweight structure, the flow of forces is visible and the enlightened care to understand what they see. Thus, lightweight structures with their rational aesthetics may solicit sympathies for technology, construction and engineers. They may help us to escape the wide-spread monotony and drabness in today's structural engineering which in turn will become again an essential part of the building culture [8].

4. Conclusion

Based on the physical properties, it was found that the highest density increase was sengon wood with vacuum pressure impregnation treatment with a 15% solution concentration (variable D) of 0.071 gram/cm^3, which was higher than the initial density. However, this does not mean that it has higher mechanical properties because it was found to undergo physical deformation after the vacuum pressure impregnation treatment. As for the review of the mechanical properties, there is the largest percentage increase in the tensile strength parallel to the fibre by 91.5% with cold-immersion impregnation treatment with 10% solution concentration (variable F). Nevertheless, the highest average increase in mechanical properties was sengon wood with vacuum pressure impregnation treatment with a solution concentration of 5% (variable B) of 9.1%. Regardless of the comparison of these variables, that the value of the mechanical properties of sengon wood as sawn wood without treatment already has the potential to be used as a construction material compared to the value of the lowest strength class of construction sawn wood. Besides sengon wood considered as a sustainable material, it is also a lightweight material, where the concept of a lightweight structure can trigger increased innovation in creating more varied architectural and structural designs, it may help us to escape the wide-spread monotony and drabness in today's architectural and structural design depicts as heavy, bulky and hard.

5. References

[1] Shah M M 2008 Sustainable Development *Encyclopedia of Ecology* vol 1 ed S E Jørgensen and B D Fath (Amsterdam: Elsevier Science) pp 3443-6
[2] Erickson L E and Brase G 2020 *Reducing Greenhouse Gas Emissions and Improving Air Quality: Two Interrelated Global Challenges* (Boca Raton: CRC Press) p 164
[3] Zainordin N and Zahra D B F 2020 *Proc. 3rd Int. Conf. on Separation Technology (Malaysia)* vol 200 (Netherland: Atlantis Press) pp 176-82
[4] Edenhofer O *et al.* 2011 *IPCC Special Report on Renewable Energy Sources and Climate Change Mitigation* (Cambridge: Cambridge University Press) p 230
[5] Frick H 1982 *Ilmu Konstruksi Bangunan Kayu* (Yogyakarta: Penerbit Yayasan Kanisius) p 292
[6] Basyaruddin, Suryaningsih A and Awali J 2019 Potensi Pemanfaatan Kayu Gelam dan Kayu Sengon dalam Dunia Konstruksi Berdasarkan Uji Kuat Lentur *Rekayasa Sipil* **13** 193-8
[7] Macdonald A J 2001 *Structure and Architecture* 2nd Edition (Oxford: Architectural Press) p 164
[8] Schlaich J and Bergermann R 2004 *Leicht Weit/Light Structures* (Munich: Prestel Pub) p 240
[9] Riyanto H D, Susi A and Ragil B W M P 2009 Kajian Sengon (Paraserianthes falcataria) sebagai Pohon Bernilai Ekonomi dan Lingkungan *Jurnal Penelitian Hutan Tanaman* **6** 201-8
[10] Swandayani T H 2015 *Pengawetan Kayu: Peningkatan Efesiensi Sumberdaya Hutan (Internet)* (Jakarta: Badan Penelitian, Pengembangan dan Inovasi Kementerian Lingkungan Hidup dan Kehutanan) (cited 2021 Jan 28) Available from: https://www.forda-mof.org/index.php/berita/post/2070
[11] Forest Products Laboratory 2010 *Wood Handbook - Wood as an Engineering Material* 100th Edition (Madison: U.S. Department of Agriculture, Forest Service, Forest Products Laboratory) p 508
[12] Gabrielli C P and Kamke F A 2010 Phenol-Formaldehyde Impregnation of Densified Wood for Improved Dimensional Stability *Wood Sci. Technol.* **44** 95-104

[13] Barly and Lelana N E 2010 Pengaruh Ketebalan Kayu, Konsentrasi Larutan dan Lama Perendaman terhadap Hasil Pengawetan Kayu *Jurnal Penelitian Hasil Hutan* **28** 1-8
[14] Malik J 2019 *Enhancing Timber Quality of Jabon Wood (Anthocepalus Cadamba) for High Quality Products by Treatment Through Densification and Impregnation with Merbau Extractives* (Melbourne: The University of Melbourne)
[15] Badan Standardisasi Nasional 2013 *Spesifikasi Desain untuk Konstruksi Kayu* SNI 7973:2013 (Jakarta: Badan Standardisasi Nasional) p 318
[16] Harrington J, Jacob M and Short C 2006 *Handbook on Structural Timber Design to Eurocode 5 (IS EN 1995-1-1) Rules Including Strength Capacity Tables for Structural Elements* (Dublin: COFORD, National Council for Forest Research and Development)
[17] Badan Standardisasi Nasional 2019 *Persyaratan Beton Struktural untuk Bangunan Gedung dan Penjelasan* SNI 2847:2019 (Jakarta: Badan Standardisasi Nasional) p 695

Acknowledgments
The results were obtained as part of undergraduate thesis research the Kalimantan Institute of Technology (ITK). This research was supported by DKM Al-Barokah PT Pamapersada Nusantara (Sangatta) and LPPM ITK (Balikpapan) as funding parties, also Forest Products Research and Development Centre (Bogor) as a support for research facilities.

IOP Conf. Series: Earth and Environmental Science 1007 (2022) 012022 doi:10.1088/1755-1315/1007/1/012022

A strategy of a masterplan of Landak general hospital: a study of architectural programming

M R Alhamdani[1,3] and D P Sari[2]

[1] Department of Architecture, Faculty of Engineering, Tanjungpura University, Jalan Prof. Dokter H. Hadari Nawawi, Pontianak 78115, West Kalimantan, Indonesia
[2] Department of Architecture, Pontianak State Polytechnic, Jalan Ahmad Yani, Pontianak 78124, West Kalimantan, Indonesia

Abstract. The plan for the development of the Landak District General Hospital is motivated by three things, namely the first from the mandate of law concerning hospitals, secondly the development of the Landak Regency in line with population growth, and the increasing demand and third the need for health services that occur in health centres. The aims of this research are to prepare of the initial stage of the Masterplan of the Landak District Hospital, for obtaining a harmonious and balanced use of regional space with the needs and capabilities of supporting urban growth and development by taking into account the harmony of the environment, the availability of a regional development plan that can function as a forum for integration for the interests of all parties. The results are the strategy of master plan development of Landak general hospital viewed from architectural programming namely activity patterns, zoning formation, space requirements, functional relationship, circulation patterns, building security and safety, landscape and open space, and parking requirements.

1. Introduction

Hospitals are part of the health system. Hospitals should constantly improve service quality following customer expectations to increase customer satisfaction [1]. Hospital as a health service facility has a mission to achieve development goals. The hospital's mission is said to be successful if the hospital can provide quality services by trying to improve the quality of services intensively and continuously and supported by the completeness of adequate infrastructure and facilities. The hospital will serve the needs of public health services, both general services and specialist services.

A consideration of the synchronization of the environment and the present architecture of healthcare services should reflect in the development of health care projects that put together best practices in energy-efficient construction, opening an original page in the evolution of architectural design in health care facilities and the general history of world design and construction [2]. Architectural programming is typically catching on as one of the primary stages of a building scheme. In practice, the procedure, driven by experts, involves the user as an information source and results in the program of needs for the building. The study proposes a Usability Briefing process model, where briefing is an active and continuous process throughout all the building phases: from pre-project, using design and construction phases to delivery and in-use. The essential values of a healthcare plan are a practical approach to

[3] Corresponding author: mridhaalhamdani@gmail.com

healthcare services, the efficient utilization of the building plus its services by the patient population serving, and in time the facility's ability to draw and preserve the best feature medical and administrative employees [3, 4]. The value of the master planning development in health care facility projects should be readily identified to those involve precisely or indirectly. The approach of how it should accomplish the needs of the future must be made as part of the Client's, Designers, Users and Facility Managers' training list to conduct the projects [5].

The plan for the development of the Landak District General Hospital is motivated by three things, namely the first from the mandate of Law Number 44 of 2009 concerning Hospitals, secondly the development of the Landak Regency in line with population growth, and the increasing demand and need for health services that occur in health centres in Indonesia. Around the coverage area, namely the surrounding health centres, so that this condition is a consideration for the development of this hospital as a referral centre for health centres and other health services by looking at the projected conditions for the coming year so that the Landak District General Hospital can prepare hospital health services in the future. The aims of this research include a) the preparation of the initial stage of the Masterplan of the Landak District Hospital, b) obtaining a harmonious and balanced use of regional space with the needs and capabilities of supporting urban growth and development by taking into account the harmony of the environment, c) the availability of a regional development plan that can function as a forum for integration for the interests of all parties.

2. Research theory and methods

2.1. Literature review

A hospital is not a "factory" for wellbeing as several people imagine; factories figure inanimate substances. It is a great deal closer to a school and faster still to a " slight settlement." In one sense, it is a momentary home for those who are nauseous, in which all their human wishes must meeting. In another feeling, it is an institute for systematic research, for each patient presents a study problem of analysis and a possible problem of therapy [6]. Hospital classification's determination is based on 3 (three) main aspects, namely service; Human Resources (HR), equipment; and buildings and infrastructure. Hospital buildings and infrastructure, as referred to in line (1), must meet the requirements for building and environmental planning and the requirements for the reliability of hospital buildings and infrastructure. The building and environmental planning requirements include [7].

a) The designation of the location and intensity of the building is following the provisions of local regulations.
b) Hospital building design.
c) Control of environmental impacts under the provisions of laws and regulations.

Health care institutions, coming to conditions with these patterns, are undertaking the following [8]:

a) Decreasing expenses due to the addition of health care into inclusive and harmonized community broad systems.
b) Decentralizing health care services. Thus, care is provided where it is required. Changing from an illness representation to a wellness model.
c) Focusing on holistic medicine and complete patient care, and integrating traditional and alternative therapies.

The hospital is not simply a building but as well a complex community institution. In considering the upcoming outlook of function-oriented hospitals and other healthcare services, the lately made concept, that is "healing environment", provides specific basis approaches to understanding various issues [9]. The constructing process of healthcare facilities, in any case, involves complex trade-offs with iterative activities such that design properties that have showing to improve building accomplishment are frequently not wholly considered. Healthcare narrative continues to describe valuable research relating the health and well-being of healthcare facility occupants to the performance of the built atmosphere [10]. Along with the increasing public demand for excellent service performance, hospitals have become a must to improve service performance. Improvement of service performance should be carried out gradually and continuously by prioritizing service improvement on elements of community satisfaction that still have the "less good" category, namely on handling complaints, service time, and elements of facilities & infrastructure [11].

Hospital design is a plan that can combine health and behaviour, the development of the design of the hospital guided by several concepts of sustainability related many researches is moving towards to an understanding of the Evidence based Design, which connects the physical environment with the patient and staff outcomes [12]. On the other hand, Sarah Hosking analogize the hospital as a workshop where people are taken to be checked and repaired before being sent off to resume their journey [13]. The aesthetic of hospital will answer the question of why the hospital is very difficult to be planned, built, operated and maintained. It consists of human comfort, well-designed and well-maintained building [14].

There are many previous studies regarding developing a hospital area master plan regarding the development of health facilities into good hospitals and their supporting facilities. The problems that arise are also diverse, resulting in different identification and analysis. The focus of the problems in previous studies also varied, ranging from the study of the perception of the use of materials to the function of the hospital's internal space to the analytical study of the relationship between selected elements of architectural design and other factors. Other studies write about the design of medical rooms that must be considered from patients' physiological and psychological needs and other studies. The results of these studies are beneficial in writing this research, especially in looking at the development and construction of existing hospitals, which were developed following the mandate of the Minister of Health Regulation Number 44 of 2009 concerning hospitals, especially in areas such as districts and sub-districts. From the way of looking at this phenomenon, this study provides concepts and recommendations in the form of architectural programming in seeing space, circulation, and zoning in the hospital.

2.2. Methodology

The approach used is the descriptive qualitative method to review the literature describing the concept of architectural programming to apply towards the case [15]. This method aims to illustrate a condition that existed at the time while the study was conducted and investigates cause and effect through the recognition of symptoms from existing problems [16]. The research stages:

a) Preparation Phase, collecting data both externally and internally from various sources to compile data related to the preparation of the Master plan for the development of the Landak District Hospital.
b) Analyse conditions both externally and internally based on data compilation and various data information held.
c) Formulation of hospital external and internal trends to obtain information for future development and development plans identifying in a hospital master plan.
d) Conduct a study of the functional program based on the Master Program by planning:
   - Space program based on work activities.
   - Functional relationship.
   - Hospital facility needs.
e) Prepare a hospital master plan in the form of an overall building plan and its stages.

3. Results and discussion

3.1. Existing data

The area of Landak Regency, which oversees 13 sub-districts by the end of 2018, has 9,909.10 km2 or 6.75% of the area of West Kalimantan Province and is the Regency with the third smallest area after Mempawah Regency and Pontianak City. Landak District Hospital Is one of the SKPD within the Scope of the Landak District Government, which was established based on Landak District Regional Regulation Number 63 of 2002. The granting of Operational Permit for Landak District Hospital stipulating by Regent Decree Number: 445/474/HK/III-2015 on 12 November 2015. Currently, the Landak District Hospital is a Type C hospital, based on the Decree of the Landak Regent regarding Class Determination and Permit for Landak District General Hospital Number.

In general, the physical condition of the Landak District Hospital needs to develop to meet the requirements of a hospital. The development carrying out by considering the existing buildings to maintain the structure of the building as much as possible and keep the hospital's operations from stopping; this can be a strength for the plan for the development of the Landak District Hospital. The Landak District Hospital consists of several buildings with a height of 1 to 2 floors. The physical buildings consist of administrative buildings and polyclinics, TB and Amero poly buildings, decontamination rooms, panel buildings, Emergency unit and CSSD buildings, RO buildings, radiology buildings, laboratory buildings, MCU buildings, physiotherapy buildings, TCM laboratory buildings, pharmacy buildings, buildings Internal Medicine inpatient, Surgical Treatment Inpatient Building, Isolation Room, doctor's office, nurse's office, security post, OK room, ICU building, ICU waiting room, Children's inpatient building, perinatology room, NICU building, bulkhead houses, workshops, morgues, generator houses, post-partum rooms, warehouses, nutrition installations, laundry installations.

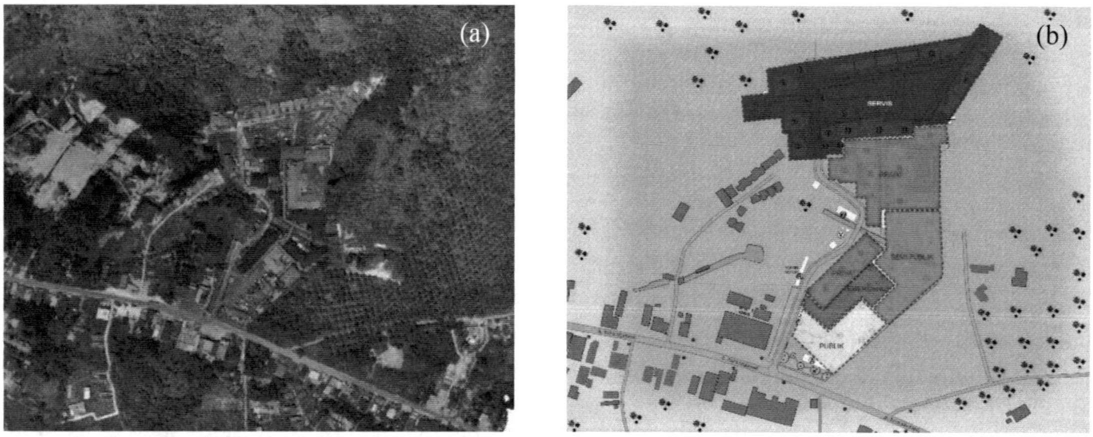

Figure 1. Satellite view (a) and zoning of Landak regional public hospital (b).

3.2. Discussion

Several previous studies that have similar topics to this research. In Melo and Souza study said to perceive a potential of flexible spaces to enrich the solutions for the architecture of health environments with functional ideas and viability of using new constructive technology [17]. Supported by statements of Hozak, Gregory and Nelson in research about relationship of hospital architecture healthcare staff, it stated that the research provided insights into architectural aspects that have not been widely studied as variables in healthcare. The partnership between the architect and the Direct Care staff can improve the improvements in the research model that can be specified more accurately, and thus more accurately determine how the architecture affects patient and staff results [18].

In the other hand, the design of external forms and internal space of hospital also begin to diversify under the powerful control of functions. Systematic design of public space should be the organic combination of functional use, mental experience and spatial experience. It is a balance of all kinds of requirements, rich expressive force and appeal of public space [19]. The architectural layout strategy developed early in the layout process has a major influence on the fate development process for healthcare facilities. Results show the impact of two architectural approaches on building development, with horizontal and vertical development over a period of. This study recommends changes during the layout procedure, creation, and allocation phases [20]. Development, improvement and implementation for the success of the green design and construction process of hospital facilities will improve the health and welfare of patients, increase the effectiveness of doctors, nurses and other hospital staff, and the life cycle of these facilities. May reduce costs [10].

The similarities between previous studies and this research lie in the topic raised, namely hospital architecture. The second equation is the research method used using a descriptive case study approach with survey techniques, secondary data, observation, and then data is interpreted narratively. Meanwhile, the difference between this research and previous research lies in the focus of the research and state of the art. The research focuses on the steps to develop the Landak hospital master plan with an architectural programming approach.

Hospital as health service facility has a mission to achieve development goals. The hospital's mission is said to be successful if the hospital can provide quality services by trying to improve the quality of services intensively and continuously and supported by the completeness of adequate infrastructure and facilities. The hospital will serve the needs of public health services, both general services and specialist services. According to Carlino in Melo [17] a public health facility destined to play a role in attending the needs of civil society. The health care services are highlighted as one among this need. In other hand the study included the relationship between the design of a patient care workflow design with staff productivity and the ability to perform the required work within patient care in hospital needed to reduce working stress [18].

From discussion above architectural programming approach for Landak general hospital masterplan is:

A. Hospital building space requirements

The need for hospital building space will adjust to the type and capacity of services and activities the hospital will provide to the community. Calculation of the size of each room in the building based on function will calculate following Architectural standards and Technical Guidelines in the Field of Hospital Facilities and Infrastructure. In a rough calculation, the total floor area of the hospital can be calculated as 80 – 110 m2/bed if the Landak District Hospital in its development is assumed to have a total inpatient capacity of 422 beds. The total floor area of the Landak District Hospital building is 33,760-46,420 m2. The configuration of the building mass grouping form by taking into account:

- Function of space in the building.
- Relationships between spaces in buildings.
- Relationships between related installations.
- Minimum area of each room.
- Zoning.
- Architecture module, structure and building materials module.

The space requirement that can be seen in table 1.

Table 1. Minimum space requirement for non-educational general hospital [21].

No	Installations	Area (m2) per bed	150 Bed
1	Administration	3 - 3,5	1.266- 1.477
2	Emergency unit	1 – 1,5	422 - 633
3	Polyclinic	1 – 1,5	422 - 633
4	Social services	0,1	42,2
5	Registration	0,2	84,4
6	Laboratory of clinic and pathology	2,5 – 3	1.055 – 1.266
7	Obstetrics and Gynaecology	1,2 – 1,5	506,4 - 633
8	Diagnostics and Radiology	3 – 4	1.266 – 1.688
9	Kitchen	2,5 – 3	1.055 – 1.266
10	Facilities officer	0,5 – 0,8	211 – 337,6

No	Installations	Area (m2) per bed	150 Bed
11	Meeting and training room	0,5 – 1	211 -422
12	Speech and Hearing Therapy	0,1	42,2
13	Housekeeping/Cleaning	0,4 – 0,5	168,8 - 211
14	Material Management	0,4 – 0,5	168,8 - 211
15	Central office	2,5 – 3,5	1.055 – 1.477
16	Purchase	0,2	84,4
17	Laundry	1 – 1,5	422 - 633
18	Medical records	0,5 – 0,8	211 – 337,6
19	Medical staff facilities	0,2- 0,3	84,4 – 126,6
20	Engineering and Maintenance	5 – 6	2.110 – 2.532
21	Kids room	0,4 – 0,5	168,8 - 211
22	Health workers	0,3 – 0,4	126,6 – 168,8
23	Pharmacy	0,4 – 0,6	168,8 – 253,2
24	Public area	1 - 1,5	422 - 633
25	Skin Treatment Room	0,1 – 0,2	42,2 – 84,4
26	Operation room	3,5 – 5	1.477- 2.110
27	Circulation	10 – 15	4.220 – 6.330
28	Inpatient Unit	– 35	10.550-144.770

B. Activity pattern and zoning

Hospital activities influence by hospital performance. Influenced by the placement of room functions that must be related or related to easy and fast access between related functions. In general, the pattern of activity in the hospital consists of the following activities:

- In Hospital Building
  The pattern of activity and circulation formed by the movement arising from hospital building activities, such as medical care activities, medical and non-medical support services, medical administration and records, services and utilities, and emergency care services.
- Outside the Hospital Building
  The pattern of activity formed from the activities that occur outside the hospital building consists of the movement of vehicles: visitors, outpatients and inpatients, doctors/hospital staff, services and emergency departments.

The zoning of hospitals has based on the risk of disease transmission, zoning based on privacy, and zoning based on service.

a) Zoning based on the level of risk of disease transmission consists of:
- low-risk areas, namely secretarial and administrative rooms, computer rooms, meeting rooms, archives/medical records.
- medium risk areas, i.e., non-communicable disease inpatient rooms, outpatients.
- high-risk areas, namely isolation rooms, ICU/ICCU rooms, laboratories, mortuary and post-mortem, and radio-diagnostic rooms.
- very high-risk areas, i.e., operating room, emergency room, delivery room, pathology room.

b) Zoning based on activity privacy consists of:
- public areas, namely areas with direct access to the environment outside the hospital, for example, polyclinics, emergency rooms, pharmacies.

IOP Conf. Series: Earth and Environmental Science 1007 (2022) 012022 doi:10.1088/1755-1315/1007/1/012022

- semi-public areas, i.e., areas that receive no direct contact with the environment outside the hospital, generally are areas that receive workloads from public areas, such as laboratories, radiology, medical rehabilitation.
- private areas, namely restricted to hospital visitors, generally closed areas, such as ICU/ICCU, surgical installations, obstetrics and gynaecology installations, inpatient rooms.

c) Zoning based on service consists of:

- Medical and Nursing Service Zone are consisting of Outpatient Installation, Emergency Room, Inpatient Installation, Intensive Care Installation (ICU/ICCU/PICU/NICU), Surgical Installation, Medical Rehabilitation Installation, Obstetrics and Gynaecology Installation, Haemodialysis Unit, Radiotherapy Installation, Nuclear Medicine Installation, Blood Transfusion Unit.
- Support and Operational Zone consisting of Pharmacy Installation, Radio diagnostic Installation, Laboratory, Integrated Diagnostic Installation, Central Sterilization Supply Dept./CSSD, Main Kitchen, Laundry, Body Retrieval and Forensics, Sanitation Installation, Facility Maintenance Installation.
- General and Administrative Support Zone consisting of Secretariat and Accounting Section, Medical Record Section, Logistics/Warehouse Section, Planning and Development Section, Internal Monitoring System, Education and Research Section, Human Resources, Procurement Section, Information and Technology Section.

C. Functional relationship

Hospital Functional Relations is the relationship between the functions of activities in providing health services that are interrelated to produce services that are according to standards and by taking into account efficiency and effectiveness factors in all fields. The Physical Building Plan of a hospital explains everything related to efforts to determine the work location of each work unit in the form of a zoning plan/group plan for allotment of space and a hospital building block plan according to the floor area and function of the building to meet the immediate needs and supporters.

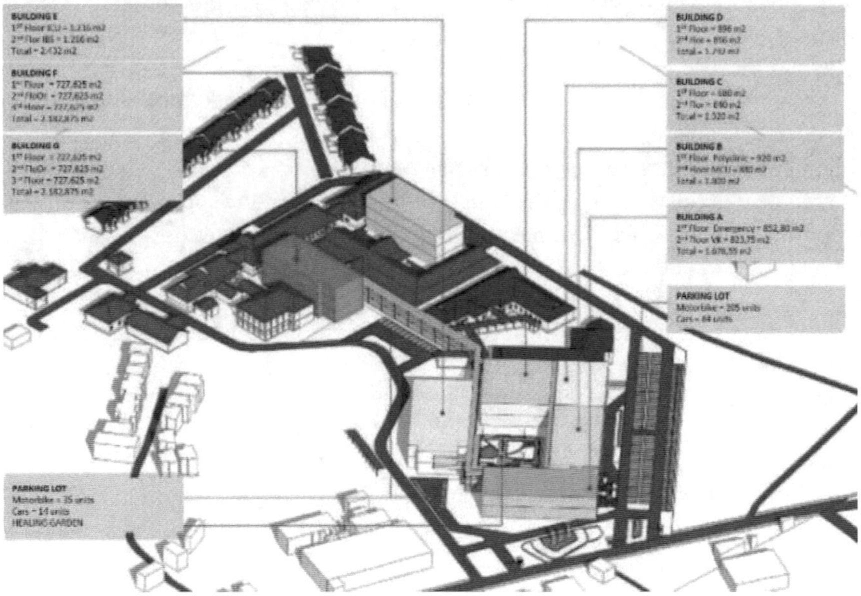

Figure 2. Axonometry of the development of the Landak district hospital.

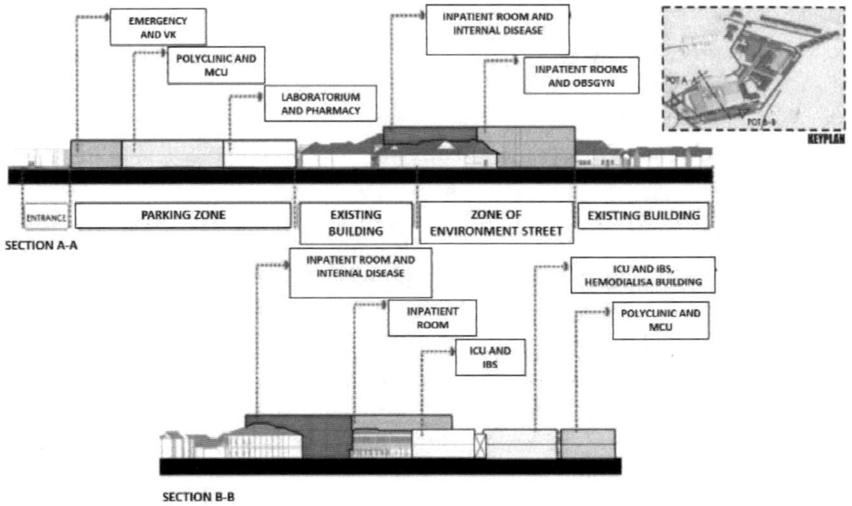

Figure 3. Section of the functional development of the Landak district hospital.

D. Hospital activity circulation pattern

Solving the problem of circulation in the building can be overcome by grouping the functions regularly and adequately. Circulation conditions outside the building regarding size, comfort, and achievement and the distance between functions must regulate pedestrians and vehicles appropriately. These aim to avoid conflicts of circulation of achievements into service functions. Certain service functions require quick and easy-to-find access, so consider:

- Place the door and its size.
- The layout of the function of the building, the distance between the mass of the building and its area.
- Circulation settings, distances, and sizes for both pedestrians and vehicles.
- The distance from the public transportation stop to the main door of the hospital location must be close and safe for pedestrians.

The planning of the circulation path to and from the building must pay attention to the following:

- Prevent cross circulation.
- The main entrance must be easily visible and accessible.
- There are adequate parking facilities and special parking for people with disabilities.
- Minimum entrance to the hospital is three doors: the main door, a particular door to the Emergency Room, and a door to the service area.

The components that make up the circulation path inside and outside the building, namely:

- Horizontal Access Outside the Building, which is closely related to the achievement/access to the hospital.
- Horizontal Access in the building, namely Hallway, consisting of corridors that are roofed and not which must provide comfort for users, specifically for the floor, non-slip building materials are used.
- Vertical Access consists of Stairs, Elevator and Ramp.

IOP Conf. Series: Earth and Environmental Science 1007 (2022) 012022 doi:10.1088/1755-1315/1007/1/012022

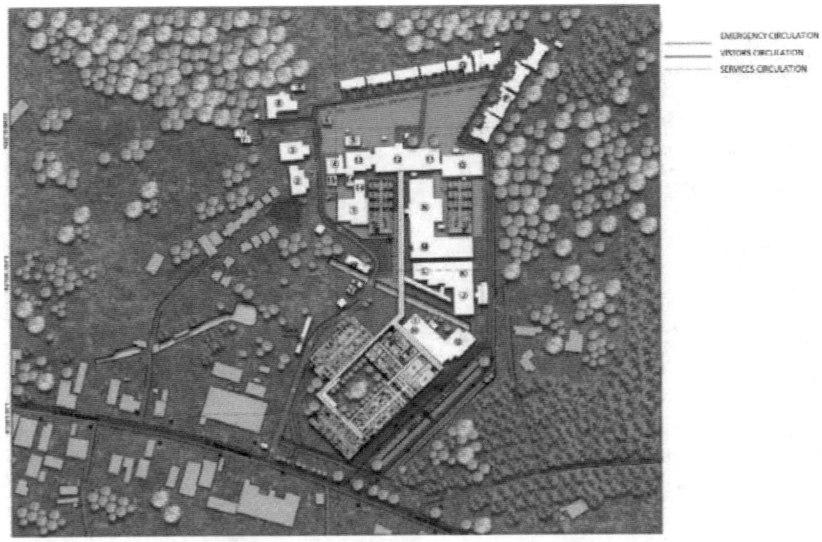

Figure 4. External circulation pattern for Landak district hospital development.

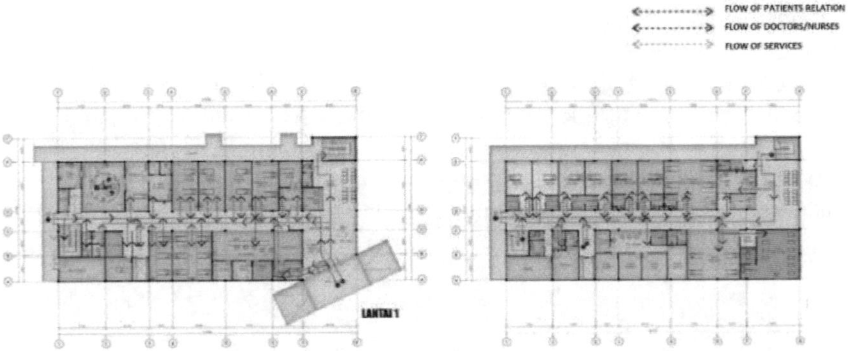

Figure 5. Internal circulation pattern of emergency room and VK building.

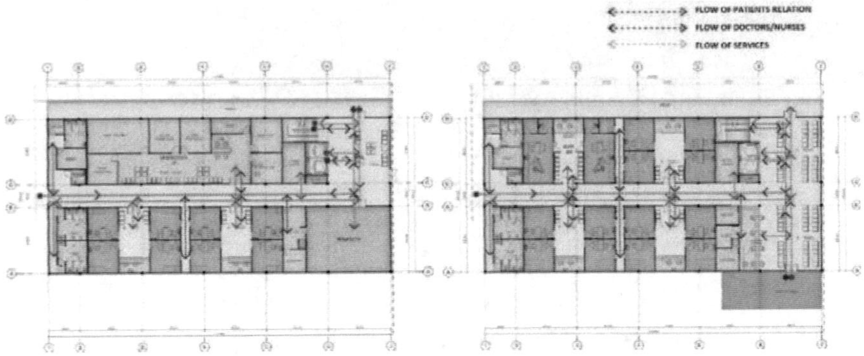

Figure 6. Internal circulation pattern of integrated polyclinic.

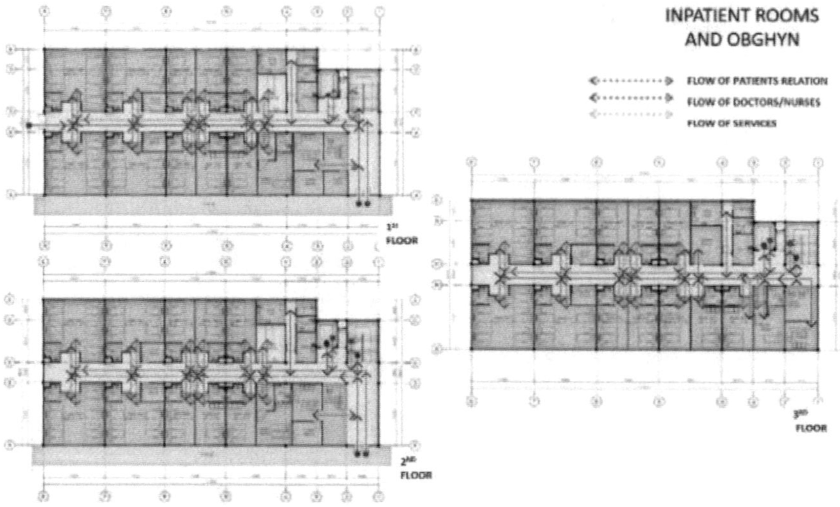

Figure 7. Internal circulation pattern of ob/gyn inpatient installation.

E. Building security and safety

Building Security and Safety are related to the movement system that takes place in the building complex. The movement in the building complex dividing into the movement of building users and access to evacuation vehicles. The movement of building users is closely related to proximity and directions to the safety of building users. Vehicle movement is closely related to the achievement of vehicles to evacuate building users and the building itself, so it is hope that the evacuation vehicle can reach as close as possible to the source of the disaster in the building or the evacuation gathering point for building users.

Figure 8. Lift, stairs, meeting point and circulation placement.

F. Landscape and open space

Conceptually, these open space functions are ideally accommodated in a core open space in the form of an inner courtyard, a courtyard surrounded by building masses. If it is not possible to place the inner courtyard, the existing open space should utilize optimally to continue providing benefits for the hospital. Functionally, this open space and landscaping functions as:

- Supply of fresh air or natural ventilation.
- Supply of sunlight or natural lighting.
- Supporting the comfort of activities in each room.
- As a rescue room (building safety).
- Reduce environmental tensions and support care and treatment efforts.

The tightness of land area leads to the most optimal design solution, namely a compact and dense cluster of building masses. The choice and placement of vegetation at the Landak District Hospital are planning as follows:

- From the front zone, a row of trees protects the building from the outside environment and the road.
- Existing vegetation that is not affected by development is maintained.
- Air filter trees, trees as protection from the parking area.
- The rest of the shade vegetation types in the meeting arrangement cover the entire open space.
- In addition to circulation and parking spaces, the ground covered vegetation is green grass of a type that is easy to handle.

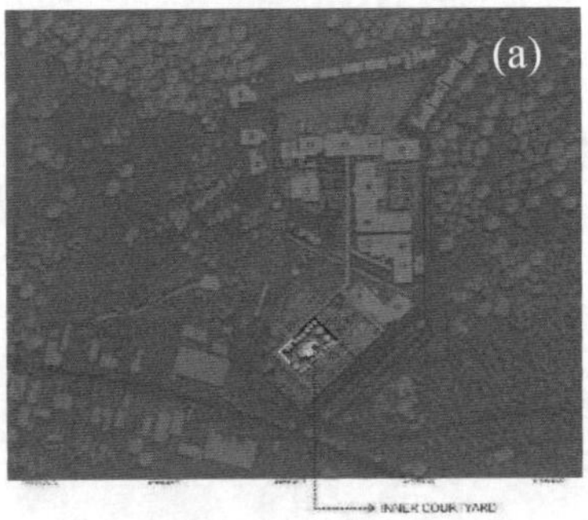

Figure 9. Inner courtyard (a) and its view (b) at the Landak district hospital.

G. Parking requirements

Based on the attachment to the Regulation of Indonesia Minister of Health Number 24 of 2016 concerning Technical Requirements for Hospital Buildings and Infrastructure, the Design and Planning of parking infrastructure in hospitals is fundamental because parking infrastructure and vehicle entrances will take up a lot of land. Assuming the calculation of parking area requirements at the hospital ideally is 37.5 m2 to 50 m2 per bed (including vehicle circulation paths) or adjusts the socio-economic conditions of the local area. Parking lots should equip with parking signs:

- The hospital building must provide a vehicle parking area with a proportional number of parking areas according to local regulations.
- Provision of parking in the yard may not reduce the reforestation area that determining.
- Parking lots must be equipped with clear parking signs.
- Besides providing a covered parking lot, the hospital building should provide a pedestrian path.
- Pedestrian paths should be safe from vehicular traffic.

Landak District Hospital is a class C hospital that is projected in 2030 to have 422 beds. Based on these data, the required parking needs reach 15,822 m2 to 21,100 m2. Some general provisions that need to consider regarding providing hospital parking areas are the separation of access and placement between motorbike parking and other motorized vehicles in planning parking pockets. In addition, parking bags are separate between employees, including doctors and paramedics, and the general public. The foremost parking with the most accessible access is the Emergency Ambulance parking.

4. Conclusion

Improving services at the Landak District Hospital will majorly impact government programs, significantly improving public health in the Landak District area. The master plan for the development of the Landak District Hospital is a mutual agreement and the result of a tiered and continuous process and stages of preparation. The development is prepared based on several assumptions, in which many parties and related agencies are involved in approving the realization of its implementation. Likewise, conditions and situations affect the development process later. So that changes in development are still possible to adjust the situation and conditions later.

However, the consistency of the plan for the development of the Landak District Hospital in the future should refer to the development master plan that mutually agrees. Corrections, improvements, and readjustments of this master plan could make before the last year of development. It is expecting that the Landak District Hospital will become a barometer of medical services with very prominent professional characteristics and rationalism. The authors gratefully acknowledge to colleagues and part who have supported this paper. Gratitude also acknowledges to Technic Consultant Service of Borneo Co., Public Health Office of Pontianak City, and Asia Raya Studio for their support and to all those who cannot write one by one.

5. References

[1] Iravanti F, Waturandang G and Aulina P Y 2019 Analisis Rencana Strategi Bisnis Rumah Sakit Harapan Jayakarta dengan Pendekatan Balanced Scorecard *Jurnal Manajemen Dan Administrasi Rumah Sakit Indonesia* **3** 85-101

[2] Bulakh I, Didichenko M, Kozakova O and Chala O 2020 *Proc. Int. Conf. on Sustainable Futures: Environmental, Technological, Social and Economic Matters (Kryvyi Rih)* vol 166 (Les ulis: EDP Sciences) p 08001

[3] Fronczek-Munter A 2017 *3rd European Healthcare Design Congr. Exhib. (London)* pp 1-13

[4] Samah Z A, Ibrahim N and Amir J S 2013 *Proc. Asia Pacific Int. Conf. on Environment-Behaviour Studies (London)* vol 105 (Amsterdam: Elsevier) pp 265-72

[5] Nawawi N M 2000 *20th Public Health Group Semin. (Stockholm)* pp 1–38

[6] Taylor L 1960 Hospitals of The Future *Br. Med. J.* **2** 752-8

[7] Oktamianti P and Pebriana A 2019 Kajian Kebutuhan Pengembangan Rumah Sakit Pemerintah Daerah Kelas B di Provinsi Sulawesi Utara *Jurnal ARSI (Administrasi Rumah Sakit Indonesia)* **5** 59-71

[8] Young J E 1996 *84th ACSA Annual Meeting and Technology Conf. Proc.(Boston)* (Washington DC: Association Collegiate Schools of Architecture) pp 566-71

[9] Nagasawa Y 2020 Global Hospitals in 2050 - A Review of the Historical Development of Hospital Building Studies from a Global Perspective *Jpn. Archit. Rev.* **3** 5–24

[10] Bilec M M *et al.* 2009 Analysis of the Design Process of Green Children's Hospitals: Focus on Process Modeling and Lessons Learned *J. Green Build.* **4** 121-34

[11] Nisa L S, Maliani, Fitriyanti S and Siska D 2020 Kajian Indeks Kepuasan Masyarakat terhadap Pelayanan Rumah Sakit Umum Daerah Pambalah Batung Amuntai Tahun 2019 *Jurnal Kebijakan Pembangunan* **15** 51–62

[12] Ulrich R P, Quan X, Zimring C P, Joseph A and Choudhary R 2004 *The Role of the Physical Environment in the Hospital of the 21st Century: A Once-in-a-Lifetime Opportunity* (Concord: The Center for Health Design) p 69

[13] Hosking S and Haggard L 1999 *Healing the Hospital Environment: Design, Management and Maintenance of Healthcare Premises* (London: E & FN Spon) p 214

[14] E Setyowati, Harani A R and Falah Y N 2013 *Proc. AMER (ABRA malaysia) Int. Conf. on Quality of Life (Langkawi)* vol 101 (Amsterdam: Elsevier) pp 189-99

[15] Creswell J W 2009 *Research Design: Qualitative, Quantitative, and Mixed Methods Approaches* 3rd Edition (California: SAGE Publications, Inc) p 260

IOP Conf. Series: Earth and Environmental Science 1007 (2022) 012022 doi:10.1088/1755-1315/1007/1/012022

[16] Sari D P and Alhamdani 2020 *1st Int. Conf. on Urban Design and Planning (Semarang)* vol 409 (Bristol: IOP Publishing) p 012028
[17] Melo M V S and Souza H A 2019 Analysis of the Perception of Users of Pre-hospital Architecture in Light Steel Frame: A Case Study in Minas Gerais, Brazil *J. Civ. Eng. Archit.* **13** 251-8
[18] Hozak M A, Nelson J and Gregory D 2016 Relationship of Hospital Architecture to Nursing Staff Caring for Self, Caring for Patients, and Job Satisfaction *Interdiscip. J. Partnersh. Stud.* **3** 5
[19] Wang N and Wang J 2017 *Proc. 4th Int. Conf. on Education, Language, Art and Intercultural Communication (Moscow)* vol 142 (Netherlands: Atlantis Press) pp 589-91
[20] Pilosof N P 2021 Building for Change: Comparative Case Study of Hospital Architecture *Health Environ. Res. Des. J.* **14** 47-60
[21] Kementerian Kesehatan Republik Indonesia 2004 *Keputusan Menteri Kesehatan Republik Indonesia tentang Standar Pelayanan Farmasi di Rumah Sakit* Nomor 1197/MENKES/SK/X/2004 (Jakarta: Kementerian Kesehatan Republik Indonesia)

Acknowledgement
The authors gratefully acknowledge to public health office of Landak Regency, the Regent of Landak Regency, and all of the parties that authors cannot mentioning for supporting this research.

The concept of dynamicity of martial arts office building design with environmentally friendly

N Kharismawardani[1] **and A Hayati**[1,2]

[1] Architecture Department, Faculty of Civil, Planning, and Geo-Engineering, Institut Teknologi Sepuluh Nopember. Kampus ITS, Sukolilo, Surabaya 60111, Indonesia

Abstract. The demand for office building for the sports branch of martial arts in Indonesia has increasingly risen in line with the rapid development of the international arena. For the administrative region, including the province of East Java, a martial arts head office is essential the administrative duties and management of the sports organisation. A martial arts office is an office to accommodate administrative and management tasks. It also requires facilities for users who mostly have an athletic background and are physically fit and well-being. Therefore, this paper aims to explore and propose architectural designs of martial arts' administrative office from finding the idea, determining the design's parameters and concept to formal design transformation. Presenting big ideas by translating the character and value of sport is applied during the design process and considering the green design approach to provide building's energy-efficient. The result proposes a building design that expresses the dynamicity of sport's value and characters. The building is also designed by reducing negative impacts on the environment and maximise energy saving.

1. Introduction

An office centre building for martial arts sports in East Java is not yet available. It requires several design criteria that provide space for administrative and management activities and represent the sport's image/value. Infrastructure Office buildings are facilities that can fundamentally stand alone and as sports facilities that can serve all groups of people [1]. Besides, the design of sports office buildings must significantly impact the needs of light, air, thermal, and others that affect the performance of athletes/users in sports buildings [2]. The martial arts office building contains offices for work and several facilities for conducting exercises, sport's matches, and galleries to attract visitors' attention. Therefore, this building has a specific design requirement: an office also be a community space. The space is organised to accommodate the public activities needed, interacting and facilitating sport's activities to selected athletes when using the facility.

The sports office has various meanings when viewed from its value. One of them is the diversity of sports, especially in martial arts, but still within the same disciplinary aspects. The office represents each sport's beliefs and standards but still in the same goal and discipline [3]. The reference study supports the design process to propose a design 'big idea and concept' by doing translation of non-architectural syntax (Martial arts sports' value and character) into the formal design (architectural syntax). As a result, the big idea is the concept of dynamicity that represents sports culture's values and has a sense of iconic

[2] Corresponding author: arina_h@arch.its.ac.id

building to its surrounding. It means a "sustainable vision" to the future generation that sports infrastructure and the office are essential in dynamic development in an environment [1]. In addition, this can be a new strategy by creating sports office buildings that can invest in physical well-being, mental well-being, individual, social and community development in the future over time [4].

To fulfil the design's concept and requirement, the building oriented to west/east facing the main street to optimise the image of a building with the surrounding has had a contrary effect that requires high costs and energy. Thus, the design approach also uses Geen Building's parameters and principles. Green building design has become a current trend as an alternative to reduce energy use in buildings. Implementing green buildings is vital to identify the environmental problems and provide efficient energy solutions related to green buildings to meet future energy needs, especially in sports buildings [5]. Therefore, this study used the green building as a general criterion and sport's values and image as a particular criterion.

2. Theory and precedent

This chapter discusses the design's ideas/innovations, criteria, and programming. The first stage of the design process is doing literacy study methods in theory and precedent studies. According to Jones, the literature searching method is part of a divergent and exploratory method that is useful for determining goals and needs in exploring situations [6].

2.1. Dynamicity and transparency

Dynamic architecture is one part of sustainable design because the building's shape follows environmental changes and can reduce energy use [7]. According to David Fisher, 2011, Dynamic architecture includes fully, partially, and statically dynamic design. Fully-dynamic is the whole building can rotate, revolve, and swivel or pivot. Partially-dynamic is a movement or change in partial specific parts & spaces, elevations, furniture, and water. And the static-dynamic represents the shape of static-dynamic in an optical illusion, form, and inspiration [8].

Transparency architecture is a visible system without being touched directly and continuously [9]. This transparency fulfils the aspects of the value of the martial arts organisation that prioritises transparency between members of the organisation [10]. A complex architectural element is produced through the concept of dynamic and transparency. The main components that are the centre of attention in the design are specific spaces, elevation, movement parts, affected by seeing sight and unobstructed. With this component, the main goal is to propose building expression that is iconic and different from the surrounding environment.

2.2. Green Building approach

Green building is an alternative to creating sustainable and energy-efficient buildings. According to Linda and Robert Veil, there are six basic principles of green architecture. These basic principles become design parameters and criteria of building design. The six principles are conserving energy, working with climate, minimising new resources, respecting users, respecting sites, and holism [11]. Several studies discuss applying the green building approach to buildings in certain climates. This study finds the wide or elongated shape of the building will be able to suppress energy to the maximum by using several technologies as a solution for energy suppression in buildings [12].

In applying the green building approach, it is necessary to pay attention to the climatic and weather conditions or the location or place where the building stands. The green building parameters according to the Green Building Council Indonesia (GBCI) [13] are:

- Appropriate land development
- Energy conservation and efficiency
- Water conservation
- Material use and selection
- Indoor comfort and health

- Building environment management

2.3. Precedent study
The precedent study discusses office buildings' design that presents the aspect of dynamicity, transparency and how the building conserves energy. The discussion about the Building of Edge Olympic, Beijiao Sports Center and Decode Urbanism Office, Olympic House, and London Aqua also focused on the ideas, criteria, and space programs.

2.3.1. Aspect of dynamicity. The Building of Beijiao Sports Center and Decode Urbanism Office is carrying out the concept of dynamic and ecologies building (Figure 1,2). The main idea is that the *architecture* intends to create a community space for activity, which enlivens its immediate vicinities. The design idea and criteria are unified all the separate spaces into one through a central "valley". In terms of space programming, this building is unique by providing the focus of the communal space being in the centre of the axis, consequently making the dynamicity convey the value of the building big idea [14].

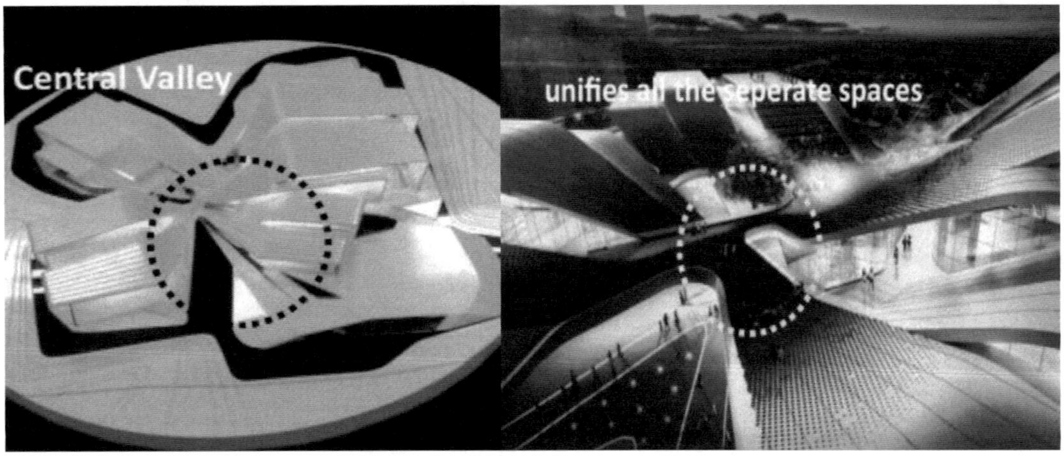

Figure 1. Dynamicity of Beijiao building [15]

Figure 2. Curved and wave shape inside and coverage building [16]

Dynamicity is expressed in spatial management inside the building design, but it can also consist of the building envelope and detail of void. The Building of London Aquatic Centre by Zaha Hadid has successfully expressed the undulating and wave motion with the fluid geometric concept in the shape of the building. The London Aquatic Centre is a particular building for matches and the swimming sports headquarters in London. The main idea of this building is dynamic and fluid geometries like water in motion. This idea was inspired by the movement of water waves which is the main character in this Building [17]. The criteria for this building are creating space surroundings and reflecting the riverside landscape of the Olympic Park. The building is planned perpendicular to the Stratford City Bridge [16] with programmed spacing.

2.3.2. Aspect of transparency. The buildings' design applied the transparency concept inside and outside. Transparency can be between spaces and concerning the surrounding environment relation. In the Olympic house building, transparency is used by providing space without skating in public areas to provide an atmosphere of openness and monitor each other from all sides [19]. Transparency in the Olympic Edge building presents skylights to give sunlight to reduce the use of electrical energy in the building (Figure 3).

Figure 3. Transparency between spaces [18,19]

2.3.3. Building made with conserving energy. There is a building with a strategy for saving more energy and making sustainable buildings. The First Building is Edge Olympic Building, built-in 2017 and located in Amsterdam, Netherlands (Figure 4). The idea of this building is to use digital infrastructure for sustainability and saving energy. The criteria for this building are using VOC-free materials, maximising the movement of the atmosphere in the room, considering the mental and well-being of users in the building by providing light openings and natural ventilation [18]. In terms of space programming, all areas on each floor are connected and the central room. In addition, there is Natural light penetrates the heart of the building.

Next is the Olympic House [Figure 5]. This building was built in 2019 and is located in Lausanne, Switzerland, as the headquarters building of the International Olympic Committee. This building carries the concept of flexibility, movement, sustainability [19]. The main idea of the building form is dynamic and undulating; it aims to commit to sustainability and multiple work styles now and in the future. The building uses photovoltaic to generate electric power by using solar cells to convert energy from the sun into a stream of electrons. Solar cells are installed to adjust building shapes that represent the symbol of the value of the concept in the Olympic Building.

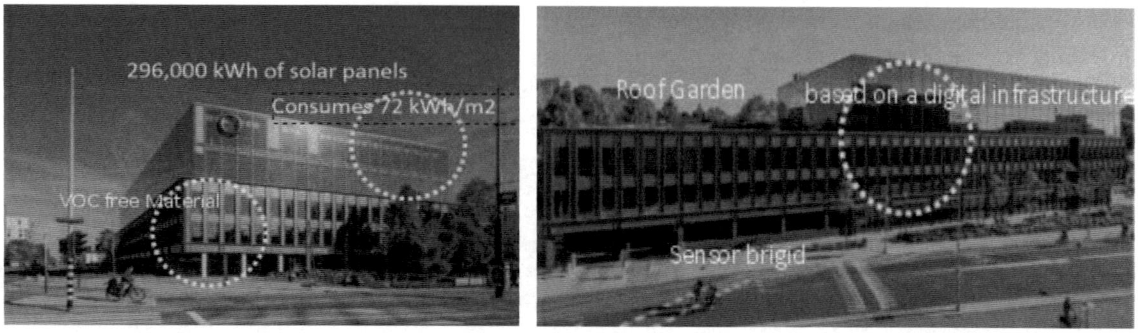

Figure 4. Edge Olympic building strategy to conserving energy in design [18]

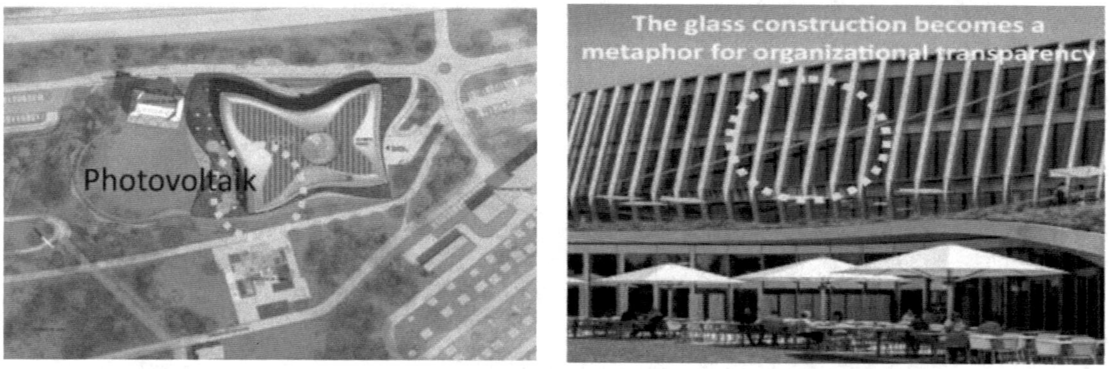

Figure 5. Strategy to conserving energy in Olympic House building [19]

Through several precedent studies, most sports office buildings or sports halls pay more attention to the form that is the centre of attention (eye-catching) and the surrounding environment. In addition, these buildings also pay attention to the sustainable effects produced in the long term so that the entire building maximises natural lighting and ventilation. The design also supports the existence of green buildings, which is a hot topic of conversation in the current era.

3. Design framework and process

This study discusses the analogy of martial arts values presented in sports office buildings as the main criteria. The goal is to offer a sports office building that has its characteristics and is different from other buildings. The design uses semiotic analogy methods that identify information based on its meaning and value, transformed into architectural syntax [20]. The processes and methods used in this research are the big ideas as the main point of the building and the study of literature related to the theory used. The way to propose the big idea is by applying the domain-to-domain transfer method. It aims as a bridge between values of non-architectural syntax (sports' value and image) into the formal design (architectural syntax) [21].

Applying the dynamic shape and transparency concept still considered the level of energy efficiency as a general criterion in today's low-energy buildings. To help determine the parameters of "green building", this study will analyse the level of energy efficiency using Edge Application tools (version 2.1.5). Further, there will also be an analysis and calculation of the "green building" level.

4. Result and discussion

This chapter will discuss the result and discussion for martial arts office buildings. This is achieved by using the design concept and overall design results of the building. The design concept describes dynamics and transparency as the main values chosen. For building design, it is explained about how the building was started until the strategy to save the energy needed and produced by the building.

4.1. Design concept: dynamic and transparency

There are three essential components in sports: athletes, coaches, and structural organisations. These three components are united in a teamwork bond that needs and supports each other [22]. In addition, the three components have a value of sport based on fairness, transparency, respect, connection, and dynamic [23]. Based on these five aspects, this research is filtered into two aspects that be explored as the focus of implementation in design.

Determination of dynamic and transparency becomes a unity representing the five aspects of the belief in martial arts. Dynamic is a condition where the whole element becomes a unity in the integrity of a martial arts movement. Transparency is a condition where there is mutual openness between martial sports organisations, including all components of athletes, coaches, and organisation members.

In the dynamics of martial arts, there are physical and natural classifications. Physically, the dynamic is the movement of the body, such as muscle changes that occur due to body movement so that it affects the form and system of the human body. In addition, the rhythm that always moves partially presents a prestigious and continuous movement. This dynamic is where the building design can continuously adapt to the surrounding environment by nature. Naturally, as an embodiment of fairness and respect, these dynamic aims to survive with the existing conditions.

Transparency is an achievement value from martial arts organisations that can monitor each other, always open access and influence each other [3]. In the world of martial arts, it prioritises awareness, no barrier, observes, and systematic. With various types of martial arts, transparency between components is presented inside and outside the building. Through collaborating these two aspects, the design uses the theory of dynamic architecture. The design chose static-dynamic by designing the building form as an optical illusion and representing the dynamicity and transparency through architectural elements.

In addition, the theory of transparency supports the translation of values. Transparency architecture shows the analogy of the value of the martial arts organisation between members of the organisation [17]. The design applied the concept to building elements, such as spatial programming, building form, shapes and facade, openings, walls, floors, ceiling, etc. Moreover, the merging of dynamic and transparency produces architectural elements in external texture control, coverage, openings, external light control, spatial continuity, circulation of a path, and entrance configuration. Static-dynamic design present partially translated from the values of the sports that is a curved shape according to the form of the trainer's muscle. The curved shape is presented through a continuous and prestigious muscle performance system.

The diagram of Figure 6 shows elements resulting from the domain transfer. It focuses on architectural elements such as facade, form, texture, shading, layout, circulation, and scale per floor elevation. It aims to observe the climate systematically, such as the kinetic facade. There is a spatial external texture control in the facade element and can experience physical or visual movement. Besides, it can also be an area that looks pure and transparent between the outside and inside.

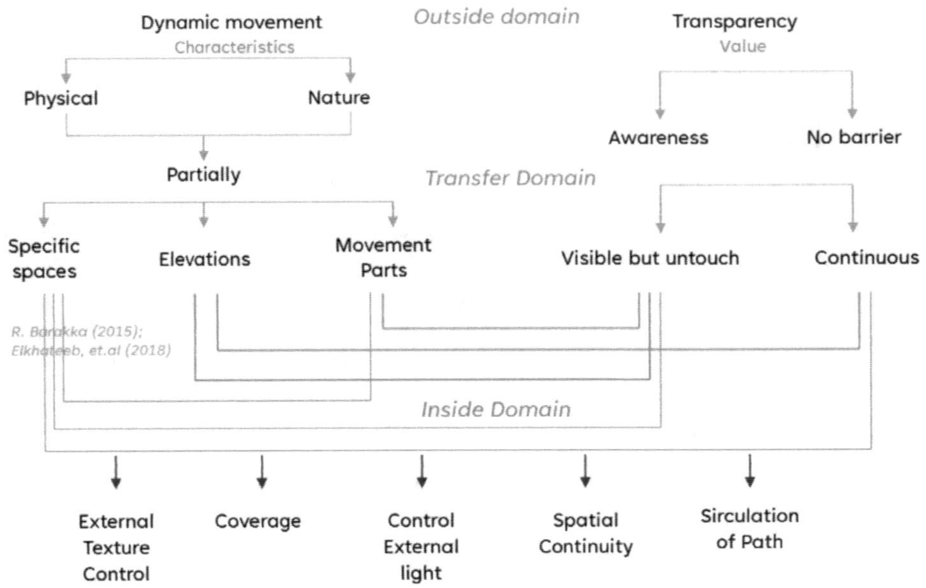

Figure 6. Diagram of domain-to-domain transfer

4.2. *Building design*

This chapter discusses implementing concept/innovation ideas in design or buildings. Concept implementation focuses on building forms, programs and diagrams, spatial quality, and materials analysed by evaluating green building parameters.

4.2.1. The journey of building shape. Building shape exploration starts with studying the semiotic analogy theory by finding information from users' and environmental parameters. The investigation starts from simple rectangular cubes and adds another one at the top with a smaller dimension (Figure 7b). The character of massiveness and cleanness is inspired by the translation of the meaning of sports values. The form and shape also consider environmental aspects, such as orientation, sunlight, radiation, and wind direction. Figure 7b shows the form exploration is integrated into spatial requirements, including spatial organisation for office and sports facilities. Offices are placed on the lower floors and sports facilities on the upper floors. The sports facilities only support light and mild martial arts exercises that do not require heavy sports equipment. The environmental views provide a positive influence to induce the spatial quality of each space for the users.

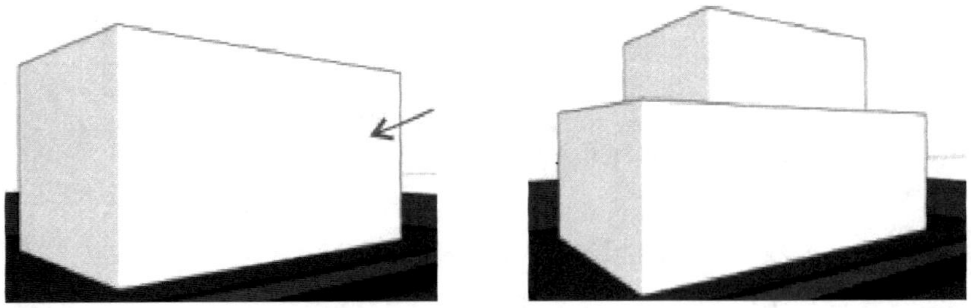

(a).Massive form (b). Divided two according to space requirements

Figure 7. Rectangular cubes form exploration

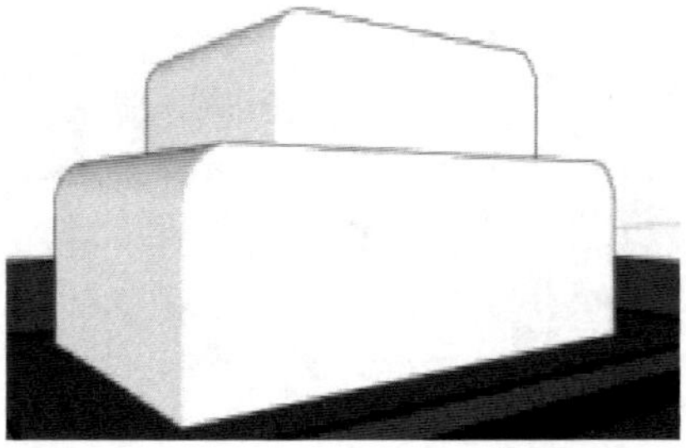

Curved side: elegance

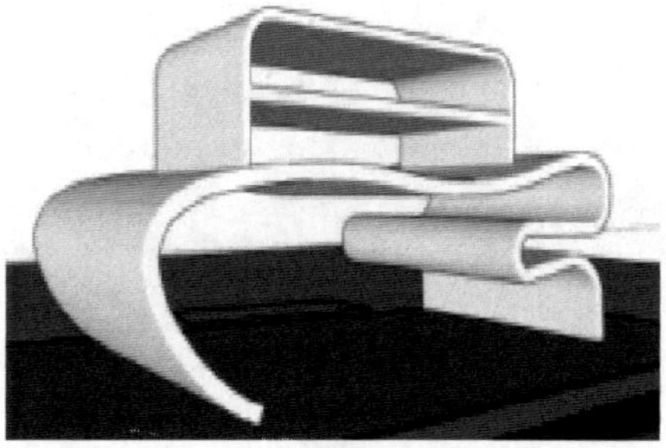

Analogy of the muscle movement in building envelop

Figure 8. The form exploration is integrated with spatial position and requirements.

Next, Figure 8 tells adding curved form at the edges of the cube building forms. The aim is to communicate elegant and flexible meaning by applying the analogy of bent joints. Further, the design presents an analogy of muscles to building a façade from the ground level to the roof. This expression shows the meaning of dynamicity and continuity. Further exploration gives a transparent side on the front-side building (9a). The transformation design is intended to be transparent between the outer and inner space. At the same time, it becomes a translation of the value of openness between organisations and the community. Protection against the amount of light entering the building due to glass on the sides, secondary skins are added. The large overhang also prevents direct light and sun radiation to balance the building position that considers positive orientation (building as the new image of the surrounding environment). A secondary skin is an undulating façade that acts as an external control light with the concept of a movement system (9b). The design is a translation of movement parts from the specific spaces in the transfer domain analogy diagram above (Figure 6).

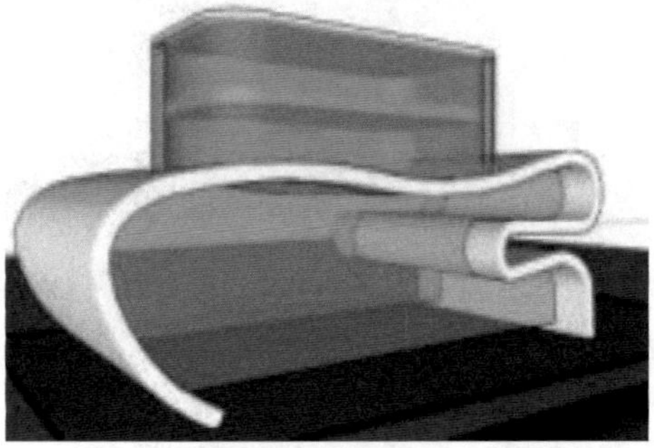

(a) Double glazing: transparent representation

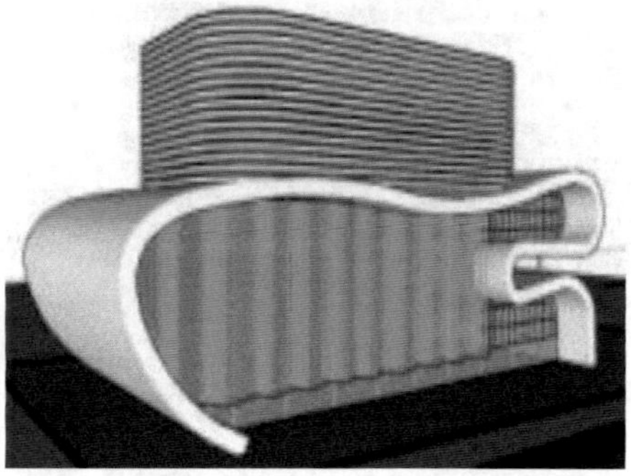

(b) Undulating façade: external control light in movement and system

Figure 9. Double glazing and undulating façade to show transparent and movement analogy

The final form of exploration is the last finishing to the façade details by combining transparent and massive building materials (figure 10). The last proposal considers the strenght and stability of building structure by changing the arch of the building shape in better proportion. The front building provides open spaces and landscape design to support the building function and meaning as martial arts, sports institutions, and organisations.

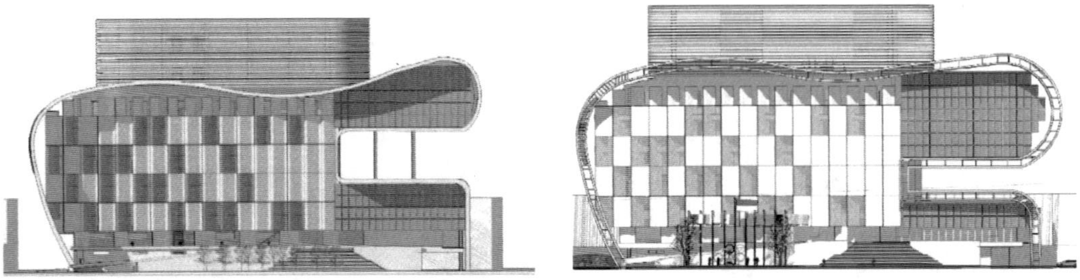

The first form exploration The final building designs

Figure 10. First design proposal and final design proposal (right) to show the dynamic and continuity representation

4.2.2. Building dynamic elements. This building has a continuous value in-building coverage as an expression of a different building from the surrounding environment (Figure 11). In addition, it can also provide a view with wavy transparency of the existing buildings in the surrounding environment. Openings in textures, materials, and shading embody value movement through textures that adapt to the surrounding environment.

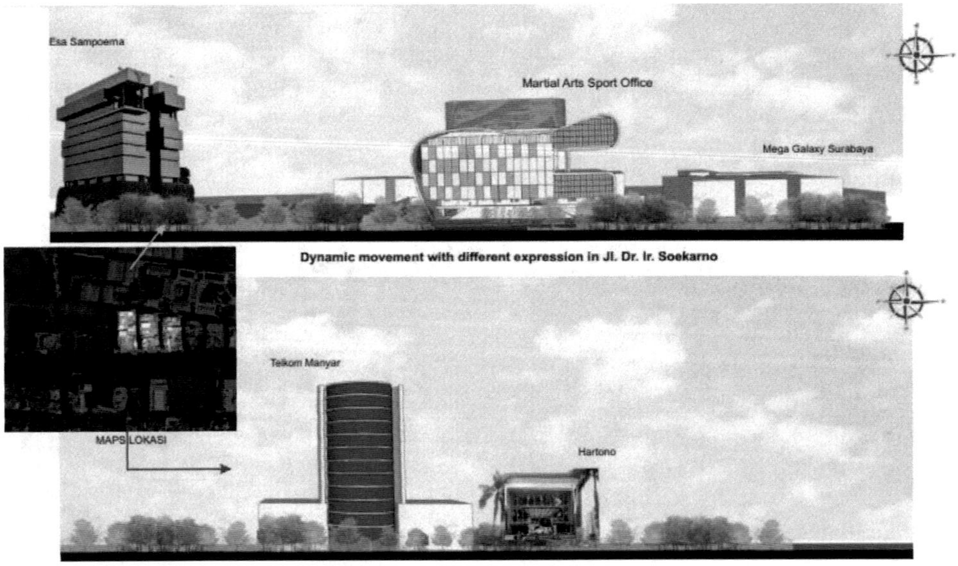

Figure 11. Street View

When viewed through a street-view image, the building is obvious with a noticeable difference in shape with the surrounding buildings. The surrounding environment near the building has a similar building expression, a square shape on each side. Thus, the design uses a curved shape and form that express dynamicity and give a balanced state to the surrounding environment. The existence of a curved shape describes a muscle working with partial movement and has the strength to provide power and withstand loads (Figure 12).

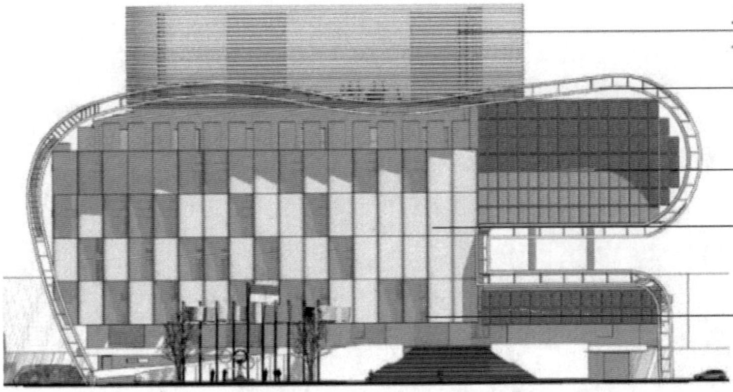

Figure 12. Detail and prestigious building coverage

4.2.3. Spatial programming and communication. The concept of visible but untouched is shown through *spatial* continuity by providing a groove in the spatial arrangement, circulation, and facade shape. The space layout is given without a gap between spaces to add the impression of transparency between spaces in the building—circulation by highlighting the differences between public and private positions to provide a view of the entire workspace (Figure 13).

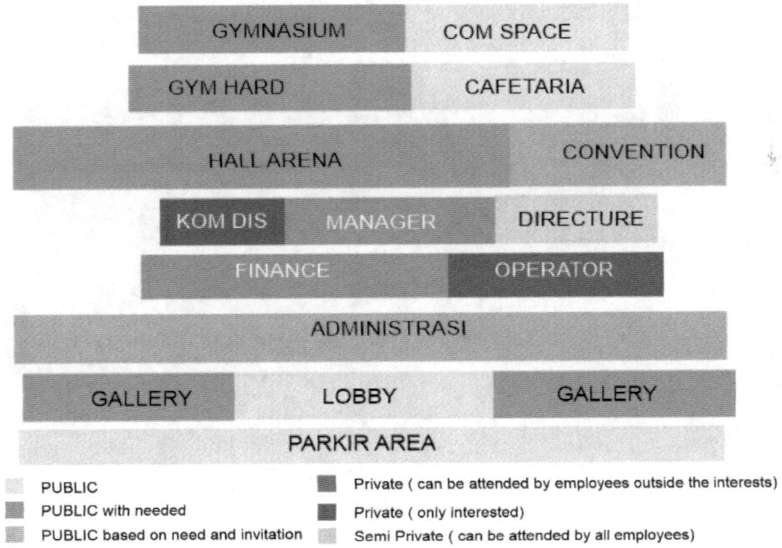

Figure 13. Dynamic space programming in each floor

4.2.4. Spatial quality. In addition, the elements of interiority such as floor, walls, openings, and ceiling show the expression of dynamic shapes and transparency. The value of seeing but being untouchable is translated through transparency between spaces in the following interior area. Figure 14 shows a multi-space that can be seen as a whole within a certain distance without a skat/barrier. In addition, spatial arrangement represents the culture of sports organizations that tends to be always transparent and dynamic. Figures 15 and 16 illustrate the administration's working area with a vast window opening. The design demonstrates the transparency between spaces so that each other can monitor each other's situation. This design is intended for public and semi-public spaces.

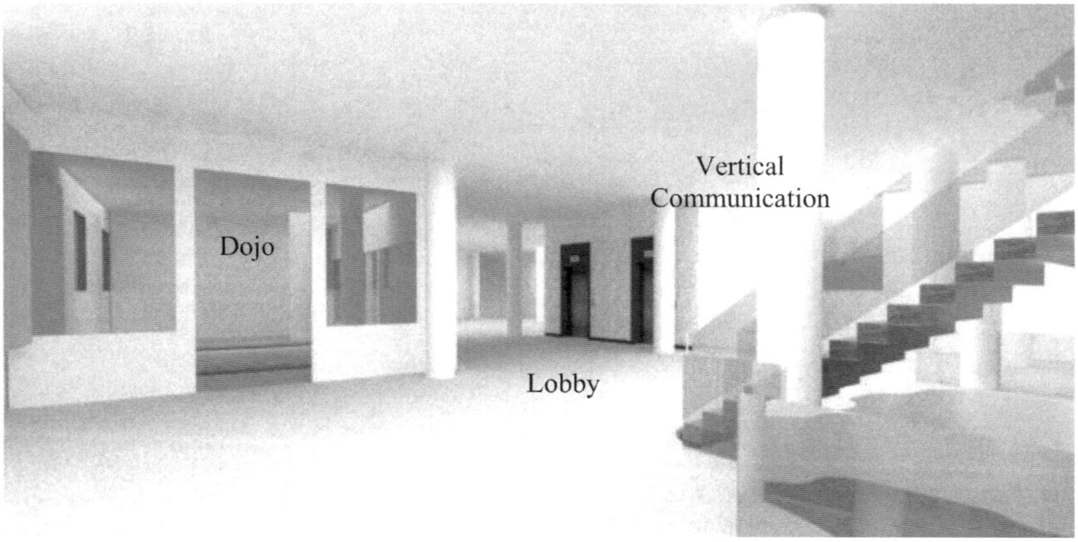

Figure 14. Multi-space in "Seeing but Untouchable"

Figure 15. Transparency between space and dynamic ceiling light source in the administration area

Figure 16. Convention Room with undulating ceiling

The sports centre area, a semi-public space in this building, carries several domain transfer results, namely the dynamic value upholding the movement in space such as light and texture. In addition, the value transparency supports the relationship between buildings by seeing the area around the building through the building wall lattice in the gym area.

The dynamic movement of the undulating ceiling and wall is in the form of a winding and uneven shape of the wall and roof (Figure 16). It gives an image of the dynamics of space that affects the atmosphere in the room both visually and soundly. This concept is given to the convention room as a meeting arena and for friendly matches and other similar events. Furthermore, the idea of sunlight movement is given to the gym room. Natural sunlight is vital for physical and physical fitness [22]. In addition, several studies have stated that the presence of natural light moving in space can provide physical and psychological comfort for space users [24]. Therefore, the gym area offers playing natural light in the room through the grille of the window (Figure 17).

Figure 17. Spot of dynamic sunlight movement in the gym area

4.2.5. Energy efficiency with the material in building. Green buildings utilize low-energy and use sources ecologically through adapted materials, resources, and conversion rates [25]. The study of building materials corresponds to three categories that are included in the green building parameters, namely energy-saving, water-saving, and energy contained. The design uses Edge Application tools to Calculate the building materials for low-energy building performance. The program also helps determine which technology, elements, and aspects are modified to maximize energy use. The buildings studied have a total sub-project floor area of 5,640 m^2.

In emphasizing embodied energy, this research focuses on the use of materials that can be low in energy by considering the shape of the building, which has become the main criteria in the design. The percentage shows that using specific materials that have been determined can reduce up to 30.83% of the total embodied energy of the building. The use of several materials in buildings, namely:

- Roof construction using In-Situ Reinforced Concrete 150 mm with glass wool insulation 25mm
- External wall with 70% curtain wall as frame lining and 30% aluminum clad sandwich panel
- Internal wall using In-Situ Reinforced Wall
- Floor Slab using In-situ Reinforced Concrete Slab 150 mm with flooring terrazzo, tile, and nylon carpets
- Windows frame using Aluminum Double Glazing

Furthermore, the energy used is up to 38.59% of the total energy needed. The materials and materials used are:

- Use of Windows to Wall Ratio 51.89% with Fairly high WWR is countered by low E Coated Glass and External Shade
- Insulation roof with glass wool U. Value 1.26 W/m^{2}K
- External shading using Combined - Shading Factor (AASF) 0.33
- Low E Coated Glass with SGHC 0.41
- LED Light Bulbs Internal and External Space in the form of artificial lighting using energy-efficient LED light bulbs in external and internal space
- Daylight sensor and light controls with using sensors based on natural resources for energy savings

The last is the emphasis on water energy, namely water saving. In this study, the results were found, namely 31.38% of total energy suppression for water use in buildings (Figure 18). The details of its use are as follows:

- Low-Flow Faucet in all bathroom 6L/min
- Dual Flush High 6L/flush & 4L/flush
- Water-Efficient Urinals in All Other Bathrooms - 3 L/flush
- Greywater treatment and recycling treatment

Figure 18. Final calculation with Edge Application

Based on the selection, the orientation of the building and the WWR significantly affect the energy consumption of the building. Thus, the total energy used per month is only 67,717.51 kWh/Month. This figure is significant compared to buildings that have not implemented a green building system. Other factors were also chosen, such as climate zones, building materials, structures used and several other things to increase the energy efficiency [26].

5. Conclusion

Based on the concept of building as image as an idea and innovation from a martial arts sport value incorporated into the building through domain transfer. Then set dynamic and transparency as the main concept. The concept makes the building different from the surrounding buildings and has the characteristics of the building itself. The idea presents the uniqueness of building design to express martial sport's dynamic and transparent value and pay attention to environmental sustainability. The object that is the focus of this research is a martial arts office building in which there are various kinds of martial arts and users but still have the same firm aspects.

With a combination of green building as a general criterion, this design reduces the total energy requirements during and after construction. Maintenance of energy efficiency, water-saving, and embodied energy analysed using low energy materials can reduce energy use in buildings. The use of the green building as a general criterion is the main factor that must be a concern in designing any building. By conducting an analysis based on green building parameters, it can reduce the use of non-renewable energy.

The existence of collaborative innovation building as an image with dynamic and transparency combined with green building is one concept that can continue to provide designers with the exploration of works and reduce the amount of energy needed. The design applies specific building materials, technology, and design arrangement that reduce excessive energy and turn into a sustainable building.

6. Reference

[1] Nassr E H and Al-Neaimi K G 2021 *The Importance of Sports Infrastructure for Residential Neighbourhoods' Centres* IOP Conference Series: Earth and Environmental Science **754**(1).
[2] Turrin M et al 2016 *Computational Design for Sport Buildings* Procedia Engineering **147**: 878–83.
[3] Marhaento P 2010 *Interpersonal Relationship Management in Sport Coaching* Process Olahraga Prestasi **6**: 57–63.

[4] Department for Culture Media and Sport 2015 *Sporting Future: A New Strategy for an Active Nation* Departament of Education and Skills 1–84.
[5] Ghaffarianhoseini A et al 2013 *Sustainable Energy Performances of Green Buildings: A Review of Current Theories, Implementations and Challenges* Renewable and Sustainable Energy Reviews **25**: 1–17.
[6] Jones J C 1970 *Design Methods: Seeds of Human Futures* **2nd** edition John Wiley & Sons Ltd.
[7] Elkhateeb A M, Fikry M A, and Mansour A A 2018 *Dynamic Building and Its Impact on Sustainable Development* Alexandria Engineering Journal **57**(4): 4145–55.
[8] Baraka R O, Shamallakh A M 2015 *Architecture Dynamic* Retrieved from http://www.share-research.org/:https://www.slideshare.net/alyaashamallakh/dynamic-architecture-gaza-own-perspective-on-dynamic-architecture. access 22 Nophember 2021
[9] Rumayang D R, Ekasiwi S N N, Hayati A 2014 *Transparansi Arsitektur dalam Proses Rancang Terminal Kampung Rambutan* Jurnal Sains dan Seni Pomits **3** No 2 2337-3520
[10] Rowe C and Slutzky R 1997 *Transparency Berlin*: Birkhauser Verlag.
[11] Vale B and Vale R 1991 *Green Architecture Design for Sustainable Future* Thames & Hudson London
[12] Masood O A I, Al-Hady M I A and Ali A K M 2017 *Applying the Principles of Green Architecture for Saving Energy in Buildings* Energy Procedia **115**: 369–82.
[13] Green Ship 2010 *Greenship Rating Tools untuk Gedung Baru* vers **1.0** Green Building Council Indonesia
[14] Vera G 2013 *Beijiao Sports Center / Decode Urbanism Office* 22 Apr 2013 ArchDaily Accessed 16 Jan 2022
[15] De Architekten Cie *Edge Olympic* [Internet] Archello 2020 [updated 2021; cited 2021 Dec 21] Available from: https://archello.com/story/75071/attachments/photos
[16] Hufton 2012 *London Aquatics Centre for 2012 Summer Olympics / Zaha Hadid Architects* 18 Aug 2011. ArchDaily. Accessed 16 Jan 2022
[17] Woods L 2018 *Drawn into Space Zaha Hadid Architectural Design* **78**(4): 28–35.
[18] Edge Tech EDGE *Olympic Amsterdam* [Image on internet] 2020 [updated 2021; cited 2021 Dec 19]. Available from: https://edge.tech/developments/edge-olympic-amsterdam
[19] Pintos P *Olympic House / 3XN* [Image on Internet] 2019 [updated 2021; cited 2021 Nov 26]. Available from: https://www.archdaily.com/919974/olympic-house-3xn
[20] Marotta A et al 2017 *Visual Images and Language in Architecture: Signifier Semiotics and Meaning Semiotics* Multidisciplinary Digital Publishing Institute Proceedings **1** (964): 1-17
[21] Plowright P D 2014 *Revealing Architectural Design: Methods, Frameworks and Tools* Revealing Architectural Design: Methods, Frameworks and Tools.
[22] Al-Rhodesly A M M A 2019 *Biophilic Approach to Architecture : Case of the Alhambra, Al-Andalus* Alexandria University
[23] Hadfield S and Hasson G 2013 *Bersikap Tegas dalam Segala Situasi* Jakarta: BIP Kelompok Gramedia
[24] Kwong Q J 2020 *Light Level, Visual Comfort and Lighting Energy Savings Potential in a Green-Certified High-Rise Building* Journal of Building Engineering **29**: 101198
[25] Bielek B 2016 *Green Building – Towards Sustainable Architecture Green Building* Applied Mechanics and Materials.
[26] Yu Z, Zhang W, and Fang T 2013 *Impact of Building Orientation and Window-Wall Ratio on the Office Building Energy Consumption* Applied Mechanics and Materials **409–410**: 606–11.

IOP Conf. Series: Earth and Environmental Science 1007 (2022) 012024 doi:10.1088/1755-1315/1007/1/012024

Light as medium for time manipulation of Start-Up-Online-Shop office building

Z R Firliani[1] **and S N N Ekasiwi**[1,2]

[1] Architecture Department, Faculty of Civil, Planning, and Geo-Engineering, Institut Teknologi Sepuluh Nopember. Kampus ITS, Sukolilo, Surabaya 60111, Indonesia

Abstract. Information and communication technology are multiplying in this era and impacting work-life balance. The discussion starts with issue related to perception of time in work life. People's perception of time can be faster or slower due to their work environment or the activities they do. Work-life balance become blurred because many workers in big city work more than they should do. Previous research compares this phenomenon with the experience living in space. Based on it, comes a question whether architectural design can substitute timer in a work environment. In this paper, we propose an approach for start-up office with the concept time manipulation using light as medium to influence people's perception of time. The goal is to influence people's perception of time and increase productivity in work environment. The theory that used in this paper discusses about relationship between architecture and time and discuss the designer's approach from design process. The design concept related to the theory then tested on human perception and applied on architectural syntax of building. Using concept-based framework with domain transfer to represent movement of time in building design through experience of space. The design prioritizes the psychology and value of time of the workers.

1. Background

The topic starts with explaining the impact of Information and communication technology on everyday life, especially in the working environment. Online-shop start-up office in this era are widely known by public and mainly develop in big cities such as Surabaya, which is the 2nd largest city in Indonesia that has green city labels. Online Shop is considered more efficient, especially in perspective of time so that it is easier for users and accessible at any time which increase profits. But because of that, the balance of work and life become blurred because many workers forget their working hours. When working, a person can become too focused-and forget to take a break. This is because the workspace is not connected to the ambient environment conditions so that the change of day and night is missed.

The main issue in this paper is the perception of time. The definition of time can also be interpreted from various scientific disciplines and perceptions. According to Tuan, time is the duration in which tension is followed by ease, so time can be described as a motion or flow [1]. The perception of time experienced by a person can be faster or slower depending on the ambient environment and activities. Sometimes we feel that time is going slow because we are bored, for example waiting for someone. It will different when we watch dramas or work for deadlines, this condition make time feel much faster. Based on it, we hypnotize that 'what if architecture can substitute of timer'. This mean that time is not only related to the number of hours. We can also know the time from the change of light and

[2] Corresponding author: nastiti@arch.its.ac.id

environmental condition. The medium for time perception which is easily visible, namely from the light media which until now can be used as timer.

Our goal is to increase the productivity of workers while also create energy efficient building. The priority is the psychological value of user. The issue of the movement of time through the medium of light that affect a person's perception. The novelty of light gives deep meaning for building, there is two precedents to compare the similar concepts such as Jean Novel in his Institut Du Monde Arabe who uses the movement of sunlight as a medium to convey meaning and experience with the concept of the pupillary diaphragm. Another example is Al Bahar Towers by Aedas in Abu Dhabi, with a kinetic facade to control light in response to the environment and give meaning to Middle Eastern culture.

The issues related to time by the movement of light can produce a preposition, namely human perception. Human perceive time feel slower or faster depending by their activities, environment ambient and experience of space. Therefore, time can be manipulated by light to create human perception. The preposition in the design of this paper can be stated that office as a time manipulation with light as a medium for time manipulation. Time manipulation concept with light to control the atmosphere in the room for worker's comfortable activities and can feel the changes in time, so that they become more productive. The purpose of this paper is expected to help readers who use time and light as concepts in designing a building to give a deeper meaning related to time and for the sustainability of the surrounding environment so that the movement of light can provide color and contrast between dark and light for time perception and energy saving.

2. Design approach and theoretical framework

Previous research about philosophy of human experience discusses the relation between architecture and human experience, especially in perception. The theory time and architecture by Erica Wickerson used as guideline to explains that time is related to activity and spatial experience [2]. Time perception, one of which can be generated through the movement of light. The movement of light can give an effect such as shadow, colour and contrast of dark and light which can visually affect one's perception of time [3]. The principles of Time and Architecture related to the light are used as guidelines, namely movement, moment and characteristics to translate concepts into architectural syntax. While the aspects that are considered to influence the concept of time manipulation with light media in the design are the facade, building form and materials that are adapted to the activities of the Start-Up Online Shop office workers in each room program.

The idea of the design is using light as a medium for perception of time with exploration through the analogy of the event of time, namely the movement of time related to the change of time from morning, afternoon, and evening in a building as a timer to manipulate time with light media. The movement of light based on the change of time give a shadow effect that is used in the design of the space program according to the activities of workers. Through the kinetic facade and material (please refer to chapter 4.1), aims to present a perception of time. With a design context that is green energy building focus on daylight and artificial light which is controlled through kinetic facade system in which concept of time manipulation for human perception of time can be translated to green energy building.

2.1. Literature studies

Previous study aims to compare and evaluate the design innovation and creativity. Table 1 shows a comparison of previous study from two precedent works, including the Institut De Monde Arabe by Jean Novel, which has the idea of a museum as a pupil diaphragm for controlling light to create deep meaning of light by kinetic facade with design criteria including Psychological priority from daylight with a flexible kinetic facade to control light with the resulting shadow effect giving meaning to the representation of eastern culture and providing quality performance for comfortable lighting and energy saving (please kindly see figure 1). The second precedent is Al Bahar Towers by Aedas with the Idea tower as expression of honeycomb geometry for flexibility facade to respond to daylight in weather extreme by movement of sunlight. The use and control of light is also for the sustainability of buildings that respond to the sustainability of the surrounding environment (please kindly see figure 2). Looking

IOP Conf. Series: Earth and Environmental Science 1007 (2022) 012024 doi:10.1088/1755-1315/1007/1/012024

at the comparison in table 1, it can be evaluated as an innovation in this design, namely Ideas Light as medium for Time manipulation with design criteria psychological priorities for time perception, Respond to Light for time manipulation in moment of user activity and movement in the building through kinetic facade and shape. Dynamic buildings as a representation of the movement of time by utilising light as a medium to provide psychological effects and building performance so that in addition to presenting time perception through spatial experience, the use of light is also used for energy efficiency so that it is in harmony with environmental conditions.

Table 1. Study of precedent [4,5].

	Institut De Monde Arabe Jean Novel, France, 1987	Al Bahar Towers Aedas, Abu Dhabi, 2012
Building Ideas	Museum as a pupil diaphragm for control light to create deep meaning of light by kinetic façade.	Tower as an expression of honeycomb geometry kinetic facade for flexibility facade to respond to daylight in extreme weather and culture.
Criteria	Psychological priority from light, flexibility, quality, efficiency energy, façade representation of eastern culture.	Psychological and building performance in response to sun movement in extreme weather.

Figure 1. Kinetic façade to control light for deep meaning and psychological priority [6].

Figure 2. Responding daylight with orientation, building form and adaptive building skin for weather extreme and respond to the eastern culture [7].

Looking the comparison on Table 1 can be evaluated as innovation in this design project. The design idea responds to Light for time manipulation in the moment of user activity and movement in the building by kinetic façade and building form through spatial experience. The light is also used for energy efficiency so that it is in harmony with environmental conditions.

2.2. Design approach

The project in this paper uses adaptive building and human perception, namely the kinetic facade to control and utilise the light that enters the building based on user activities in the building [8]. The design focus on light to control natural light and artificial light to create perception of time. Time is very meaningful for Start-Up Office because it is a commercial building so that with the medium of natural light it can be explored how problems related to light in certain natural conditions such as morning, afternoon, evening which we call the "event of sky" and in conditions of changing light seasons can still be seen for perception of time. The orientation and shape of the dynamic building plays a very important role in this research because the position of the incoming light that is used will give the effect of colour and shadow on the space inside as well as with a dynamic shape and kinetic facade to represent the movement of time and have an effect on controlling light so that the value and meaning of time manipulation can be used from the psychological side also from green building. Human perception of time based on the theory used using the principles of movement, moment and characteristics. Movement of time give reminder for work activity “feel fast & slow” and it means with continuous stage of movement give dynamic perception of time because it relation with dark and light called shadow. Because the problem of workers is that they often forget time so that it is not balanced with life, this design will also be grouped with the human perception approach so that the time manipulation concept is adjusted to the activities in the space. Moment is also related to the movement of light can be capture visualisation by sequence “event of sky” so can give experience through by space. Characteristics of light such as light travels in straight line, while also can be bent and refracted give effect for human perception and experience of space to remind something which is also influenced by the ambient environment to form the atmosphere. This approach applied to the kinetic facade as an adaptive building skin to control light, both natural light and artificial light on the concept of time manipulation and energy saving efforts can be achieved.

3. Design method

This design started from the exploration of time manipulation as a response the issues related to problems in working life by deductive thinking. The design process uses concept-based framework proposed by Plowright with the big idea as a hypothesis. This framework that has a design process stage, socio-cultural, hypothesis, refine, mapping, propose moment, arrange elements, and proposal [9]. Concept-based framework design process phase to guide the design from problem seeking to problem solving using several design methods. At the socio-cultural stage with exploratory thinking activities on problem seeking regarding the issues raised, namely time and light using the stating objective method by Jones, namely exploring problems and situations related to the theory of time and light in buildings, especially in the office [10].

The design process starts with the big idea of "Time Manipulation" based on design problems at work and the theoretical approach of "Time and Architecture" which is focused on light [2]. In this phase, it examines the state of the existing environment related to information of lighting and the ambient environment of the site, so that with the issue of time and light that has been explored by controlling the light to create time perception based on user activity in the building. Exploration related to light as a medium for time manipulation and reviewing user activities in the Start Up Office so as to obtain the hypothesis that big idea of time manipulation is related to the main goals of increasing work productivity and also by conducting a response to site method to analyse the existing conditions related to the light, activity situation in the site location. This exploration resulted in a preposition that was obtained based on a theoretical study of time and light, namely human perception of feeling slow or fast are affected by their activities, environment ambient and experience of space, so time can be manipulated by the light to create human perception. In this phase, light utilisation is divided into two categories, namely daylight and artificial light, each of which has a different role in the implementation of time manipulation.

Using domain transfer, we translate outside domain into architecture syntax (Please refer to Figure 3). The idea as a design concept is transferred to generate form and develop criteria which later be determined by a proposed moment and detailed at the stage of array elements into a proposal as a design solution. There is a relationship between time and light with their attribute to get some keywords as propose moments. There is three propose moments that translate to architectural syntax, it is movement, moment, and characteristic. After translating the proposed moments to target domain like building form, façade, material and program, the next step is to decide judgement criteria. Several design criteria were obtained to be applied for building design. There are three design criteria, first is building response to change of time (morning to evening) and ambient environment. Second is building responding climate for energy saving and representation movement of time. The last criteria is building as a reminder for user.

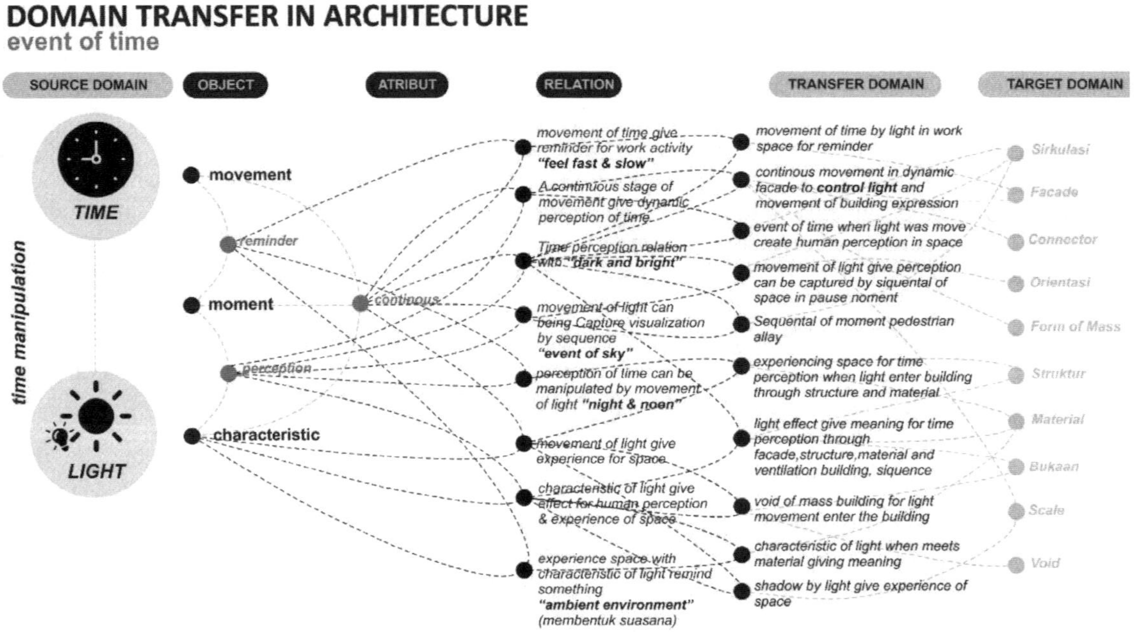

Figure 3. Domain to domain transfer.

The next phase is Refining, which transforms the concept of time and light from outside the architectural domain to the inside domain architectural syntax so that in the translation there is a relation that connects the concept to the architecture of the building. The activity reduces the control concept for judgment criteria, namely the building must represent the movement of time with light media so that users can feel the experience of space on time perception. The criteria generated in this phase are that the building must represent the movement of time by light as medium for time perception. Daylight is controlled through building openings and kinetic facade to present a space atmosphere based on changes in time from morning to evening. Incoming light is set through the kinetic facade according to the characteristics of light that can be bent and reflected so as to give a shadow effect to spaces, such as workspaces, corridors, and communal spaces. Meanwhile, artificial light serves to provide an atmosphere so that time feels fast so that there are no reminders of day and night, such as in meeting rooms and display rooms. As for the outdoor space, use materials and vegetation for the movement of light so that it gives a dark and bright effect and adjusts the seasons such as the dry and rainy seasons, for example on the rooftop there are natural wood and glass materials that function to refract light and give a colour effect to the space.

The results of the judgment criteria will produce a proposed moment office as a time manipulation with Light as medium for Time Perception. From the proposed moment, it is detailed into the arrangement element stage to explore architectural elements that fit the concept, namely how the moment can be arranged as elements which mostly use natural materials and building systems to support the concept of time manipulation of light media so that theoretical principles can be implemented. on buildings to translate concepts into architectural elements including building programs. Then at the final stage, the proposal is a whole design building as the coordination of all the previous steps.

4. Result and discussion

4.1. Building form

This paper discusses about exploration about data mapping with site analysis. Form data mapping on site location of building, one from some aspect is sun path to explore the orientation zoning and building form. There is asset and constraint that used to make design response. Figure 4. shows a site analysis of the surroundings so that the concept of time manipulation based on the height of the building, the orientation towards the building can support the concept of time perception and as an environmental response through green buildings that utilise light. Exploration of mass building design as a design alternative by movement and change of daylight and shading by surrounding.

The circular shape with the presence of voids was chosen because in this analysis the shape of the building is made dynamic so that it can be seen from various sides as well as the height and orientation of the building aiming for a futuristic concept if there are surrounding buildings that are taller and block light from entering the building so that the building concept can be sustainable. The circular shape also responds to the movement of light so that it is well distributed and as an axis for manipulating time with light media. But there is one problem for futuristic perception in south side of building site area. On the future the is some possible to building surrounding have the same height with this office project so can interrupt the time manipulation concept with daylight control. So, from this problem, for sustainable concept use the artificial light and building program on the south side don’t need too much daylight. The concept of time manipulation applied on the display room as building program. Display room have the concept like mall building to make user forget the time and enjoy buying or watching the product without feel bored. So even the south side there is building with the same high as the office building, the concept won’t interrupted by them.

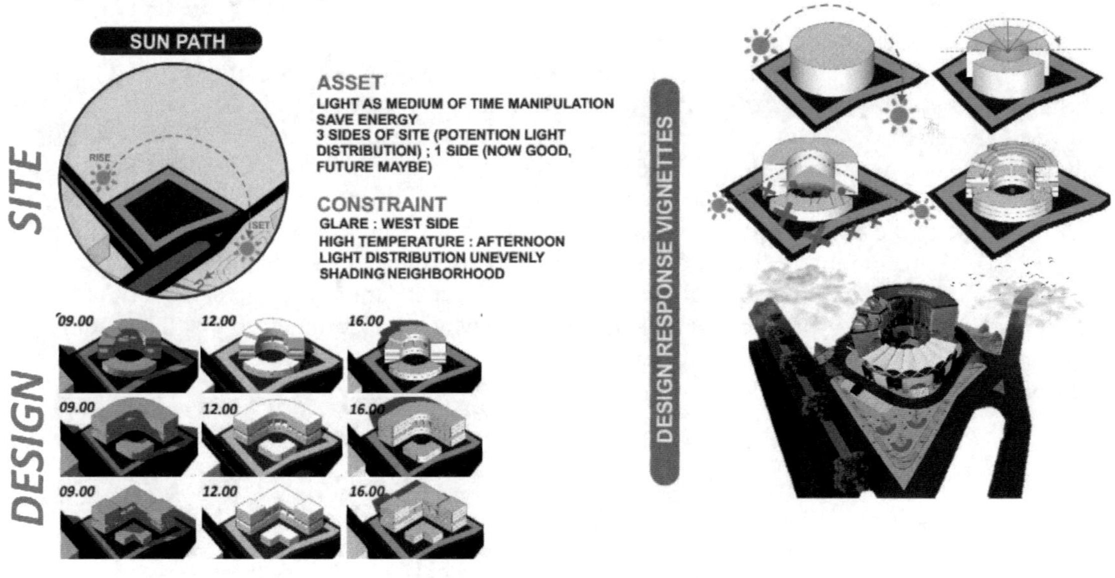

Figure 4. Site analysis.

The next step is exploring the orientation building and the position of the openings that effect the movement of light when entering the building mass. Furthermore, determine the main circulation of building based on surrounding analysis to represent the concept so that the building is in harmony with the surrounding environment. The relation generated from the proposed moment is through the kinetic facade as an architectural element of the implementation of the movement to control light so that it can create a moment according to the user's activities in the office so that the movement of light between morning and evening can give a different perception and can be used as a reminder because it gives a shading effect. which can affect a person's perception of time which we can see in Figure 5.

Figure 5. Building orientation.

Figure 5 also shows that the orientation of the building and the position of the openings greatly affect the movement of light entering the building. Due to the hot sun on the west side, with the orientation of the building facing the northwest and dynamic shape on the west side, it uses a kinetic facade as vertical shading for sunscreen and controls the incoming light so that the room inside gets sufficient light and time manipulation can be set appropriately according to space function, you can see on Figure 6. When the light is controlled by the kinetic facade so that it gives a shadow effect according to the movement of time from morning to evening then it causes the user to have a perception of time to feel that time is not felt while working and is not saturated. The effect is also a reminder for the user so that there is a balance between live and work. While on the east side using a balcony that forms like a flow for time interpretation and also serves to dispel the heat from the east and for cross ventilation. So even though there are openings on the east side and open areas, the concept of time manipulation can still be applied.

The facade of the building that uses the kinetic facade vertically to control the light that enters the building so that it can form a shadow as an interpretation of time, so that the big idea of the building as a timer can be achieved in order to achieve live and work are balanced and increase work productivity. In the picture there are also details of the kinetic facade which are designed so that the shading pattern is in accordance with the concept of time manipulation so that the movement of time can be seen through the shading produced by daylight.

Figure 6. Kinetic facade.

4.2. Diagram and program

Figure 7. shows the diagram of the concept of a building related to the characteristics of light and movement for program building and circulations. When Light enters the building, it gives a dark and bright effect which give colour to the building and give person's perception of time. We can see on

Figure 8. that the movement of light and shadow in the workspace as reminder which gives space experience. This room uses the wave ceiling and sensor technology of lighting to visually and functionally control the atmosphere, so the user can perceive time. The theoretical principles can be implemented through the characteristics of light when it enters the building to give a dark and bright effect so that it can give colour to the building which will give a person's perception of time.

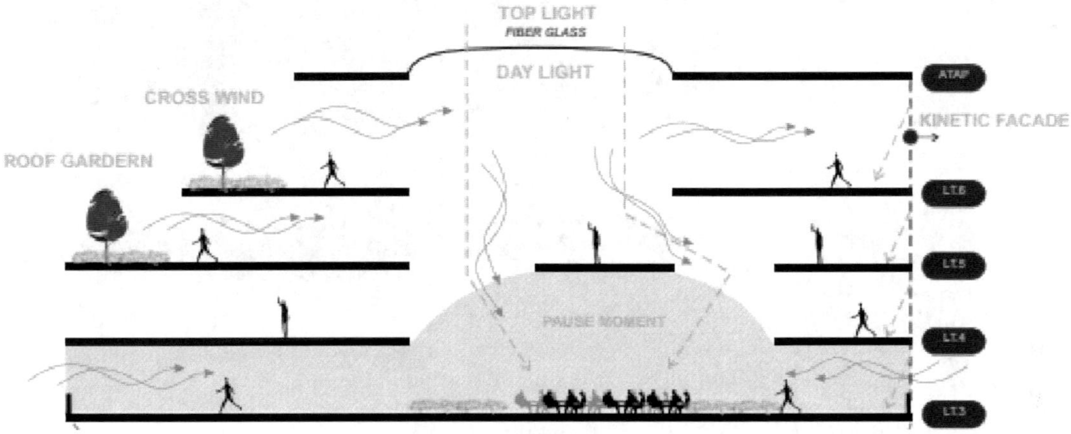

Figure 7. The section plan of light concept.

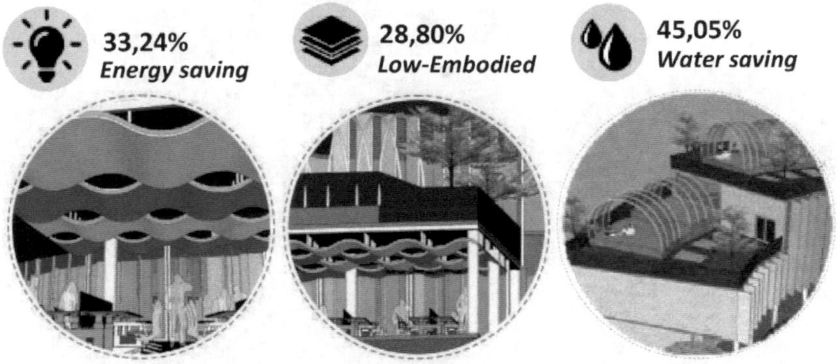

Figure 8. Working space and roof garden.

4.3. Building material

Figure 9. shows the perception of time through movement's expression in space. movement of time as a timer by light in workspace for reminder with shadow by light give experience of space. On the rooftop there is a glass material combined with ACP wood material that has been coated with termites. When the rainy drops on the glass, it can be refracted and give a certain colour to the room inside as a moment of sky so that the moment can be captured from the Sequential roof garden. In addition, on the rooftop there is also a garden as an effort to save energy by utilising rainwater to be distributed to plants in the building by using edge building tool for green building design.

IOP Conf. Series: Earth and Environmental Science 1007 (2022) 012024 doi:10.1088/1755-1315/1007/1/012024

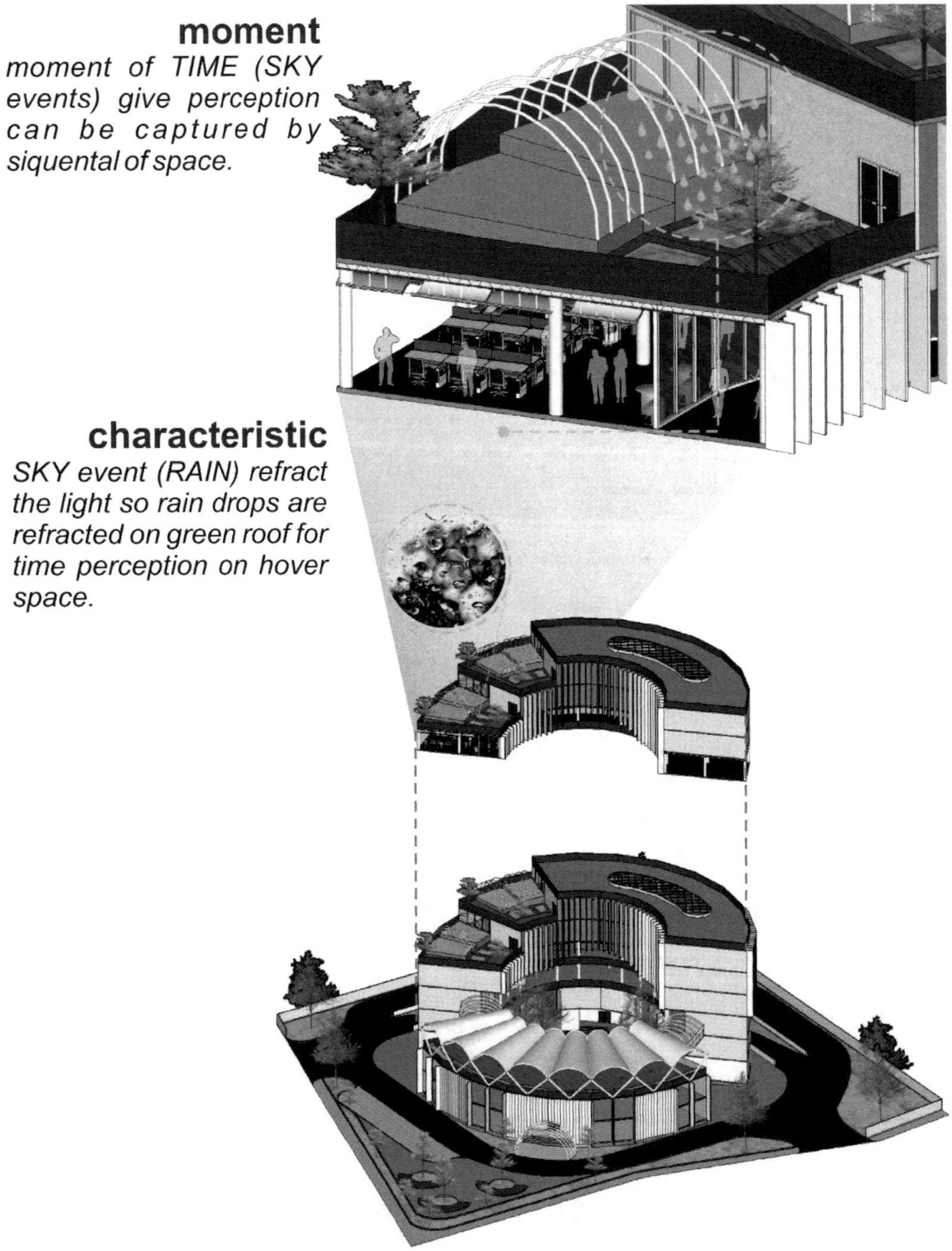

Figure 9. Building material in roof garden.

4.4. Building system

Figure 10. shows the building system as response to support green-building concept. As with green roofs, plants are selected according to their characteristics to respond to the climate in Surabaya and also to shade visitors so that at certain moments, such as in the evening, one can be reminded by time and by the materials used as a response to the building's environmental conditions. So that the building system coordinates to help realize the concept of time manipulation and green building.

IOP Conf. Series: Earth and Environmental Science 1007 (2022) 012024 doi:10.1088/1755-1315/1007/1/012024

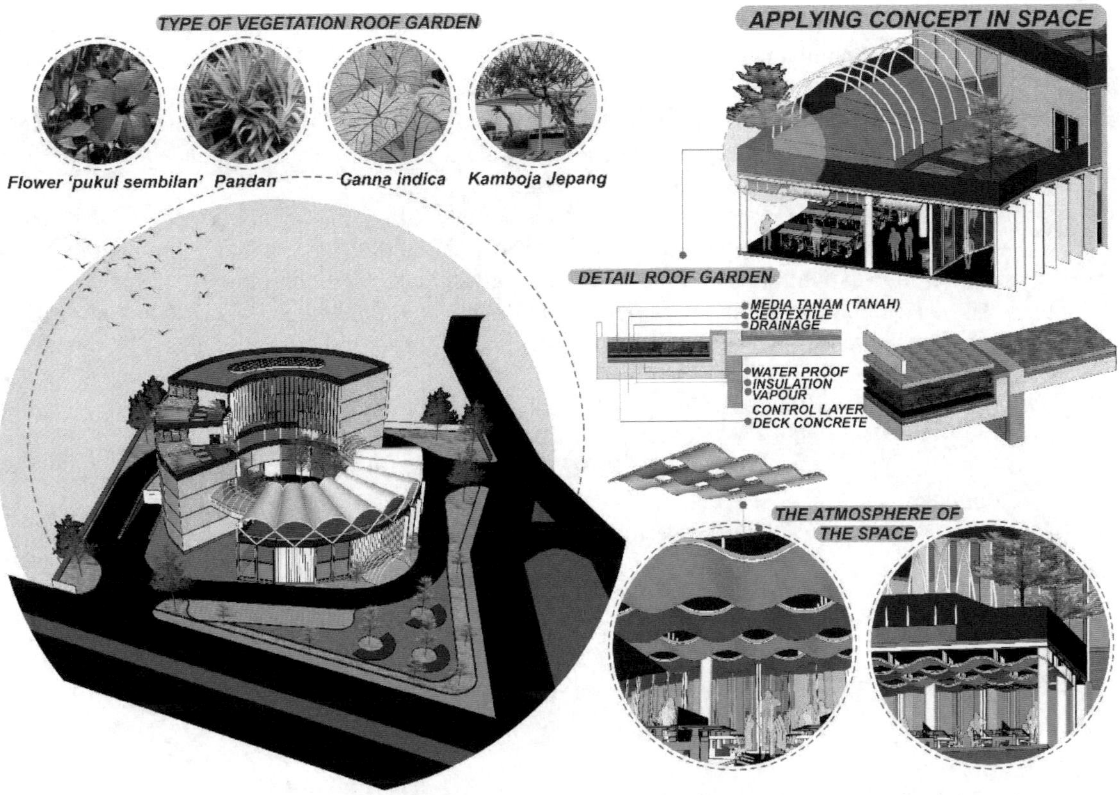

Figure 10. Green building system.

5. Conclusion

From the discussion, it can be concluded that the goal is to increase work productivity while keeping live and work balance. The process of designing the Online Start-up Office with the concept of time manipulation is based on framework proposed by Plowright which departs from the big idea of time manipulation using time and light as a guideline to realise the concept. Using method Domain Transfer for translating outside domain to architectural syntax. The guidelines are transformed using the transfer domain tools into architectural elements of syntax. Light as a medium for perception of time can give a color effect to the building which produce light and dark so that the design goal to increase work productivity while keeping live and work balance are achieved. The advantage of this design is that it can utilise light and control light with a kinetic facade so that it can save energy and a timer for both building users and the surrounding environment as well as buildings that have characteristics that make them different from the surrounding buildings. Concept of Time manipulation with light as medium make buildings have unique characteristics and are beneficial because time for commercial building is very important. With the selection of environmentally friendly materials can make the building green and environmentally friendly. Time manipulation by utilising light as a medium for time perception becomes a design innovation so as to make buildings have unique characteristics and are beneficial for users and the surrounding environment.

6. References

[1] Tuan Y F 1977 *Space and Place: The Perspective of Experience* (New York: University of Minnesota Press)
[2] Wickerson E 2017 *The Architecture of Narrative Time* (United Kingdom: Oxford United Press)
[3] Lobell J and Kahn L I 1979 *Between Silence and Light: Spirit in The Architecture of Louis I. Kahn* (Boston: Shambhala Publications, INC)

[4] Novel J 1987 *Arab World Institue* Ateliers Jean Nouvel http://www.jeannouvel.com/en/projects/institut-du-monde-arabe-ima/ (accessed February 19th 2021)
[5] Attia S 2017 *Evaluation of adaptive facades: The case study of Al Bahr Towers in UAE.* GlassonWeb https://www.glassonweb.com/article/evaluation-adaptive-facades-case-study-al-bahr-towers-uae (accessed February 19th 2021)
[6] Schielke T 2014 *Light Matters: Mashrabiyas-Translating Tradition into Dynamic Façade.* ArchDaily https://www.archdaily.com/510226/light-matters-mashrabiyas-translating-tradition-into-dynamic-facades?ad_medium=gallery (accessed February 19th 2021)
[7] Ciento K 2012 *Al Bahar Towers Responsive Façade.* ArchDaily https://www.archdaily.com/270592/al-bahar-towers-responsive-facade-aedas (accessed February 19th 2021)
[8] Kopec D and Kanaani M 2016 *The Routledge Companion for Architecture Design and Practice* (New York: Routledge)
[9] Plowright P D 2014 *Revealing Architectural Design: Methods, Frameworks and Tools* (London: Routledge)
[10] Jones J C 1970 *Design Methods* (New York: John Wiley & Sons)

IOP Conf. Series: Earth and Environmental Science 1007 (2022) 012025 doi:10.1088/1755-1315/1007/1/012025

Shape exploration through airflow simulation for reducing dust in apartment design

R T Maharani[1], S N N Ekasiwi[2], FX T B Samodra[2, 3]

[1] UPN Veteran Jawa Timur, Faculty of Architecture and Design, Jalan Rungkut Madya No.1, Surabaya 60294, Jawa Timur, Indonesia
[2] Architecture Department, Faculty of Civil, Planning, and Geo-Engineering, Institut Teknologi Sepuluh Nopember. Kampus ITS, Sukolilo, Surabaya 60111, Indonesia

Abstract. Dust is a particle in the air which at certain concentrations can be harmful to human health. Therefore, in designing buildings, we need to consider forms that can reduce dust exposure. The design process, which uses a force-based framework, was carried out to design an apartment located in a very dusty industrial area. Shape exploration is done using simulation software. The process of exploring the basic form of an apartment with a courtyard type is realized by forming a surface that is easy to flow through by wind, namely by making a sloping surface, so that the wind easily removes dust, so it does not enter the room.

1. Introduction

Sick building syndrome (SBS) affects the health of humans who are directly exposed to these conditions. Its influence on health problems is either physical disorders, like asthma, allergies, or headaches, as well as mental disorders [1], [2]. Dust is one of the small particles, which can cause SBS. There are three types of dust: 110 micron-inhalable dust, 10-micron inhalable dust, and total dust, which comprises all airborne particles [3]. This investigation focuses on resolving inhalable dust produced by high concentrated dust pollutant areas to buildings.

Air movement is needed to transport dust from one location to another. The minimum speed is 5.5 m/s and we can see the dust concentrate from observing the surface [4]. When the wind strikes, the mass of the structure will affect to building shape. The mass of the building has a different effect on the distribution of dust. This is caused by the surface of the building which can become a dust cushion. That must be considered in designing building shapes [5].

Because the amount of dust sticking to the surface is distributed by the wind, the surface can react to dusty conditions [6]. As a result, the consideration is in surface material because the rough texture will easier collect the dust than carpets [7]. Nevertheless, many materials now can be self-clean utilizing titanium dioxide which is recommended as an antibacterial and cleaning agent. This connection aims to use UV light to make it easier to remove the dust [8]. As illustrated in Figure 1, the shape of the building surface determines the amount of dust carried by the airflow. The flat, box-shaped surface, for starters, has a large dust concentration on the floor [9]. Second, the trap's surface catches dust in the room and because the inclined surface lacks symmetrical shapes, dust does not stick to it [10]. Finally, due to changes in momentum, the curve's curvature has the power to change the particle line [11]. Deep exposure to dust also modifies the body colour of the surface. You'll need a self-cleaning substance to

[3] Corresponding author: fxteddybs@arch.its.ac.id

fix this problem, as indicated above. The design of an apartment is a dwelling with a typology or characteristic that includes floor uses, single or double corridors, balanced proportions, or a terrace. To control air pollution [12], the design considers the site environment typically and the climatological profile, such as the wind type switched on for air exchange. The living space should be placed on top to minimize direct sunlight.

Green is commonly used as a treatment for reducing dust in existing housing, such as an apartment. The tree is used to remove pollutants from the air [13]. For example, in Stefano Boeri's Vertical Forest or traditional desert houses with a compact arrangement with a terrace, the pollutant filter is made up of green elements, and terraces or balconies help ventilate the building. This paper reported the investigation of building surfaces, which are considered to be the building shape to minimize dust.

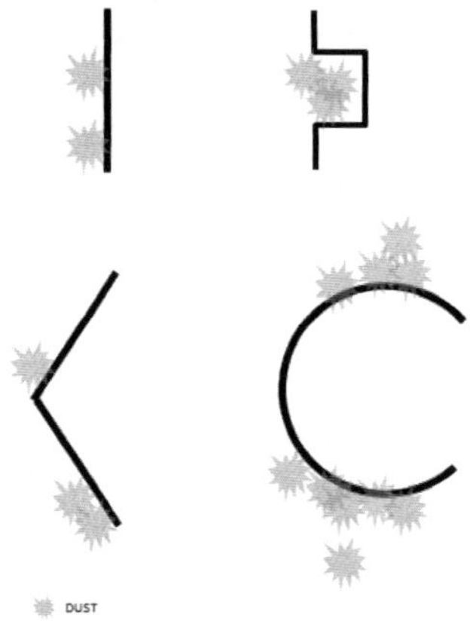

Figure 1. Dust sticking to the surface

2. Methodology

As shown in Figure 2, the design process employs a force-based framework [14]. Before proposing a preliminary design, we elaborate on forces. Regarding that framework, the forces include context, culture, and needs, which in turn formulate design criteria and first propose form to cope with the criteria. Secondly, the proposed form should be refined and followed by assembly through exploration and evaluation in a loop process to create the design concept for proposing a preliminary design. This paper focuses on identifying the force for knowing the specific problems using simulation methods with Autodesk Flow Design. The software assists in determining the wind direction, speed, and pressure in the face. The software was previously utilized to learn about performance-based design in tall building envelopes [15]. In this simulation, the wind flow has two important roles. Firstly, the ability to fly dust. The second is sweeping dust on the surface of the building.

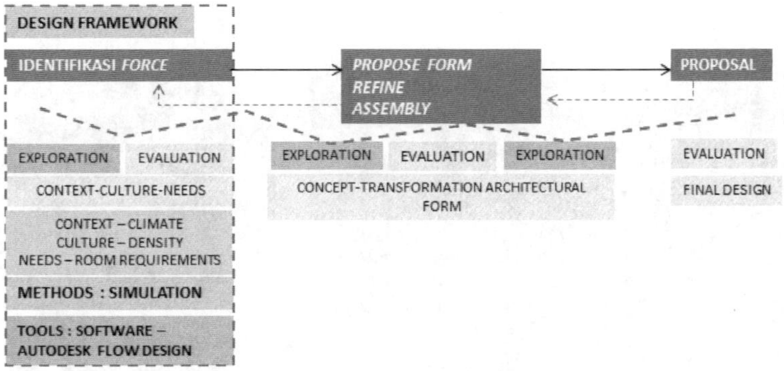

Figure 2. Design framework and methodology (adapted from Plowright [14])

3. Result and discussion

We consider three factors when exploring the force. First, the context focuses on environmental conditions such as solar movement, rain precipitation, wind speed and direction, temperature, and daylight conditions. Afterward is the culture. It includes the density of the site. It will perform an impact on the dust spread. Lastly is the need, which corresponds to room analysis for knowing the building volume for an apartment. The primary action in identifying the force is site selection. The building site is in Cilacap, an industrial city in Central Java with a medium-high PM10 dust concentration level [16]. Nonetheless, the location meets the requirement for an apartment building. The closeness to the supporting facilities, public transportation, and easiness accessible, are the requirements [12]. The three aspects of forces are as follows:

3.1 The Context

There are two site options in this selection stage. Both are in the area that meets the location requirement for apartment buildings. The SWOT (Strength, Weakness, Opportunity, and Threats) analysis was conducted. It is based on the criteria as mentioned in Table 1. The first one is in *Rinjani* street (S1) and the second one is in *Tentara Pelajar* Street (S2). Both are located on the main road, which can be easily accessed as mentioned in Figure 3. Regarding the analysis, the (S2) is selected as the site for the program development. That is because there is a distance between Industrial and site so there is an advantage for reducing dust itself. Furthermore, the site location is near the public transportation with the result people can be easy to access the site and the area is a developing area so that in the future it will become the new city district and can be the driving force for the apartment market.

Figure 3. Site selection options.

Table 1. Site analysis.

Site	Strength	Weakness	Opportunity	Threats
(S1)*Rinjani* Street	Close to the commercial and residential area	Very closed to Industrial area. It is about 1-2 km, high level dust pollutions.	People can take advantage their living-working distance	High level pollution
(S2)*Tentara Pelajar* Street	Close to the commercial and residential area	Far distance from central city is about 10 km	the location in the developing city area can be the driving force for the prospective residences.	Closeness to the main road might be harmful from vehicle traffic

The microclimate is one of the force factors which influence the criteria in determining the building form. The climate condition of the location is analysed based on Meteoblue data 2019. The wind speed and direction are identified using the wind rose analysis. (Figure 4). The temperature data shows the mean maximum 28 -30 degrees C, while the mean minimum 21-23 degrees C respectively [16]. Concerning solar movement, the surface facing West and East are very critical to gain solar radiation.[17]. The months with low precipitation can be critical for dust spread.

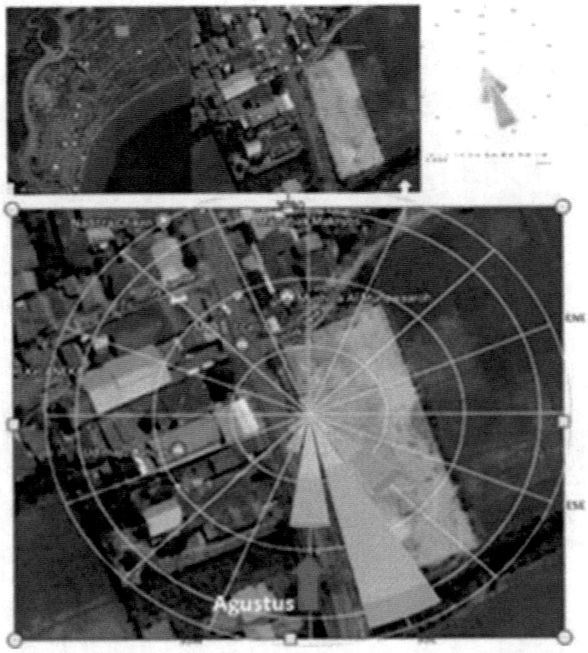

Figure 4. Site location (adapted from Meteoblue [16]).

The average wind speed ranges from 1 km/h to 19 km/h (0.3m/s to 6 m/s), according to the wind rose analysis. Apart from that, the highway at the west side, close to the site can be the source of dust in those areas. As a result, the southwest side is a critical point for dust spread. Concerning to that situation, a surface analysis from various angles is required because each surface can produce a different response [9]. High airspeed levels will transport more dust particles, but they will also aim in surface cleaning. The surfaces should allow the wind to freely flow, so that dust does not become sticked. The concave surface should be avoided because dust will cling to it and make cleaning difficult. During the design process, you may choose to use a planar, tilted, or convex surface. Nonetheless, the convex surface catch high wind speed. It probably causes uncomfortable for the residences. Moreover, it causes difficulty to control dust spread because the particle's momentum will change according to the first Newton Law [11].

The dust will easily stick to the plane surface, either vertical or horizontal one. It can be worse if the surface has niches. This situation makes the wind blow the dust, to press it on the surface plane, and difficult to blow it up. Therefore, some experiments are conducted in this research, on the plane shape that can respond to the direction of wind flow. Table 2 shows the analysis of dust spread on the planar surface, tilt surface, concave, and convex surface, with incident wind direction of 15 degrees, 45 degrees, and 90 degrees to the surface.

Table 2. Surface analysis.

Surface Shape and Wind Direction			Wind Analysis	Dust Analysis
15°	45°	90°		
1st Experiment on Planar Surface 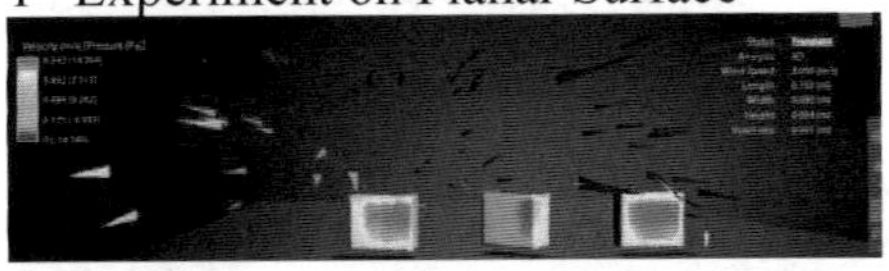			The highest airspeed and pressure are found at angles 15 degrees and 90 degrees to the surface. The wind speed 5 m/s creates speed and pressure on	On the vertical plane surface, the dust will concentrate downward or turn sideways. But if high wind speed

Surface Shape and Wind Direction			Wind Analysis	Dust Analysis
15°	45°	90°		
			the surfaces of maximum 6m/s, 14 Pa. and the minimum speed 0 m/s with -4 Pa	occurs, the dust will be removed. Furthermore, dust also can move into backside, which is caused by the turbulence effect [6], [9].
2nd Experiment on Tilt Surface 			Most of the area with the highest air speed and pressure on the surface occur at angle of the wind 15 degrees from the surface. At a wind speed of 5 m/s, the velocity and pressure on the surface of the plane are: maximum: 6.5 m/s with 13 Pa minimum: 0 m/s with - 17 Pa	On a tilt surface, the dust cannot be easy to stick because of the tilt angle. The design should be avoiding the angle 0 and 15 degree because the dust can be easy to pile up in that angle. [10] [18].
3rd Experiment on Convex Surface 			At convex surface obtains the highest airspeed and pressure if the wind direction is 90 ° from the surface. The 5 m/s wind speed resulted the maximum speed 10 m/s with 5 Pa, and the minimum speed of 0 m/s with - 9 Pa.	In convex surface, there is a changing of momentum that impact to wind speed acceleration, so that the dust is difficult to control [11].
4th Experiment on Concave Surface 			At concave surface is found that the highest air speed and pressure occur at the angle 15° and 90°. If there is a wind with 5 m/s, the speed on the surface will be 6 m/s with 15 Pa and 0 m/s with -19 Pa.	In concave surface, the dust is trapped in this shape so that it needs special treatment for removing the dust itself. [9].

3.2 The Culture

There are three elements should be included in culture analysis. They are the land use, regulation and the density of the surrounding. The local building regulations for Tritih Kulon area includes 60% Building Coverage, 40% Green Coverage, 1.8-2.1 Floor Area Ratio, and 60% Basement Coverage respectively. From these rules, the limitation of buildable area is 2460 sqm out of the 4100 sqm total area, the building level allowed to 3 storeys. The coverage area is shown in Figure 5.

IOP Conf. Series: Earth and Environmental Science 1007 (2022) 012025 doi:10.1088/1755-1315/1007/1/012025

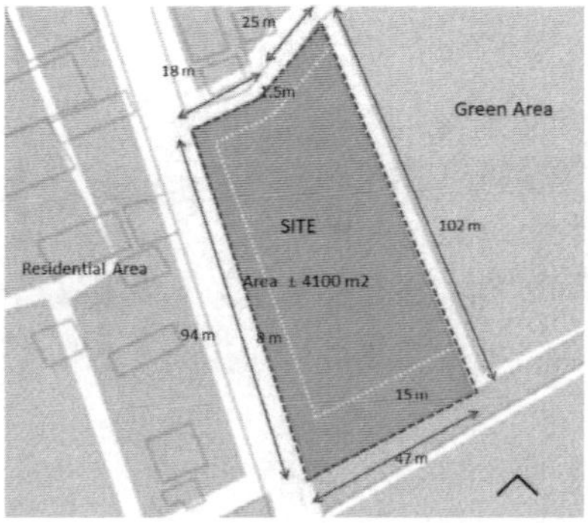

Figure 5. Site boundary.

The culture that influences dust spread is about density area, in which dust can quickly come to the building in low-density areas but not in high density areas [19]. The site is vacant land bordered by areas of different densities. The North-West side is a densely populated and commercial area, while the relatively low-density areas in the South, West, and East are perceived as low-density areas. So, it might influence dust dispersion because the wind flows are unobstructed.

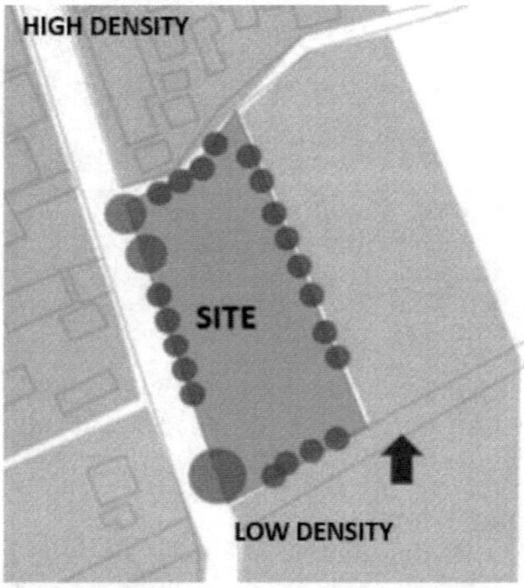

Figure 6. Density area.

3.3 The Needs

The process for selecting the number of units in an apartment is counted by efficiency tower calculation. Many developers strive for getting 80% efficiency. Based on the assumption, the tower efficiency is counted by Semi Gross Area (SGA) divided into total Gross Floor Area (GFA) at times 100%. SGA also can get from total GFA Unit adding to GFA corridor. So that, if SGA is 3444 sqm, the GFA unit will get 2755.2 sqm and GFA corridor is 688.8 sqm because GFA corridor is 20% of the total GFA. The standard of two-bedroom (2-BR) is 73 sqm and three-bedroom (3-BR) is 90 sqm. The percentage of 2-BR is 60%, while that of 3-BR is 40%. This assumption produces the total unit apartment that is 35. Several tower forms and configurations are created based on those results. Afterward, the experiment of wind flow and dust spread due to tower compositions is conducted. Table 3 demonstrates the results. The results indicate that the apartment mass configuration should regard carefully to the wind direction. So that, if dust is carried by the wind with a minimum of 5.5 m/s [4], the shape and position should be in the right position to get this proper wind speed. Moreover, the awareness of the unwanted windspeed should be taken into consideration.

Table 3 Building shape analysis.

Building Shape	Analysis
1st Experiment: Three-tower composition with Planar Surface 	 The wind will be more exposed in these shapes, and the distance in between the masses will create the possibility of a venture effect [20]. In which, it will result in a higher wind speed. Due to this effect, the inner side is still susceptible to dust exposure. The following is the result of surface air speed and pressure with maximum 7 m/s at 2.4 Pa and minimum: 0 m/s with a pressure of -3 Pa
2nd Experiment: U-Shape with Planar Surface 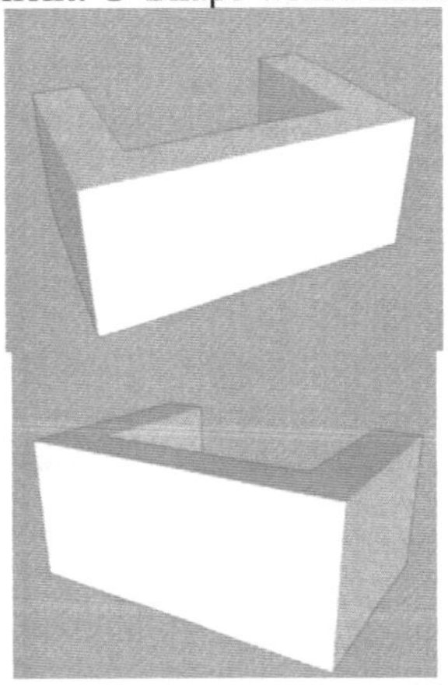	 Due to the corner effect, the inner side is still susceptible to dust exposure [20][6]. The following is the result of surface air speed and pressure: Maximum: 6.5 m/s at 16 Pa Min: 0 m/s with a pressure of 0 Pa
3rd Experiment: Two-tower composition with I-shape and Planar Surface 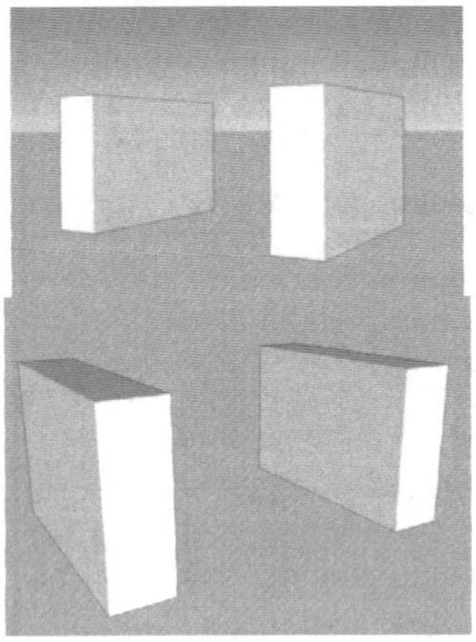	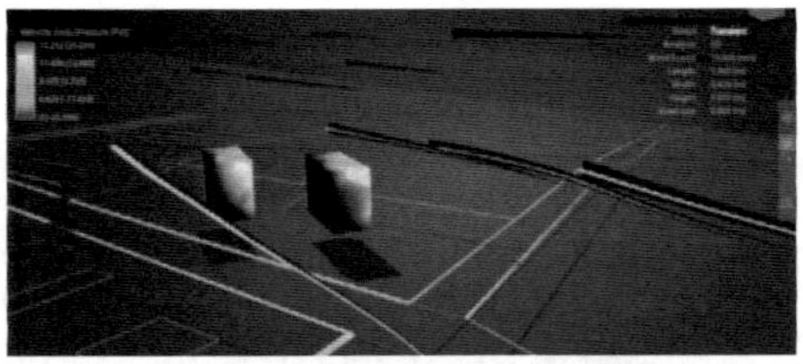 The surface in between the buildings results varying pressures. This configuration causes the wind to shift from high to low pressure. Due to this effect, the inner side is still subjected to dust exposure [6, 20]. The result of surface air speed and pressure: With a pressure of 25 Pa, the maximum speed is 13 m/s, while the minimum is 0 m/s with a pressure of -23 Pa
4th Experiment – Single Tower with Convex Concave Surface	

Building Shape	Analysis
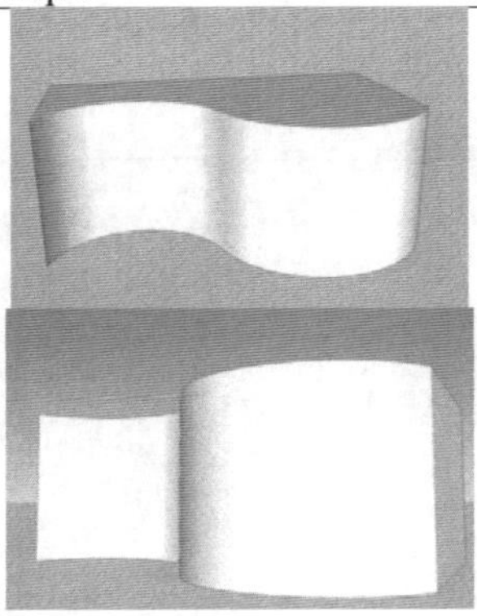	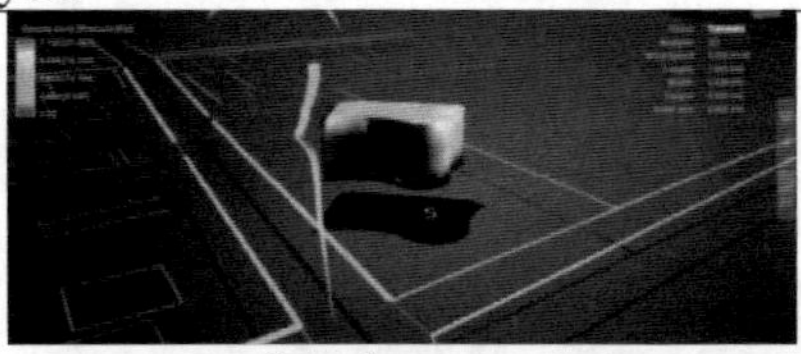 The concave surface has more exposure to the wind. As a result, the dust sticks easily. In contrast to a convex surface, the concave surface has more opportunity of dust blowed by the wind. [9, 11, 20]. The result of surface air speed and pressure: 7.7 m/s with a pressure of 21 Pa Min: 0 m/s with a pressure of 0 Pa
5th Experiment – Single tower with courtyard and planar surface 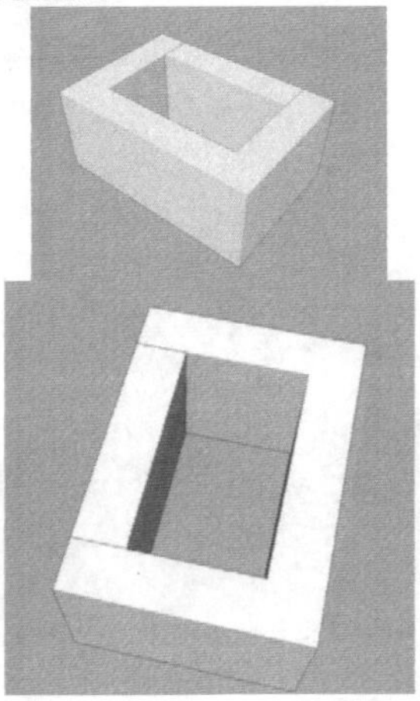	 The courtyard type performs wind flow above the building. It prevents dust from entering the courtyard area. However, in certain dimension of the courtyard, the wind flows into the building. It depends on the height to width ratio. [21, 22]. The courtyard type indicates the best control of the outside strong wind. The inside openings are needed to supply the natural ventilation of residential buildings [23]. The result of surface air speed and pressure: Max: 6 m/s at 13 Pa Min: 0 m/s with a pressure of -5 Pa

Courtyard type was chosen, with certain consideration, based on the analysis in Table 3. For an apartment design, the designers often offer balconies. If it is the case, the balcony forms a niche, in which dust is easily stuck. Therefore, added elements to the external façade, which are exposed to the wind direction must be considered. Elements, like screen should be considered for the function of reducing dust. The balcony position is in a protected area and is not directly exposed to wind from the road. Furthermore, the analysis of the opening and ventilation requirement for each room is needed to know where the room is that have a dust spread potential from outside the site. In Table 4 are the relation in-between room, opening schedule, ventilation system, opening types, surface, and the possibility of dust spread.

Table 4 The room programming.

Room	Opening Schedule (hours)	Air Requirement	Opening Type	Surface	Dust	Synthesis
Outdoor Area (carpark)	24	Natural	Direct	Floor – wall - roof	yes	The wall panel on the South-West side to act as a wind barrier.

IOP Conf. Series: Earth and Environmental Science 1007 (2022) 012025 doi:10.1088/1755-1315/1007/1/012025

Room	Opening Schedule (hours)	Air Requirement	Opening Type	Surface	Dust	Synthesis
Security Area	12	Mechanical and Natural	Window , door	Floor – wall - roof	yes	 The opening is located on the opposite side of the wind direction
Drop off Area	24	Natural	Direct	Floor – wall - roof	yes	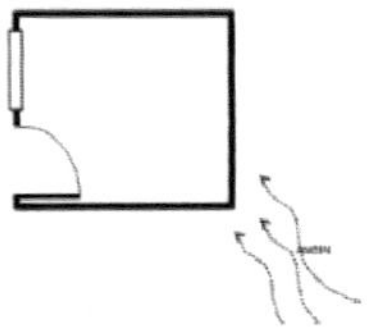 The drop-off area is located on the 'sterile' side for dust exposure 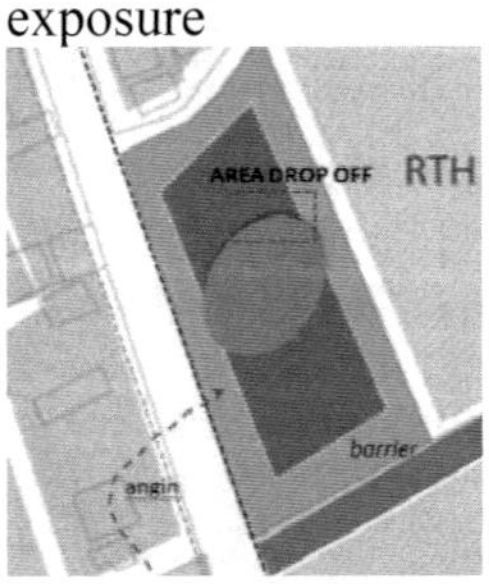
Living Room	8	Mechanical and Natural	Door, Window	Floor – wall - roof	yes	To set a screen or panel on the external side (balcony) can be a dust-reduction zone.
Kitchen	8	Mechanical and Natural	Window	Floor – wall - roof	yes	
Dining Room	8	Mechanical and Natural	Window	Floor – wall - roof	yes	

Room	Opening Schedule (hours)	Air Requirement	Opening Type	Surface	Dust	Synthesis
Bedroom	8	Mechanical and Natural	Window	Floor – wall - roof	yes	
Bathroom	0	Mechanical	Door	Floor – wall - roof	no	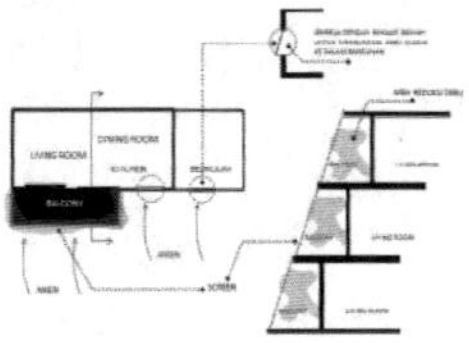
Balcony	24	Natural	Direct	Floor – wall - roof	yes	

From Table 4, it is seen that the niches such as terrace or balcony will be located in the side in which protected from exposure to the wind coming from the highway. A screen or panel on the balcony must be set for reducing dust from outside. Such a design, unfortunately, makes a massive and closed impression on the design. This condition will give other consequences, especially in airflow requirements. A design with specific building distances should be applied, for maintaining densities and allowing airflow. Landscape elements, like trees, shrubs, or wall barriers should be located for minimizing dust [24] Moreover gap composition should be created with regard to openings position which are suitable for air requirements. Concerning warm humid conditions, the opening configuration models differ from the usual cracks used for buildings in desert climates, like in Middle Eastern traditional homes [25]. A crack-like opening is more suitable for desert climates, to prevent dust exposure, which is characteristic of that hot-dry climate. Unlike in the hot-humid climate in Indonesia, in which higher speed airflow is needed, a consequence sufficient opening area is required. It would probably conflict with the need to avoid dust.[24].

4. Conclusion

The courtyard type is proposed for apartment design. The courtyard is recognized as a dust-free zone, while the external facade becomes a dust barrier. As a dust barrier, the external facade is designed with a sloping/tilt surface to allow dust to move easily, with the help of the wind flow. The dust does not adhere to the surface. Wind direction, geometry, and the position of the building surface are factors that is considered in placing the apartment activities. In addition to inclined surfaces, convex surfaces can also respond to dust flow. However, it has to be considered its placement on the overall building shape, so that the flow of dust does not cause discomfort in the indoor space due to the dust.

5. References

[1] Shamsadin A S 2019 Allergic asthma and sick building syndrome *Arch. Asthma Allergy Immunol* **3** 1 pp 001-002

[2] Utami D L and Nasri S M 2018 The evidence of sick building syndrome (SBS) among oil gas refinery officers *KnE Life Sci.* **4** 5 p 566

[3] Mody V and Kajhete R 1990 *Dust Control Handbook* (New Jersey: Noyes Publication)

[4] Chen Q 2006 *Wind in Building Environment Design Sustainable Urban Housing in China*, Dordrecht: Springer Netherlands, 9 pp. 100–115

[5] University of Idaho 2017 *Creating Microclimates - Control of Wind* (Idaho: Community Planning and Design)

[6] Aristodemou E et al 2018 How tall buildings affect turbulent air flows and dispersion of pollution

within a neighbourhood Environ. *Pollut.* **233** pp 782-796
[7] Mukai C, Siegel J A and Novoselac A 2009 Impact of airflow characteristics on particle resuspension from indoor surfaces Aerosol *Sci. Technol* **43** 10 pp 1022-1032
[8] Barakat M A and Kumar R 2016 *Photocatalytic Activity Enhancement of Titanium Dioxide Nanoparticles* (Springer)
[9] Yu Y, Kwok K C S, Liu X P, Zhang Y 2017 Air pollutant dispersion around high-rise buildings under different angles of wind incidence *J. Wind Eng. Ind. Aerodyn.* **167** pp 51-61
[10] Hefny M M and Ooka R 2009 CFD analysis of pollutant dispersion around buildings: Effect of cell geometry *Build. Environ.* **44** 8 pp 1699-1706
[11] Chrummer C A 2018 *Aerodynamics at the Particle Level* (Santa Cruz: University of California)
[12] NSW Department of Planning and Environment 2015 *Apartment Design Guide- Tools for Improving the Design of Residential Apartment Development*. (New South Wales: NSW Department of Planning and Environment)
[13] Sett R 2017 Responses in Plants Exposed to Dust Pollution *Hortic. Int. J.* **1** 2
[14] Plowright P D 2014. *Revealing Architectural Design: Methods, Frameworks and Tools*. London: Routledge
[15] Khallaf M and Jupp J 2017 Performance-based design of tall building envelopes using competing wind load and wind flow criteria *Procedia Eng.* **180** pp 99-109
[16] Meteoblue, Simulated historical climate & weather data for 7.67°S 109.04°E Meteoblue. https://www.meteoblue.com/en/weather/historyclimate/climatemodelled/-7.67N109.042E3_Asia%2FJakarta (accessed Mar. 15, 2019).
[17] Alfata M N F, Hirata N, Kubota T, Nugroho A M, Uno T, Antaryama I G N, Ekasiwi S N 2015 Thermal comfort in naturally ventilated apartments in Surabaya, Indonesia *Procedia Eng.* **121** pp 459-467
[18] Qasem H, Betts T R, Müllejans H, AlBusairi H and Gottschalg R 2014 Dust-induced shading on photovoltaic modules: Dust-induced shading on photovoltaic modules *Prog. Photovolt. Res. Appl.* **22** 2 pp 218-226
[19] Liu S, Pan W, Zhao X, Zhang H, Cheng X, Long Z, Chen Q 2018 Influence of surrounding buildings on wind flow around a building predicted by CFD simulations *Build. Environ.* **140** pp 1-10
[20] Wellington District Plan 2000 *Design Guide for Wind* (Wellington City District Plan)
[21] Defiana I, Samodra F T B and Setyawan W 2018 Night ventilation at courtyard housing estate in warm humid tropic for sustainable environment *IOP Conf. Ser. Earth Environ. Sci.* 126 p 012029
[22] Samodra F and Irvansyah 2021 Decoding space syntax on the translation of site cluster permeability into building courtyard *IOP Conf. Ser. Earth Environ. Sci.* **738** 1 p 012041
[23] Maharani R T, Ekasiwi S N N, Samodra F T B 2019 Minimizing dust in low rise housing design through surface: A literature review *Jurnal Sains dan Seni* **8** 2 p 7
[24] Mediastika C E and Scanlon T 2001 Pagar Untuk Mengurangi Instrusi Polusi Debu Halus ke Dalam Bangunan *Dimensi* **29** 1 p 7
[25] Alkhaidi A 2013 Sustainable Application of Interior Spaces in Traditional Houses of the United Arab Emirates *Procedia-Soc. Behav Sci* **102** p 288-299

Acknowledgments
This research is part of Excellent Primary Research of Higher Education, Penelitian Unggulan Perguruan Tinggi No. 933/PKS/ITS/2021. This authors gratefully acknowledge this financial and technical support.

AUTHOR INDEX

Institute of Physics
Dirac House, Temple Back
Bristol BS1 6BE UK

ISSN: 1755-1307
ISBN 978-1-7138-5449-4